DIAGNOSTIC IMAGING OF INFERTILITY

DIAGNOSTIC IMAGING OF INFERTILITY

ALAN C. WINFIELD, M.D.
Associate Professor,
Department of Radiology and Radiological Sciences
Vanderbilt University Medical Center
Nashville, Tennessee

ANNE COLSTON WENTZ, M.D.
Professor,
Department of Obstetrics and Gynecology
Center for Fertility and Reproductive Research
Vanderbilt University Medical Center
Nashville, Tennessee

WILLIAMS & WILKINS
Baltimore • London • Los Angeles • Sydney

Editor: Timothy Grayson
Associate Editor: Carol Eckhart
Copy Editor: Megan Barnard Shelton
Design: Joanne Janowiak
Illustration Planning: Lorraine Wrzosek
Production: Anne G. Seitz

Accurate indications, adverse reactions, and dosage schedules for drugs are provided in this book, but it is possible that they may change. The reader is urged to review the package information data of the manufacturers of the medications mentioned.

Printed in the United States of America

Library of Congress Cataloging-in-Publication Data

Winfield, Alan C.
Diagnostic imaging of infertility.

Includes index.
1. Infertility, Female—Diagnosis. 2. Diagnostic imaging. 3. Generative organs, Female—Radiography. I. Wentz, Anne Colston, 1940– . II. Title. [DNLM: 1. Hysterosalpingography. 2. Infertility, Female—diagnosis. 3. Ultrasonic Diagnosis. WP 570 W768d]
RC201.W55 1987 618.1′780757 86-9072.
ISBN 0-683-09148-4
Printed at the Waverly Press, Inc.

87 88 89 90 91
10 9 8 7 6 5 4 3 2 1

Foreword

It is a pleasure to write a foreword to *Diagnostic Imaging in Infertility*, which will be a valuable addition for the education of every reproductive endocrinologist and those radiologists who work as part of an infertility team. A hysterogram furnishes information unavailable from any other source that may help in the diagnosis and management of every infertility patient, whether primary, secondary, or due to repeated miscarriage problems. The congenital anomaly of a septate uterus, which causes miscarriages, can only be diagnosed by a hysterogram, and many have been completely missed by laparoscopic viewing. Other congenital anomalies, fistulae, or diverticula require this type of diagnostic procedure. One needs only to review the chapter titles—"Congenital Anomalies of the Uterus and Tubes," "Diethylstilbestrol Exposure In Utero," "Intrauterine Synechiae"—to realize thc importance of this technique.

In addition to these well-recognized applications for hysterography in gynecology, the editors have included a chapter on male infertility. This is an area that is lacking in the education of most reproductive endocrinologists. The chapter on "Sonography in Gynecologic Infertility" is crucial at this time for anyone who is involved in ovulation induction, be it in anovulatory patients or in normally ovulating patients for in vitro fertilization. The chapter on "Sonography of Abnormal Early Pregnancy" comes as a logical sequela. Those who have been interested in inducing ovulation and pregnancy are obviously eager to determine the normalcy of that pregnancy. It is important to determine this as rapidly as possible in order to reassure the patient in a case of a normal development or prevent unnecessarily prolonged hope in the case of an empty sac syndrome. Patients with infertility of any etiology have a 20–30% miscarriage rate, most of which is associated with an empty sac. The most rapid way of identifying this abnormality is by detection of an absent fetal heart at 8 weeks.

It is important that the radiologist be aware of history, physical findings, and clinical concerns, and it is equally essential for the reproductive endocrinologist to be familiar with the best technique for demonstrating the suspected lesion in collaboration with the radiologist. The teamwork between the gynecologic specialist and radiologist is essential, and both should be present at hysterosalpingography for appropriate evaluation and diagnosis of the problem. *Diagnostic Imaging in Infertility* should help to reinforce these statements and therefore should improve our teaching and training of reproductive medicine. Doctors Winfield and Wentz are to be congratulated on this effort.

Georgeanna Seegar Jones, M.D.
Professor
Department of Obstetrics and Gynecology
Eastern Virginia Medical School
Norfolk, Virginia

Preface

This book has been conceived to fill a need for those physicians involved in the management of the infertile couple. During the course of the diagnosis and therapy of these patients, numerous imaging techniques are needed. We have tried to bring these modalities together into one text, useful to the radiologist, gynecologist, and urologist both in the care of these patients and in the organized instruction of their students and colleagues.

It was apparent to us in the development of this text that the information included must be presented in a framework of multiple specialties. The gynecologist, urologist, and radiologist represent a team contributing to the care of the infertile patient in such a manner that the totality of their participation may well exceed the sum of the individual parts. The radiologist is more than an imager; he/she is a concerned and feeling participant in diagnosis and management. With this in mind, we hope that the chapters presenting a clinical overview of infertility are useful. Practitioners in infertility need to understand the basis of the imaging techniques and their utilization in patient management. Radiologic concepts and technical factors of importance have therefore been included.

Much more of this book is devoted to the female partner of the infertile couple, not because of any sense of relative importance but rather because the imaging techniques play a larger role in the diagnostic investigation of the female. Ultrasound has become a significant tool in the management of this special group of patients, and will probably expand its usefulness in the future. Radiologic imaging of the male partner, although less frequently needed, is of tantamount importance where indicated. The forthcoming role of new modalities of imaging is uncertain. There is, however, no question but that imaging techniques will continue to develop to answer the needs of our patients.

It is apparent that close cooperation and knowledgeable communication between involved physicians are necessary in the management of our patients. If this book can convey this concept to our colleagues and residents, it will have fulfilled our goal.

Alan C. Winfield

My collaboration with Dr. Alan Winfield has been of tremendous value, both to me and particularly to my patients. The infertility evaluation is a team effort requiring coordination between several specialties, and the infertility specialist must be in a position to manage the diagnostic workup of the infertile couple. This includes hearing the history, evaluating temperature charts, performing postcoital tests, performing and interpreting the endometrial biopsy, and, importantly, being involved in the evaluation of intrauterine and tubal function. These principles were initially taught to me by Dr. Georgeanna Seegar Jones, who helped me to understand the importance of performing my own studies, interpreting them, and using the results to develop a management plan. Accordingly, for the past 6 years, Alan and I have performed hysterosalpingograms together, working with different techniques, contrast media, and methods to detail the pathology encountered. Therefore, this text is a personal expression of what we think is important in patient care, particularly that there be close communication, collaboration, and cooperation between those involved in the diagnostic and therapeutic aspects of infertility.

Anne Colston Wentz

Acknowledgments

A book such as this requires the input, cooperation, and efforts of many people. It is difficult to adequately acknowledge all of the help received in this task. Needless to say, major impetus has come from so many of our past teachers and associates as well as the impact of a long line of patients.

Several of the illustrations found in this text were graciously offered to us by our colleagues, both in this country and abroad. Our thanks are extended to T. A. Baramki, N. R. Dunnick, L. Ekelund, S. Howards, H. W. Jones, Jr., M. Lancet, P. Porch, Jr., R. Rosenblatt, and W. Stern.

Art work was created by Bonnie Norman with skill and imagination. John Bobbit supplied the photography with his customary devotion to excellence. We are grateful to them both.

Secretarial support as well as unending patience was supplied by Carolyn Cooper, Beverly A. Steele, and Vernice C. Dunlap and was received with appreciation. The Editorial Section of the Department Radiology and Radiological Sciences at Vanderbilt University Medical Center, in the person of Holly J. Pelton, was of invaluable help.

George Stamathis and Carol Eckhart of Williams & Wilkins have guided the development of this text from concept to fruition. Their encouragement and meaningful suggestions have reduced our burdens considerably and we are grateful.

Finally, we would thank Barbara and Dennis for their patience, tolerance, encouragement, and support. Their ability to share our time during the creation of this manuscript made the task a much lighter one.

ACW
ACW

Contributors

Arthur C. Fleischer, M.D. (Chapters 10 and 11)
Associate Professor of Radiology and Radiological Sciences, Assistant Professor of Obstetrics and Gynecology, Vanderbilt University Medical Center, Nashville, Tennessee.

Fred K. Kirchner, Jr., M.D. (Chapter 12)
Associate Professor of Urology, Vanderbilt University Medical Center, Nashville, Tennessee

Murray J. Mazer, M.D. (Chapter 12)
Associate Professor of Radiology and Radiological Sciences, Vanderbilt University Medical Center, Nashville, Tennessee

Wayne S. Maxson, M.D. (Chapter 10)
Assistant Professor of Obstetrics and Gynecology, Center for Fertility and Reproductive Research, Vanderbilt University Medical Center, Nashville, Tennessee

David A. Nyberg, M.D. (Chapter 11)
Assistant Professor of Radiology, University of Washington, Seattle, Washington

Glynis A. Sacks, M.D. (Chapter 11)
Assistant Professor of Radiology and Radiological Sciences, Vanderbilt University Medical Center, Nashville, Tennessee

Contents

1 Clinical Aspects of Infertility and Its Evaluation

Infertility affects an estimated 15% of couples or about one in every seven marriages. The total number of infertile couples has increased in the past decade, due to both a larger population and an increase in sexually transmitted disease. The number of babies available for adoption has decreased because of the widespread use of effective methods of contraception and the availability of abortion techniques to limit unwanted pregnancy. More and more couples are seeking diagnosis of their infertility problem, and with recent advances in therapeutic technology, they have a high expectation of a satisfactory solution. This has caused an increased demand for the services of physicians specializing in the diagnosis and management of male and female infertility. In response to the need, subspecialty training in reproductive endocrinology and infertility, certified by the American College of Obstetricians and Gynecologists, has been developed. However, practitioners in not just one but several specialties and disciplines must work together to improve the outcome for the infertile couple. Hysterosalpingography and ultrasonography bring the radiologist in close contact with these patients. The further evaluation of the couple requires the coordinated involvement of urologists, andrologists, geneticists, nutritionists, and many others. Cooperation and communication between the gynecologist and the radiologist are exceedingly important in the diagnostic approach to both members of the infertile couple.

An infertility evaluation should be instituted after no more than 1 year of unprotected intercourse, and earlier if either partner suspects that a fertility problem may exist. Approximately 25% of couples desiring pregnancy will conceive in the first month of unprotected intercourse, 60% in 6 months, 75% in 9 months, and 90% in a year. Since only an additional 5% will achieve pregnancy in the next 6 months of exposure, there is no reason to delay evaluation, particularly if the woman is over 30 years of age. Increased coital frequency will improve the chance of pregnancy: only 16% of couples having intercourse less than once a week achieved pregnancy in less than 6 months, whereas 83% of couples having intercourse four or more times a week achieved pregnancy during that time.

The infertility evaluation should be efficient and thorough. There are two primary goals: (*a*) to determine the *etiology* of the couple's infertility and (*b*) to give a *prognosis* for future fertility. Until recently, even with optimal management, only 50–60% of couples consulting a physician or clinic for infertility could expect to achieve pregnancy. These statistics are made invalid by the rapid evolution of the technique of in vitro fertilization with embryo placement directly into the uterus, which circumvents some of the causes of infertility. Nevertheless, a careful evaluation to determine the reason for infertility and to provide an appropriate management plan is still important to maximize the chances for pregnancy. The methods of diagnosis have not changed, only the statistics associated with ultimate pregnancy rates. It must be recognized that not all couples who are infertile will have the financial, emotional, and cultural resources to undergo in vitro fertilization. In addition, some patients will decide not to undergo studies essential for a thorough evaluation. The infertility workup is, after all, elective, but the limitations of an incomplete investigation should be explained. For this reason, an organized, efficient, and thorough approach is essential, both to minimize the time needed for a diagnosis to be made and to maximize the ultimate chance of pregnancy. However, a satisfactory result will be achieved only when the patient understands the goals of the evaluation.

The etiology of the infertility and the prognosis for future fertility are both established by taking into consideration three main areas: (*a*) the age of the woman, (*b*) the duration of infertility, and (*c*) the medical factor (or etiological diagnosis) responsible for the infertility.

The age of the wife, not the husband, has impact upon the prognosis for infertility. The age of maximal fertility in the female is approximately 24 years, with only 1 of 60,000 births occurring past the age of 50. The fecundity or fertility capacity of a female of any particular age is difficult to analyze because there are numerous variables, such as intercourse frequency, but fecundity appears to decrease with increasing age. For example, at age 25, 75% of conceptions occur in less than 6 months of exposure, while after age 40 the number drops to 23%. Fecundity obviously begins to decrease after the age of maximal fertility, but a significant difference is found only after age 35. However, from psychologic and physiologic standpoints, it is important to institute an infertility evaluation without delay in a woman over 30 years old and to understand that the prognosis for fertility decreases over time.

The duration of exposure to pregnancy also helps to determine the prognosis for future fertility. A couple with long-term infertility clearly has a cause to be identified and an excellent chance of having the etiological diagnosis made. However, the longer the couple has been married without achieving pregnancy, the less the chance of pregnancy, partly because of advancing age but also because failure to conceive over many years implies a serious problem.

The third major area to be considered, crucial to prognosis, is the medical factor or etiological diagnosis responsible for the infertility.

The following six factors are the most likely causes of infertility; all must be evaluated in a thorough investigation.

1. Central or ovulatory factor—Ovulation of an oocyte, which is difficult to establish, must occur. Regular menstrual cyclicity implies but does not prove that the ovarian processes essential to ovulation are occurring.
2. Male factor—The male partner must have sperm that can penetrate and fertilize the oocyte.
3. Mucus or cervical factor—Abnormalities of the cervix can act as a cervical barrier to fertility.
4. Endometrial-uterine factor—The fertilized oocyte, or embryo, must implant in the uterus.
5. Tubal factor—The fallopian tube(s) must allow transport of the ovum to the uterine cavity.
6. Peritoneal factor—Pelvic adhesions, endometriosis, or other problems in the peritoneal cavity can cause a physical or mechanical block to fertility.

After the initial history and physical examination of both partners, a definite plan of evaluation, with a predictable time frame of investigation, should be established for the couple. A routine screening infertility evaluation should near completion in three or four office visits. Since two or more factors have been found to be operative in fully 35% of all infertility cases, all factors should be evaluated even if an abnormality that is found in one area appears to explain the infertility.

Five of the six factors above can be evaluated in one ovulatory menstrual cycle. If, by history, the wife seems to be ovulating, as judged from her description of regular menses and characteristic premenstrual symptoms, she is asked to keep a basal body temperature chart, which is useful to help with the timing and interpretation of the ensuing tests (Fig. 1.1). If the history suggests that she is not ovulatory, this diag-

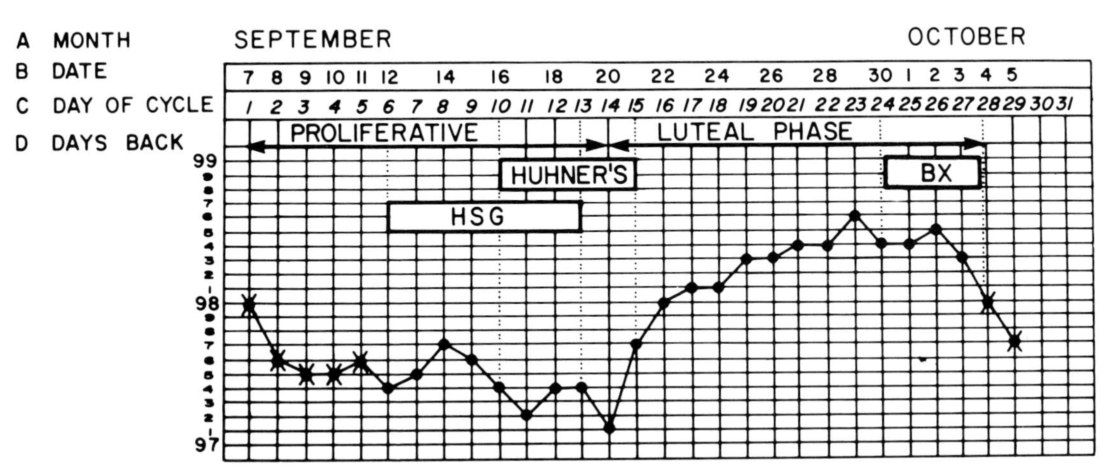

Figure 1.1. Example of basal temperature chart during a typical ovulatory cycle. The optimal times for hysterosalpingography (*HSG*) postcoital semen examination, and endometrial biopsy (*BX*) are noted. Times of active menstruation are designated.

nosis must be clarified by further gynecological and endocrine testing to determine the etiological cause and the correct treatment. If there are no contraindications, medications for ovulation induction may be used to re-establish an ovulatory pattern, and the evaluation proceeds.

The basic infertility workup is accomplished by scheduling a postcoital test for evaluation of sperm survival in cervical mucus as close before the time of anticipated ovulation as possible. The hysterosalpingogram may be accomplished at the same time, if it can be established that ovulation is impending but has not yet occurred. For practical purposes, a basal body temperature that has not risen and a clear, watery cervical mucus, both observed on or before cycle day 14, suggest that ovulation is impending; if there is any question it is better to cancel the hysterographical procedure and do it in the next cycle. Before preparing the patient for hysterosalpingography, the cervical mucus is aspirated from the endocervical canal into a 1-ml syringe that is capped and saved for microscopic evaluation later. The next two procedures, the endometrial biopsy for evaluation of the endometrial pattern and a semen analysis, can be accomplished in the same menstrual cycle, just before the onset of menstruation. In retrospect, if the temperature chart has a biphasic pattern characteristic of an ovulatory cycle, then five of the first six infertility factors could have been, and hopefully were, investigated in that menstrual cycle. The ovulatory factor is evaluated by observation of a biphasic basal temperature chart, and confirmed by endometrial biopsy; the male factor by sperm count and postcoital test; the endometrial uterine factor by the hysterogram and the endometrial biopsy; and the tubal factors by hysterosalpingography. Since the chance of pregnancy seems to increase in the 3–4 months after a hysterosalpingogram has been performed, it is useful to wait for three to four cycles before scheduling a diagnostic laparoscopy for evaluation of the pelvic or peritoneal factor. This time can be beneficially used to correct any abnormalities discovered in the basic infertility evaluation.

Central or Ovulatory Factor

The presence or absence of ovulation ordinarily can be surmised from history. Breast tenderness, dysmenorrhea, weight gain or bloating, fatigue, and irritability are all associated more frequently with ovulatory than anovulatory cycles. The symptomatology described in the history that supports an ovulatory cycle is due to progesterone output in the luteal phase and can occur in the absence of the physical release of an oocyte from the follicle. This requires formation of the stigma, an area of dissolution of the membrane of the follicle, which is induced by a complex interaction of hormones, enzymes, and prostaglandins; the follicles can produce progesterone even if the ovum is entrapped and not released. Absolute proof of the physical act of ovulation requires either a pregnancy or identification of the oocyte. For practical purposes, any woman not taking hormone medications, and who is having menses every 30 ± 4 days, can be considered to be presumptively ovulatory.

The basal temperature chart is a useful adjunct in the infertility evaluation. It cannot establish that physical ovulation is occurring, but it can aid in the scheduling and interpretation of tests. The follicular or proliferative phase of the cycle occurs between cycle day 1 (the day of menstruation) and the occurrence of ovulation, which is indirectly related to the rise in basal temperature. The luteal or secretory phase of the cycle begins after ovulation and extends to the next menses. The basal temperature is low, usually less than 97.6°F, before ovulation, and rises 0.6–0.8°F after ovulation under the influence of progesterone secretion, which causes a mild hyperthermia. Follicular phase shortening, late ovulation, luteal phase shortening, and delayed menses are easily ascertained from the chart. Proper timing for postcoital tests, hysterosalpingogram, and endometrial biopsies can be decided and pregnancy diagnosed early. That ovulation may have already occurred is suggested by an elevated temperature on the chart. For example, if a hysterosalpingogram had been scheduled for cycle day 15, and the patient's basal temperature increased on day 12 and remained elevated, then the hysterosalpingogram should be cancelled to avoid performing it in the postovulatory period.

The patient should be asked to record daily temperatures only for the limited time necessary to complete the infertility evaluation. She should be instructed to take her temperature orally, at the same time each morning, before she has indulged in any activity, including getting out of bed or brushing her teeth. For individuals who work nights and sleep days, the temperature chart can still be valuable: the patient can take her temperature after 3 or 4 hours of uninterrupted sleep, when her temperature is likely to be at its basal level. All happenings of interest should be recorded on the chart, including late nights, alcohol consumption, sickness, intercourse, bleeding, cramping or spotting, and so forth. However, the keeping of a temperature chart should not be open-ended, and the chart itself should only be used for scheduling and interpretation of tests.

Mucus or Cervical Factor

The cervical mucus must maintain and promote sperm survival and motility, but is only of importance for conception in the periovulatory period; therefore, the postcoital test (PCT) must be properly timed. During the follicular phase, rising estrogen levels cause the cervical mucus to become copious, watery, acellular, and conducive to sperm motility. The term "spinnbarkeit" is used to describe the ability of the mucus to be drawn out into or spin a thread that can reach 10–12 cm long in the periovulatory phase. The effect of postovulatory progesterone output is to reverse all estrogen effects on the mucus, which becomes thick, scant, viscous, cellular, and a barrier to sperm movement. Patients may be referred with a history of "hostile mucus" simply because the PCT was performed in the midluteal phase or prior to day 10 of the cycle, when the cervical mucus is physiologically inadequate to promote sperm motility.

The PCT is done within 24 hours of intercourse and is accomplished by aspiration of the cervical mucus, which is examined microscopically. The quantity and quality of the mucus is recorded, and an assessment made of the number and motility of sperm present. If abnormalities are detected, it may be necessary to perform in vivo and in vitro penetration tests to identify and diagnose the cause of the mucus that is acting as a barrier to sperm.

It may be convenient for the patient to schedule both the hysterosalpingogram and the PCT for the same day, because both tests must be done before ovulation. The patient may be scheduled for a PCT in the gynecologist's office to be followed later by the hysterosalpingogram, in the radiology suite; we have found it convenient to aspirate the cervical mucus immediately before performing the hysterosalpingogram. The capped syringe will maintain the mucus for examination later.

Male Factor

Although a great deal of information about the male can be obtained from the PCT, a semen analysis is still essential. Since up to 40% of infertility diagnoses are attributed to various problems of the male partner, the quantitative measurements obtained from a sperm count, including volume, count/milliliter and total count, percentage of normal and abnormal forms, and percentage motility, are helpful in determining areas requiring further evaluation in the male. An increased number of white blood cells in the ejaculate may indicate infection. Head-to-head or tail-to-tail agglutination or "sticking together" of sperm may suggest the presence of antisperm antibodies. The infertility specialist will be alerted to take a careful history and to guide the investigation along particular lines by abnormalities found in the routine sperm analysis. If a completely "normal" semen analysis is found, attention is then directed to evaluation of the cervical mucus, to ascertain its quantity and quality. The possibility exists that the mucus may be acting as an immunological barrier to sperm penetration, particularly if other tests of sperm function have indicated their ability to swim through donor or bovine mucus. Finally, the capability of sperm to fuse with and enter a hamster oocyte may be assayed as an analogy to fertilization in the thorough evaluation of the male factor.

Uterine-Endometrial Factor

Local abnormalities of the uterus, including polyps, submucous myomata, chronic endometritis, and retained products from a prior pregnancy may prevent implantation or placentation. Congenital uterine abnormalities, particularly the arcuate and bicornuate anomalies, are not ordinarily associated with impaired fertility, and only 25% of patients with a bicornuate uterus have any problems during pregnancy. The incidence of Asherman's syndrome or endometrial sclerosis appears to be increasing, and intrauterine synechiae may cause both secondary infertility and recurrent abortion (see Chapter 7). An unusual endometrial lesion is the deranged pattern caused by an infectious process, and culture for ureaplasma and/or chlamydia may be necessary.

Luteal phase inadequacy is a term that refers to either an inadequate output of progesterone from the corpus luteum or an inadequate endometrial response to progesterone stimulation, and is another category of endometrial factor evaluation during the basic infertility investigation. Its incidence in women with primary infertility is less than 5%, but it can be found with increased frequency in certain clinical settings: (*a*) 35% of patients undergoing the induction of ovulation with clomiphene citrate consistently manifest an inadequate endometrial development; (*b*) approximately 35–50% of patients with frequent early miscarriages have an inadequate progesterone output; (*c*) women over 35 are diagnosed to have endometrial inadequacy more frequently than younger women; (*d*) ovulating women with hyperprolactinemia are more likely to have defective corpus luteum function; and (*e*) women who jog or undergo strenuous athletic conditioning secrete less progesterone.

An adequate output of both estradiol and progesterone from the corpus luteum is needed to convert the proliferative endometrium into a secretory pattern, which then undergoes a predictable and reproducible change from ovulation to menstruation. In luteal phase inadequacy, an inadequate hormonal stimulation, or the inability of the endometrium to respond to hormones, results in a deranged endometrial pattern that is easily detected by taking a timed endometrial biopsy within 1–2 days of the expected menstrual period. The biopsy, taken from the anterior or posterior wall of the fundus or upper corpus, should produce a piece of tissue about 1.5 cm long. After fixation, microscopic sections are prepared and stained, and the histological pattern is "dated" according to established criteria. An endometrial biopsy taken 2 days before the next period should have a histological dating compatible with secretory day 26, because the first day of menstruation is arbitrarily called day 28. Biopsies that date early by 2 or more days lag expected development and are called "out-of-phase"; they are suggestive of a defective progesterone output. The site of the biopsy is rarely detected by hysterosalpingography. However, the scar may extend to the myometrium, and may occasionally be observed as an indentation if a hysterosalpingogram is done in the follicular phase of the next cycle.

Tubal Factor

Hysterosalpingography provides detailed information about the uterine cavity and tubal patency; some forms of intrapelvic disease may be detected, particularly if areas of calcification are seen or if lack of tubal mobility suggests fixation and adhesive disease. The Rubin test, in which CO_2 is introduced through the cervix, is nonspecific, gives misleading information, and has technical difficulties: bilateral tubal patency and normality of the cavity cannot be readily ascertained, and tubal spasm cannot be differentiated from occlusion.

The hysterosalpingogram should be performed prior to ovulation, when the oocyte is in prophase of Meiosis I and relatively radioresistant, before resumption of meiotic maturation. No increase has been documented in the incidence of spontaneous miscarriage or congenital anomalies in patients in whom a hysterosalpingogram was performed in the cycle of conception. Importantly, all hysterograms should be performed after cessation of menses so as not to introduce blood in retrograde fashion into the peritoneal cavity, to avoid the development of endometriosis and perhaps infection.

Hysterosalpingography may be therapeutic in some women since the incidence of conception increases during the subsequent 3 or 4 months. A delay before proceeding to diagnostic laparoscopy may further the chance of pregnancy and avoid the need for general anesthesia and a surgical procedure. Also, the time can otherwise be used to follow up and evaluate any other problems suggested by abnormalities in the other basic tests.

Peritoneal Factor

A pelvic or peritoneal cause for infertility refers to any physical or mechanical pelvic finding that can provide a barrier or block to any of the processes required for fertility. Pelvic adhesions, endometriosis, and postsurgical scarring limiting tubal mobility are the most common findings, and are not necessarily diagnosed radiographically. A normal hysterosalpingogram has been documented in 35–60% of patients found to have endoscopic or operative evidence of endometriosis. The hysterosalpingogram tends to underdiagnose peritubal adhesions and endometriosis, and overdiagnose cornual occlusion (see Chapter 2). Endometriosis may be suspected if "convoluted oviducts," which have an appearance like corkscrews, are observed, but there are no other pathognomonic findings. Peritubal adhesions may be suspected if poor tubal mobility is seen, but this interpretation is very subjective. The introduction of contrast medium into the uterine cavity may cause cornual spasm that may not be relieved by antispasmotics; at subsequent laparoscopy done under general anesthesia, cornual patency may be observed because the anesthesia has relaxed the spasm. Overall, when endoscopy is performed in patients in whom no reason for infertility has been found and a thorough evaluation of the areas discussed above has been done, about 20–25% will be found to have intrapelvic pathology, such as endometriosis or pelvic adhesions. Therefore, diagnostic laparoscopy is an essential part of the evaluation of the infertile couple.

Other Considerations

An additional factor, with a less obvious relation to fertility, must be considered; this is the psychogenic aspect. Infertility imposes a strain on any marriage, and the anxieties and frustrations expressed by a couple with infertility are poignant. With continued failure to conceive, a woman tends to blame "poor timing," and puts such pressure on her husband to perform that ultimately the stress can lead to impotence or ejaculatory disturbances. The onset of the menstrual period becomes a time of profound depression. Recording daily temperatures can wrongly become a guide for choosing the day for intercourse, resulting in "sex-on-schedule" and a bizarre and unsatisfactory sexual relationship. The menstrual cycle becomes a vicious cycle. The unsuspecting, unaware, or insensitive hysterosalpingographer may find himself or herself in the middle of this milieu of anxiety, guilt, and emotional distress. An explanation for years of infertility may be diagnosed at hysterosalpingography, and immediately precipitate a barrage of questions about etiology, treatment, and results. For these reasons, it is imperative that the hysterosalpingographer have a thorough working knowledge of the field of infertility, be cognizant of the history and previous evaluation of the patient, and have the patient's present clinic and basal temperature chart available for review. The physician involved must be sensitive to the needs of the couple. It is crucial that he or she have insight into both the infertility evaluation and the psychological and emotional distress that infertility can cause.

Finally, what can one expect for the couple in whom a thorough infertility evaluation, including diagnostic laparoscopy, has been completed without finding the etiologic reason for the infertility? Recent studies suggest that fewer than 5% of couples fail to show some etiological factor associated with infertility; the 10–20% incidence of unexplained infertility quoted in the earlier literature is too high an estimate in view of the additional information made available through pelvic endoscopy, testing for immunological causes, and improved evaluation of the male. The so-called normal infertile couple is a misnomer; these patients are obviously not normal, yet they are among the most poorly managed patients in terms of diagnosis-oriented therapy; since nothing specific was found to correct, the tendency is to try a little of every approach to help them achieve pregnancy. However, recent studies suggest that their prognosis may be better than previously thought. In one study of couples with unexplained infertility, 60% of the women found to be normal at laparoscopy conceived within 36 months (Rousseau et al., 1983). On the other hand, if conception has not occurred by 36 months after laparoscopy, then the couple is very unlikely to achieve pregnancy. For these couples, with truly unexplained infertility, a therapeutic alternative exists: in vitro fertilization and embryo transfer results in an acceptable rate of pregnancy.

Success in the diagnosis and management of infertility is achieved if an etiological reason for the infertility is found. A systematic approach to establishing diagnosis and prognosis provides for the couple a realistic view of what to expect. If therapy is based upon the discovered etiology, then the prognosis for future fertility can be determined, and the goals of an infertility evaluation will have been accomplished.

SUGGESTED READINGS

Akin A, Elstein M: The value of the basal temperature chart in the management of infertility. *Int J Fertil* 20:122, 1975.

Aral SO, Cates W Jr: The increasing concern with infertility. *JAMA* 250:2327, 1983.

Collins JA, Wrixon W, Janes LB, Wilson EH: Treatment-independent pregnancy among infertile couples. *N Engl J Med* 309:201, 1983.

Corson SL (guest ed): Diagnosis and treatment of infertility: An invitational symposium. *J Reprod Med* 18:113, 1977.
Drake TS, Grunert GM: The unsuspected pelvic factor in the infertility investigation. *Fertil Steril* 34:27, 1980.
Duff DE, Fried AM, Wilson EA, Haack DG: Hysterosalpingography and laparoscopy: A comparative study. *AJR* 141:761, 1983.
Goldenberg RL, Magendantz HG: Laparoscopy and the infertility evaluation. *Obstet Gynecol* 47:410, 1976.
Guzick DS, Bross DS, Rock JA: A parametric method for comparing cumulative pregnancy curves following infertility therapy. *Fertil Steril* 37:503, 1982.
Jones WR: Immunologic infertility—fact or fiction? *Fertil Steril* 33:577, 1980.
Leridon H, Spira A: Problems in measuring the effectiveness of infertility therapy. *Fertil Steril* 41:580, 1984.
Philipsen T, Hansen BB: Comparative study of hysterosalpingography and laparoscopy in infertile patients. *Acta Obstet Gynecol Scand* 60:149, 1981.
Rosenfeld DL, Seidman SM, Bronson RA, Scholl GM: Unsuspected chronic pelvic inflammatory disease in the infertile female. *Fertil Steril* 39:44, 1983.
Rousseau S, Lord J, Lepage Y, Van Campenhout: The expectancy of pregnancy for "normal" infertile couples. *Fertil Steril* 40:768, 1983.
Snowden EU, Jarrett JC II, Dawood MY: Comparison of diagnostic accuracy of laparoscopy, hysteroscopy, and hysterosalpingography in evaluation of female infertility. *Fertil Steril* 41:709, 1984.
Steinberger E, Rodriguez-Rigau LJ: The infertile couple. *J Andrology* 4:111, 1983.
Verkauf BS: The incidence and outcome of single-factor, multifactorial, and unexplained infertility. *Am J Obstet Gynecol* 147:175, 1983.

2 Techniques and Complications of Hysterosalpingography

A normal female reproductive tract is essential to fertility, and several approaches to investigation are used during the routine evaluation of the infertile couple. Healthy and patent oviducts facilitate conception, and a normal intrauterine surface, without polyps, submucous myomata, septa, or scars allows the normal process of implantation to occur. An endocervical canal without diverticula, a lower segment without postoperative scars or defects, and a normal internal cervical os may be important to a successful pregnancy. The hysterosalpingogram is useful in evaluating all of these areas.

Hysterosalpingography is indicated early in the investigation of the infertile female, particularly in those with a history of previous abdominal surgery, known episodes of pelvic inflammatory disease, prior postpartum infection or cesarean section, and/or palpable adnexal pathology compatible with hydrosalpinges or endometriosis. Hysterosalpingography is also indicated in women with previous sterilization who are requesting reversal of tubal ligation. The examination is essential for all patients prior to tubal surgery, whether or not laparoscopy has been done, because hysterosalpingography yields information not elucidated at laparoscopy. The appearance of the endocervical canal, the uterine cavity, and the fallopian tube lumina cannot be appreciated at laparoscopy, and some findings may be missed at hysteroscopy.

Hysterosalpingography has several advantages as a diagnostic test performed early in the investigation of infertility. It provides immediate information about serious problems such as tubal occlusion and abnormalities of the uterine cavity. General anesthesia is not required, and the procedure can be performed rapidly and is ordinarily well tolerated by the female patient. On occasion, a mild analgesic may be needed (1). A prostaglandin synthetase inhibitor given an hour or so before the procedure may be of benefit. Some clinicians use a paracervical block to decrease discomfort from dilatation of the cervical os, but this is not useful to decrease pain induced by a cramping uterus and seems otherwise unnecessary. Since so little advance preparation is required, and since hysterosalpingography is not invasive, the procedure is extremely cost effective.

A major limitation of hysterosalpingography is its inability to demonstrate peritubal disease with consistency, and to evaluate the status of the fimbria of the tubes. Lack of movement of the tubes during traction may suggest fixation by peritubal adhesions, but this observation is neither sensitive nor specific, and is highly subjective. A convoluted, corkscrew appearance of the salpinx or marked crowding or bunching of the tube suggests broad ligament scarring and shortening of the fimbria ovarica. These findings have been proposed to indicate pelvic endometriosis or inflammatory peritubal disease (2). However, confirmation of such suspicions usually requires endoscopy.

Hysterosalpingographic Technology

INSTRUMENTS FOR HYSTEROSALPINGOGRAPHY

Various instruments have been designed and other methods adapted to the performance of hysterosalpingography. The most common of these include: (*a*) the Kidde cannula (Fig. 2.1); (*b*) the pediatric Foley catheter; (*c*) the Malmstrom vacuum apparatus (Fig. 2.2); (*d*) the Spackman cannula; (*e*) the Harris Uterine Injector; (*f*) the Fikentscher and Semm portio cervix adapter for hydropertubation; and (*g*) the Jarcho-type cannula (Fig. 2.3).

The ideal instruments for hysterosalpingography should be easily applied, avoid uterine and cervical trauma, provide maximum delineation of the uterine cavity, be unaccompanied by cervical leakage, have no added discomfort due to instrumentation, and allow patient maneuverability for the performance of oblique films without dislodgement of the instrument. We have found the Kidde cannula to be a simple, safe, rapid, and easy-to-learn technique. Its inherent flexibility permits ready positioning under most circumstances (Fig. 2.4).

RADIOGRAPHIC INSTRUMENTATION

Appropriate hysterosalpingography requires modern, well-designed radiographic equipment to produce an optimal examination yielding all pertinent information while exposing the patient to the least possible amount of radiation. It is mandatory that the examination be done with fluoroscopic control. Equipment design utilizing an undertable tube reduces radiation exposure to the operator. Reduction of the object–film distance, maximal collimation, a small focal spot, and a movable grid are all necessary to reduce radiation exposure and enhance patient imaging. Optimal kVp for visualization of the iodinated contrast agent should be in the 75–85 kVp range.

The image may be recorded on spot film, 105-mm camera film, or videotape. The modality chosen for imaging is not as important as the meticulous technique required to produce optimal radiographic images. Attempts to do hysterosalpingography without fluoroscopic observation are fraught with risk and should be discouraged.

Selection of an appropriate film–screen system is necessary. To achieve optimal imaging with reduced radiation, fast intensifying screens and moderate detail film are suggested if a cut-film technique is to be used. The rare-earth intensifying screens are of considerable value in this regard.

Fluoroscopic observation and radiographic exposures should be minimized to those necessary to complete the examination. Although oblique views are sometimes of help and delayed radiographic examination is occasionally warranted, if the examination can be satisfactorily completed with one or two exposures the amount of radiation is similarly reduced.

Finally, it is important to remember that radiographic equipment may change with time. It is mandatory that equipment be monitored on a regular basis for evaluation of radiation output. Radiation exposure to the pelvis should be reduced to a minimum. Radiation dosage to the gonads will vary dramatically, depending upon the number of radiographic exposures, duration of fluoroscopy, equipment, film–screen combination, and the like. Familiarity with the equipment and the procedure will shorten the examination and reduce the radiation exposure. A recent survey published by the American College of Radiology (3) estimated that a typical hysterosalpingogram exposed the ovaries to radiation dosage of 590 mrad (5900 mGy).

CHOICE OF CONTRAST AGENTS

Contrast material used for hysterosalpingography may be either water or oil soluble. As a matter of interest, the first contrast agent introduced to the uterine cavity was a paste-like suspension of bismuth, used by Rindfleish in 1910 in a 21-year-old woman suspected of an ectopic pregnancy (Fig. 2.5) (4). The oil-soluble agent Lipiodol, composed of poppy seed oil with 40% iodine, was first used in hysterosalpingography in the mid-1920s and represented a satisfactory contrast agent. Concern developed over the possible effects of oil embolization to the lungs and the persistence of the iodized oil contrast within the peritoneal cavity due to lack of reabsorption. Attention turned to the use of water-soluble agents because of their rapid absorption, improved visualiz-

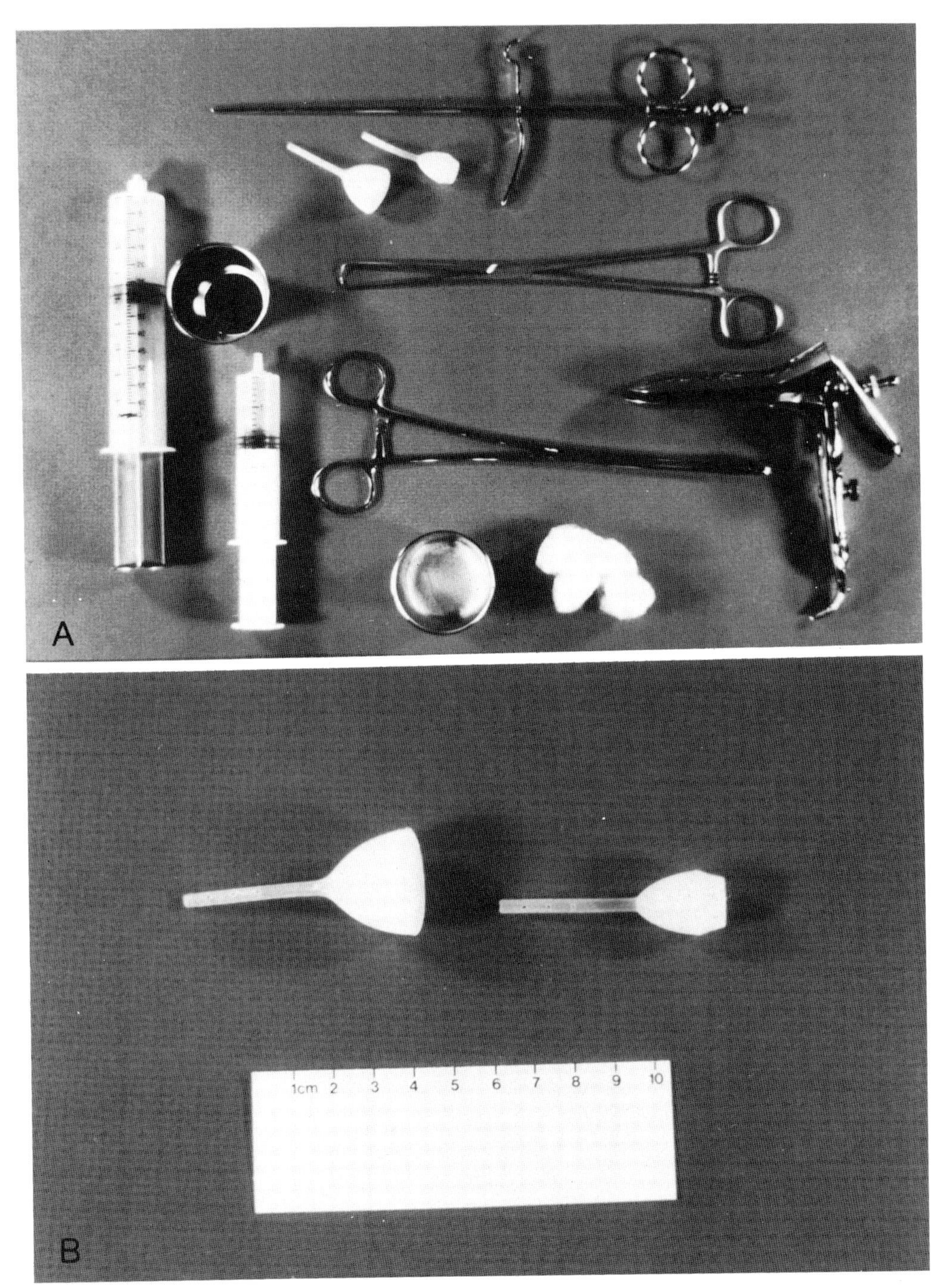

Figure 2.1. *A:* Contents of a hysterosalpingogram tray at Vanderbilt University Hospital. Kidde cannula and handle are the routine instrumentation. *B:* Kidde cannulas are available in two sizes. The smaller cannula is generally used. The tips may be shortened by scalpel if necessary.

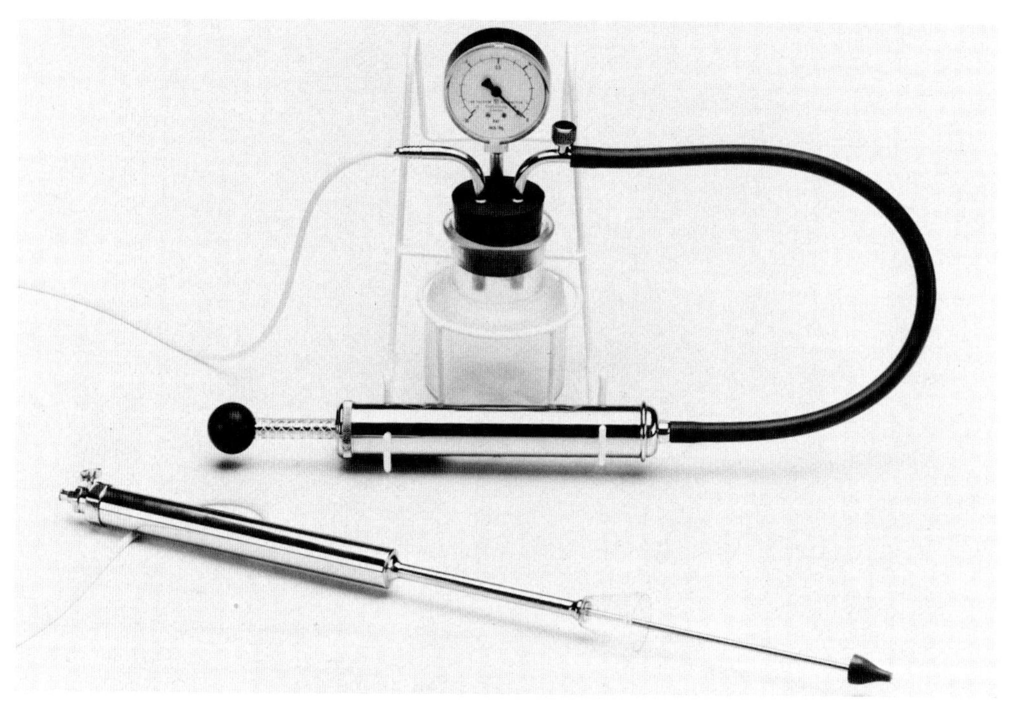

Figure 2.2. Malmstrom vacuum apparatus.

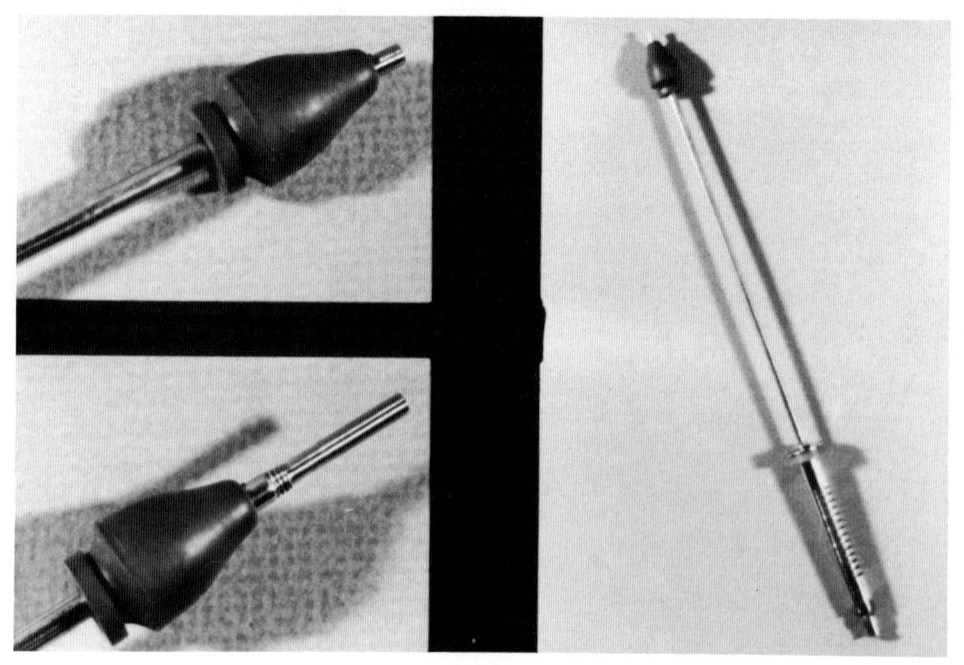

Figure 2.3. *Right:* Jarcho application device with hard rubber tip. The tip is movable. *Left upper:* Preferred position for rubber tip on metal rod. *Left lower:* Tip has moved, inappropriately exposing a long segment of metal.

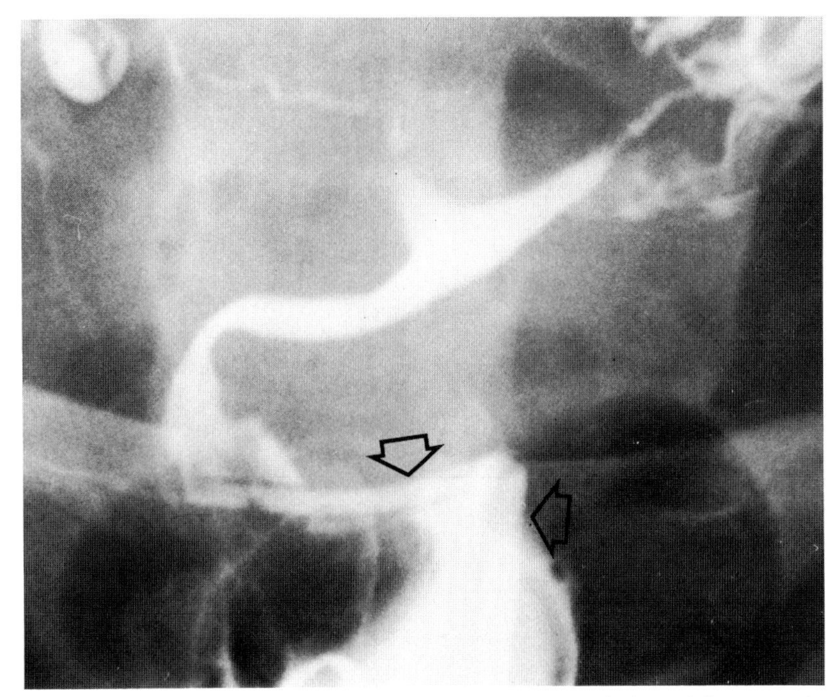

Figure 2.4. Kidde cannula (*open arrows*) in place. Note the angulation of the cannula, permitting introduction of contrast despite unfavorable alignment of the cervical os.

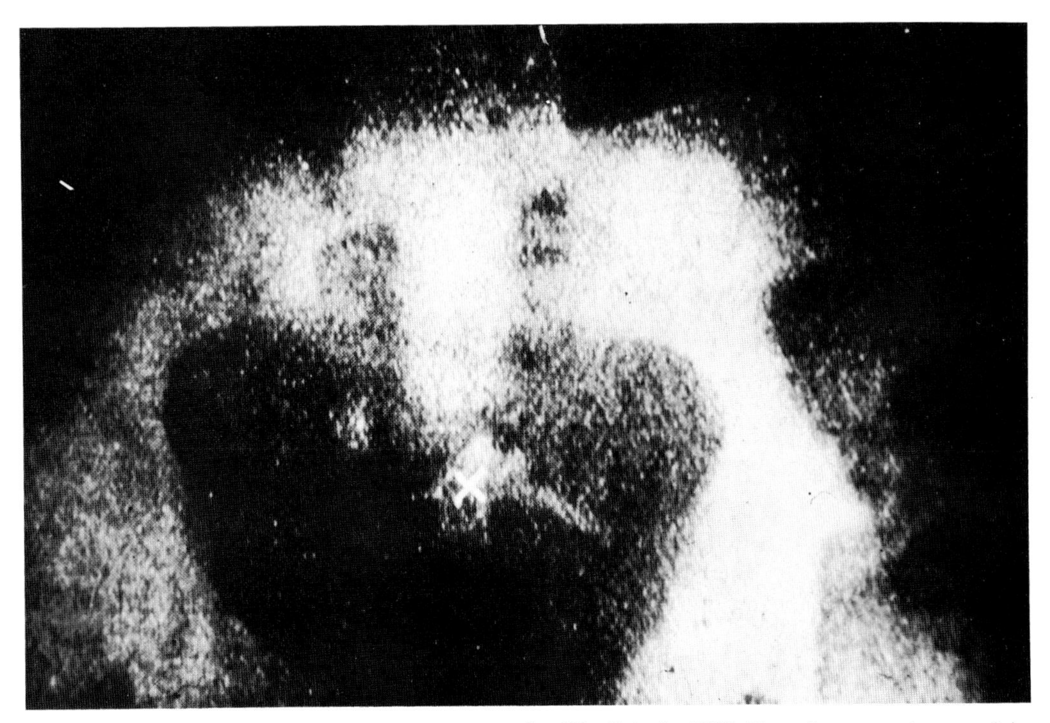

Figure 2.5. Earliest known attempt at hysterography (Rindfleisch, 1910). Bismuth suspension was introduced into the uterine cavity and radiographs were obtained. The *X* marked the faint visualization of the uterine cavity. (Courtesy of David B. Spring, San Francisco, California.)

ation of detailed observation in the uterine cavity and fallopian tubes, and a perceived increase in the level of safety. The oil-soluble agents continued to be used, and in the 1950s, Ethiodol was introduced as an ethyl alcohol ester of poppy seed oil with 37% iodine (Fig. 2.6) (5).

Numerous agents have subsequently been developed (6). Prior to 1981, Salpix (sodium acetrizoate and polyvinyl pyroladone, Ortho Pharmaceutical) was the agent of choice of most hysterosalpingographers. When manufacture of this agent was discontinued, several other contrast agents were used with satisfactory results. A frequently used alternative, Sinografin (Squibb), had physical properties quite similar to Salpix but appeared to cause an increased level of postprocedure discomfort. Work in our laboratory suggested that methylglucamine iothalamate (Conray 60, Mallinckrodt Pharmaceutical) was a satisfactory agent, demonstrating reduced patient discomfort, ease of injection, rapid resorption, and excellent visualization (7). It seems likely that methylglucamine as the contrast cation is probably superior to the sodium salts of iothalamate or diatrizoate. The nonionic low osmolar agents have been tested and, although equally satisfactory, seemed to offer no advantages. When these newer agents become commercially available, cost factors are expected to be a discouraging factor. Currently the methylglucamine salt of diatrizoate (Hypaque 60, Winthrop Breon; Renografin-60, Squibb Pharmaceuticals) is another widely used and efficacious agent.

The optimal contrast agent should be sufficiently viscid to be introduced easily and with control. Rapid resorption is deemed an advantage. The iodine content is the most important factor for optimal visualization of the uterus and tubes. Excessively high iodine concentration may mask small filling defects or irregularities within the uterine cavity. Further, the more concentrated agents are probably more irritating to the peritoneal cavity. Dilute contrast agents, on

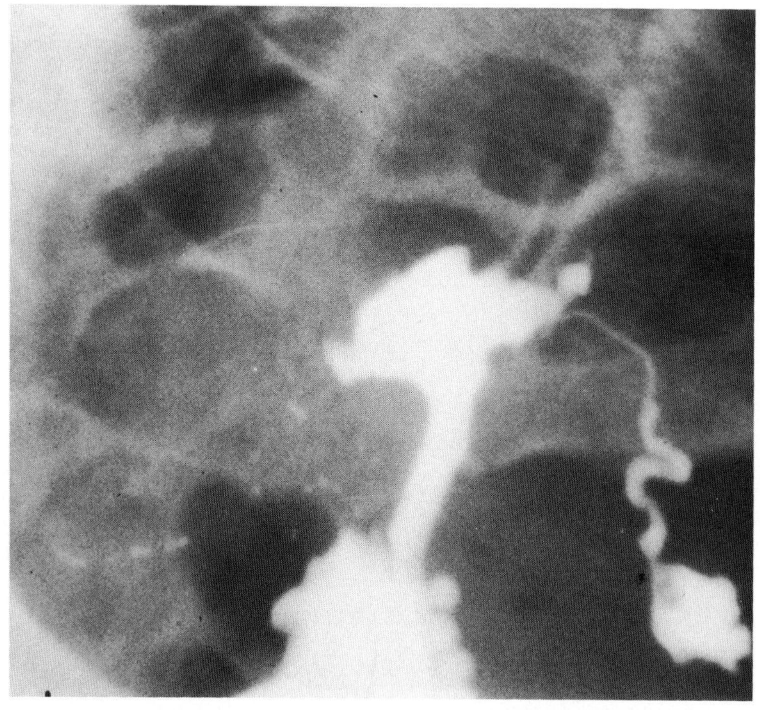

Figure 2.6. Hysterosalpingogram utilizing Ethiodol, an oil-soluble contrast, in a patient with uterine cavity deformities secondary to maternal exposure to diethylstilbestrol during gestation. There is less mucosal detail and a tendency toward droplet formation.

the other hand, make visualization difficult, particularly as regards the fallopian tubes. The role of viscosity in pain production is uncertain; data in the literature conflict (6, 7). Evidence has continued to mount that the oil-soluble contrast agents are as safe as the water-soluble media. The use of Ethiodol is accomplished with less discomfort than the use of a water-soluble contrast medium. Moore (8) compared the oil-base contrast Ethiodol to the aqueous-base contrast Salpix and found that the amount of pain associated with early steps during the procedure was identical for each medium. Pain experienced later was markedly different, with virtually no discomfort identified on spread of Ethiodol in the peritoneal cavity, in contrast to the frequently severe pain experienced as the Salpix contacted the peritoneal surface.

CONTRAST STUDIES IN CONTRAST MEDIA–REACTIVE PERSONS

A history of allergy to iodine, or of a previous reaction to contrast media, is occasionally found in an infertility patient in whom hysterosalpingography is indicated. Since intravascular intravasation is common with hysterosalpingography, the procedure cannot be considered risk-free. Appropriate measures should be taken before the procedure to prevent an anaphylactic or other allergic reaction.

Even with a past history of iodinated contrast reaction, it is impossible to predict an adverse reaction with accuracy. Evidence shows that 75% of patients with past experience of contrast reaction will tolerate administration of contrast media with impunity (9). Nonetheless, careful consideration of the indications of the procedure should be reviewed before undertaking the administration of any contrast agent.

Accumulating data demonstrate an allergic-like basis for these adverse contrast reactions. Demonstration of antibody formation (10) and histamine liberation (11) during such episodes has led to the practice of premedicating with corticosteroids those patients considered at high risk for such a reaction. An appropriate regimen uses 150 mg/day of prednisone, or its equivalent, orally, in divided doses, the day preceding and the day of the examination (12). It is prudent to have an intravenous infusion in place and an anesthesiologist available prior to the onset of contrast administration. Such precautions can permit performance of the hysterosalpingogram with relative safety for the patient. We have, at the time of this writing, not experienced a major reaction in our patient population.

Incidentally, there is some evidence suggesting that the low osmolar (nonionic) group of contrast agents may have a lesser incidence of reactions in high risk patients (13). When these agents become available, their use in this situation may be reasonable.

The Procedure of Hysterosalpingography

TIMING OF HYSTEROSALPINGOGRAPHY

Hysterosalpingography is performed during the follicular or proliferative part of the cycle, after menstruation has ceased and before ovulation has occurred. This "window" betwen cycle days 7 and 14 is chosen to avoid potential problems. For example, if hysterosalpingography is performed after ovulation, an early fertilized oocyte might be "blown out" of the tube in a retrograde fashion, leading to the possibility of ectopic gestation. Hysterosalpingography performed late in the secretory phase might dislodge secretory endometrium that is about to break down and desquamate; this could occlude the tubal ostia, and force menstrual blood in retrograde fashion into the peritoneal cavity. This is thought to predispose to the formation of endometriosis, and also may activate pelvic infection. Finally, preovulatory hysterosalpingography eliminates any possibility of radiation to a fertilized egg, and the exposed oocyte is still at a relatively protected stage. The final process of oocyte maturation, beginning with the reinitiation of Meiosis I, occurs with the luteinizing hormone surge; hysterosalpingography with its necessary radiation exposure should be avoided during the luteal phase when the oocyte is in a less radiation-resistant condition, and therefore should be done before ovulation has occurred.

It is extremely important that the hysterosalpingographer be knowledgeable in both the technique of the procedure and the history and physical findings of the patient. The examination needs to be tailored to the patient's clinical status. The patient with a history of recurrent early fetal wastage is more likely to have a uterine than tubal abnormality, so close attention will need to be paid to the appearance of the uterine cavity and internal cervical os. A close working arrangement between the gynecologist and radiologist is needed to assure optimal patient care and efficient utilization of time. When possible, interpretation of the films is done with both specialists in attendance.

PERFORMING THE PROCEDURE

Before bringing the patient into the room and preparing her for hysterosalpingography, it is important for the operator to check the sterile instrument tray. In particular, the selected cannula should be inspected and then filled with contrast medium to eliminate air bubbles. All stopcocks and valves should be turned, and the availability of ancillary equipment and drugs verified. It is always disconcerting for the operator, and potentially detrimental to the patient, to require the use of an unavailable sound, dilator, tenaculum, different-sized acorn tip, oil-base contrast medium, Foley catheter, or glucagon. Such needs invariably arise in the middle of the procedure, so a moment spent checking the instrument tray beforehand will save time and avoid aggravation during the procedure.

Knowledge of the internal pelvic anatomy is essential before performance of hysterosalpingography. The position of the uterus, whether anteflexed or retroflexed, should be ascertained, and it is helpful to know whether or not the uterus is mobile. The absence of adnexal masses or cul-de-sac abnormalities should be verified before the procedure is accomplished.

Following bimanual examination, which can be performed on the x-ray table, the patient is placed in some variation of lithotomy position. The "frog leg position," with the buttocks on several folded towels and the heels resting on the table, is convenient, because stirrups are not required and the patient can move easily for oblique or other special views. Knee rests supporting the bended knee or stirrups for the heels may be employed, but depending on the equipment used these may prevent moving the patient for oblique views.

Once the lithotomy position has been assumed, visualization of the perineum with an operating light, a head lamp, or even a gooseneck lamp should be accomplished. Good lighting is needed to visualize the cervix and the external cervical os, and to identify possible vaginal abnormalities such as may be seen with a diethylstibestrol-exposed patient.

A speculum is used for visualization of the cervix. The speculum may be metal or plastic, and double bladed, with or without a side opening. The disadvantage of a double-bladed metal speculum is that the speculum must be disarticulated and removed before the films are taken. The plastic speculum may remain within the vagina during the procedure. However, some patients complain that the plastic vaginal speculum is uncomfortable, and operators unfamiliar with its use may have considerable difficulty removing it from the vagina without causing uncomfortable distention. Some plastic specula have an attached light, of considerable advantage in visualizing the cervix and its environs. A side-opening bivalve speculum, a favorite of some practitioners, may be removed without being disarticulated and can be easily reinserted in the vagina should the need arise.

Following cervical visualization under adequate lighting, the cervix is cleaned with a disinfectant. An iodine solution, not iodine soap, is entirely satisfactory, with one precaution. Excess solution should be removed with a dry swab to eliminate its retrograde passage into the uterine cavity and tubes, since iodine may be a sclerosing solution.

If a postcoital test is to be performed, aspiration of cervical mucus is accomplished after the cervix has been wiped with a dry swab, and before disinfectant has been applied. The cervical mucus can be aspirated using a tuberculin syringe (without needle) that can be capped for later examination of the mucus. The operator will often perform this simple operation before hysterosalpingography. Importantly, two tests may thus be accomplished on one day, saving time and cost for patient and physician.

Following cervical visualization, mucus aspiration, and disinfectant preparation of the cervix, the method for instillation of contrast medium is applied. If the Kidde cannula with the flexible polyethylene tip is used, a tenaculum must be placed on the anterior cervical lip. The single tooth tenaculum is convenient, causes minimal spotting and bleeding when removed, and is associated with only a minor cramp with application. A sufficient "bite" of anterior cervical lip is essential, because a single tooth tenaculum may tear through cervical tissue if a great deal of traction is applied. A relatively large piece of anterior lip is grasped with a single tooth tenaculum; if a double or triple tenaculum is used, less of a bite is necessary, but bleeding and discomfort are likely to be enhanced. The Kidde cannula, which has already been filled with contrast medium and has the syringe attached, is next applied to the external os. The cannula is locked in place, the speculum is removed as necessary, and the patient positioned for fluoroscopic observation.

We find it appropriate in many cases to bring the husband into the room, to support his wife and to observe the remainder of the procedure. A TV monitor is positioned so that all can view the contrast material as it outlines the uterine cavity and fallopian tubes.

Under fluoroscopic monitoring, the contrast is injected slowly and with even pressure. At this point, it is unlikely that significant traction on the tenaculum is needed. The contrast flows through the endocervical canal and fills the uterine cavity. About 1–1.5 ml of contrast medium may be introduced into the uterine cavity before significant cramping occurs. The operator must be ready to take a "spot" film to record any intrauterine pathology, because the contrast fills this small (potential) space rapidly, even with a very slow injection. The operator next watches the contrast begin to fill the fallopian tubes. Abnormalities are noted, and the first interval film is taken when either the first abnormality is observed or contrast begins to spill from the fimbriated ends of the tubes. Excessive contrast will obscure internal architecture of the uterine cavity, and will not allow optimal delineation of tubes. We strive to reduce the total number of films taken, and to perform the procedure with rapidity to lessen the total radiation exposure to patient and personnel. Once fallopian tube filling has been accomplished, it is valuable to move the uterus using the combined tenaculum and cannula. This allows evaluation of tubal mobility and may cause the fimbriated parts of the tubes to move away from the pelvic sidewall.

Carbon dioxide may be introduced after the contrast material has been used. This is primarily of value when tubal patency is in doubt. Occasionally spill from one tube may be such as to mask determination of patency of the contralateral tube. Fluoroscopic observation of carbon dioxide passage may resolve this question. The carbon dioxide may also enhance visualization of intrauterine abnormalities. This technique does introduce gas into the peritoneal cavity, and is therefore associated with an increase in pain perceived to be in the shoulder, the result of irritation of the diaphragm by the intraperitoneal gas. Absorption of carbon dioxide occurs relatively rapidly, so the shoulder pain is self-limited and ordinarily disappears within 12 hours.

When abnormalities are encountered, modifications of the technique must be accomplished to allow further delineation of the perceived abnormality. For example, the finding of intrauterine filling defects or possible synechiae may necessitate turning the patient, introducing carbon dioxide, or changing some aspect of the technique to allow a better view of the area in question. The same applies to delineating tubal abnormalities, and it may be necessary to use oblique films. The simple trick of asking the patient to cough may help disperse contrast material. Occasionally, elevation of the hips, or briefly assuming a sitting position may help with delineation. Flexibility of technique is useful, which is why we do not use knee rests or stirrups. Significant stenosis of the external cervical os may impede the introduction of the cannula, tempting undue pressure to force the cannula into the canal. Dilatation may be appropriate, under these circumstances. We have been successful with a simple 7 or 8F angiographic dilator for both dilatation and injection of the contrast agent (Fig. 2.7).

Delayed films may add additional information, but the need for further films should not be routine. For example, dispersion of loculated contrast may be identified, which indicates tubal patency that may not have been documented earlier. Uni- or bilateral tubal occlusion should almost always suggest the need for a delayed film.

Cornual occlusion presents an interesting challenge. Because of the muscular arrangement in the uterine cornua, uterine contraction can result in occlusion of the tube. Although this is spasm of uterine musculature and not of the tube, it may prevent the passage of contrast material into the salpinx, and give the impression of an anatomic occlusion. Before making the diagnosis of cornual occlusion, an attempt should be made to relax the spasm. The general anesthetic halothane can efficiently relax uterine musculature, but is not usually available in the setting desired. A number of drugs have been employed for this purpose, including atropine, nitrogly-

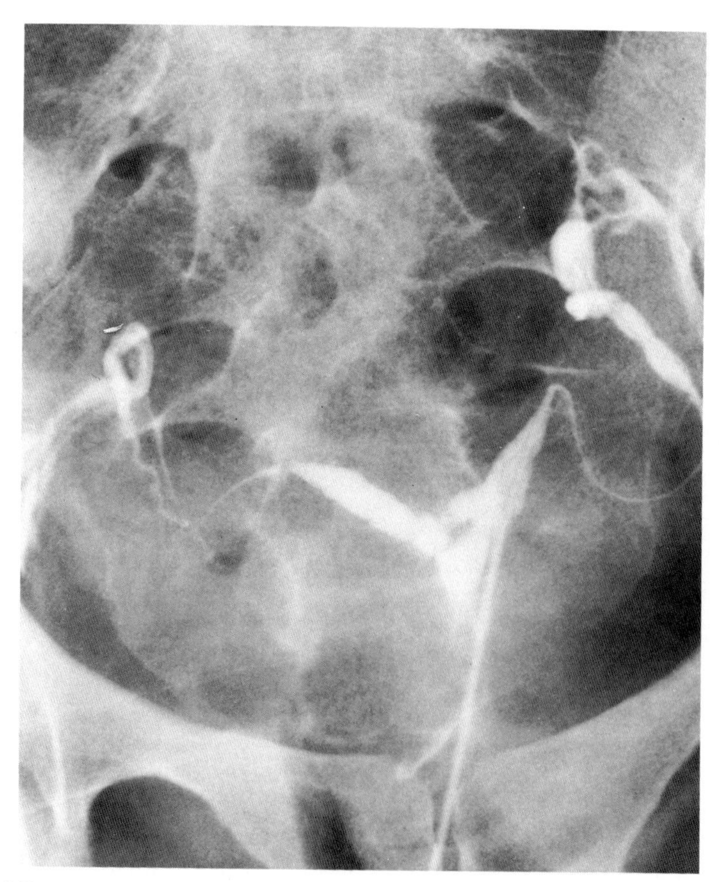

Figure 2.7. Use of vascular dilator for cannulation in patient with cervical stenosis.

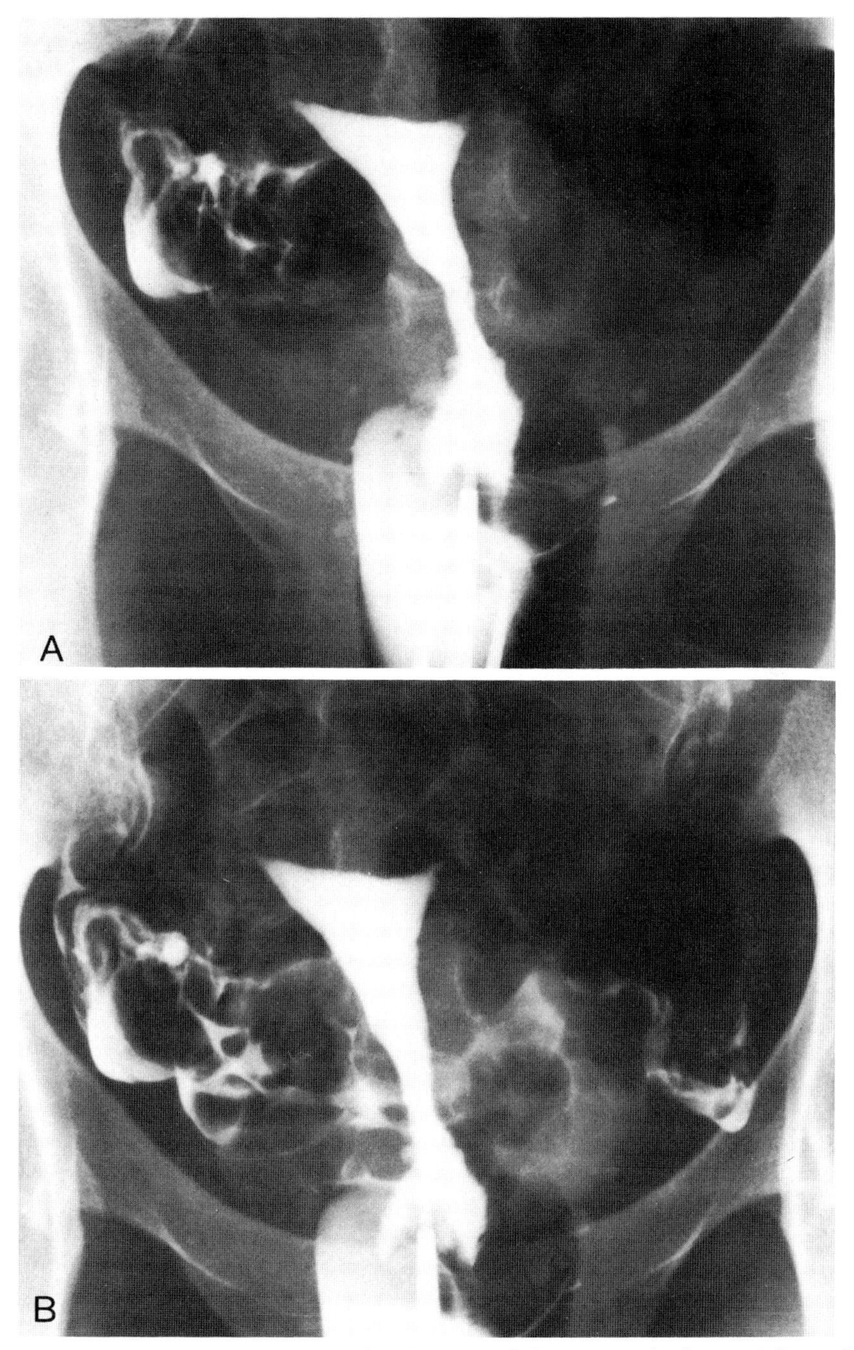

Figure 2.8. *A:* Cornual occlusion (*left*) secondary to spasm. *B:* Same examination as *A*. Several seconds following the intravenous injection of glucagon, there is ready filling of the left fallopian tube.

cerin, various analgesics and narcotics, tranquilizers, and isoxsuprine; glucagon appears to be the most promising drug available. Glucagon has been investigated and found to be a moderately efficient agent in inducing relaxation of tubal spasm (Fig. 2.8). This straight-chain polypeptide is produced normally in the pancreas, and, when given parentally, causes an increase in blood glucose and a relaxation of smooth muscle. One-third of those patients presenting with cornual occlusion were demonstrated to have patent, normally filled fallopian tubes after the intravenous admininstration of 1 mg of glucagon (14).

FOLEY CATHETER METHOD FOR HYSTEROSALPINGOGRAPHY

A pediatric Foley catheter has been used to advantage in hysterosalpingography, primarily because it eliminates the need for use of a tenaculum on the external os, minimizes cervical laceration or bleeding during the procedure, avoids some associated discomfort, and allows more maneuverability than a rigid metal cannula (Fig. 2.9).

The preparation of the patient does not differ from that previously described, and the sterile instrument tray is similar, with the addition of a number 8 or 10 Foley catheter. A uterine dressing forceps or sponge stick is used to introduce the catheter tip into the external os. The tip ordinarily slides easily, finding its own direction; there is usually no need to grasp the cervix with a tenaculum to straighten the canal. The catheter proximal to the balloon has enough rigidity to allow the tip to enter the uterine cavity, and only those patients with severe stenosis due to a previous traumatic procedure may require dilatation or the use of a silver wire probe to enable placement. Following insertion, the balloon of the Foley catheter is inflated with 1–1.5 ml of water, and downward traction is used to occlude the uterine side of internal cervical os. The balloon must be adequately distended to avoid catheter expulsion into the endocervical canal during the procedure, but not so much as to cause undue discomfort to the patient. The vaginal speculum can then be removed, and radiographic contrast mate-

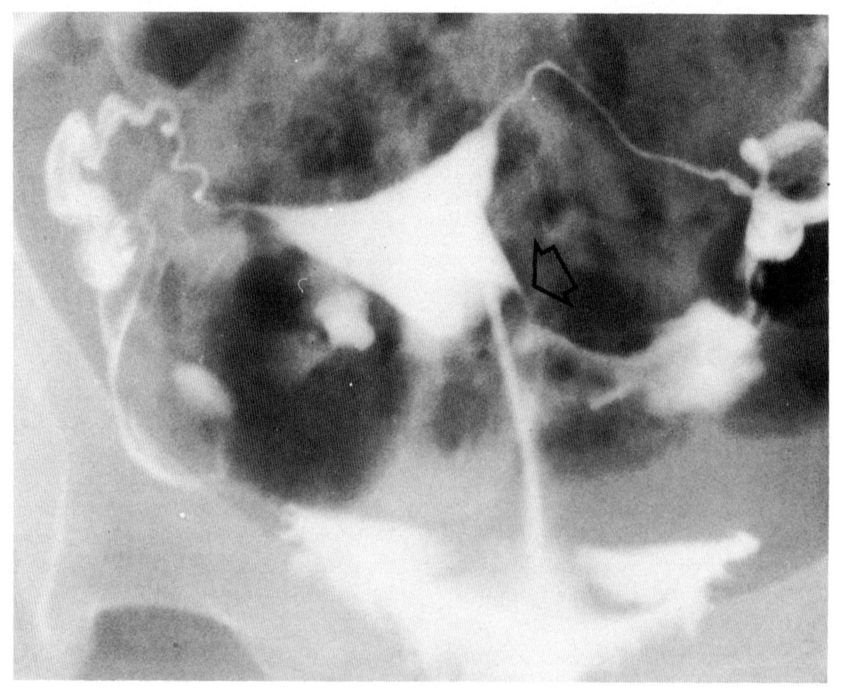

Figure 2.9. Hysterosalpingogram utilizing a pediatric Foley catheter. Balloon (*arrow*) occludes lower portion of the endometrial cavity and prevents any filling of the endocervical canal.

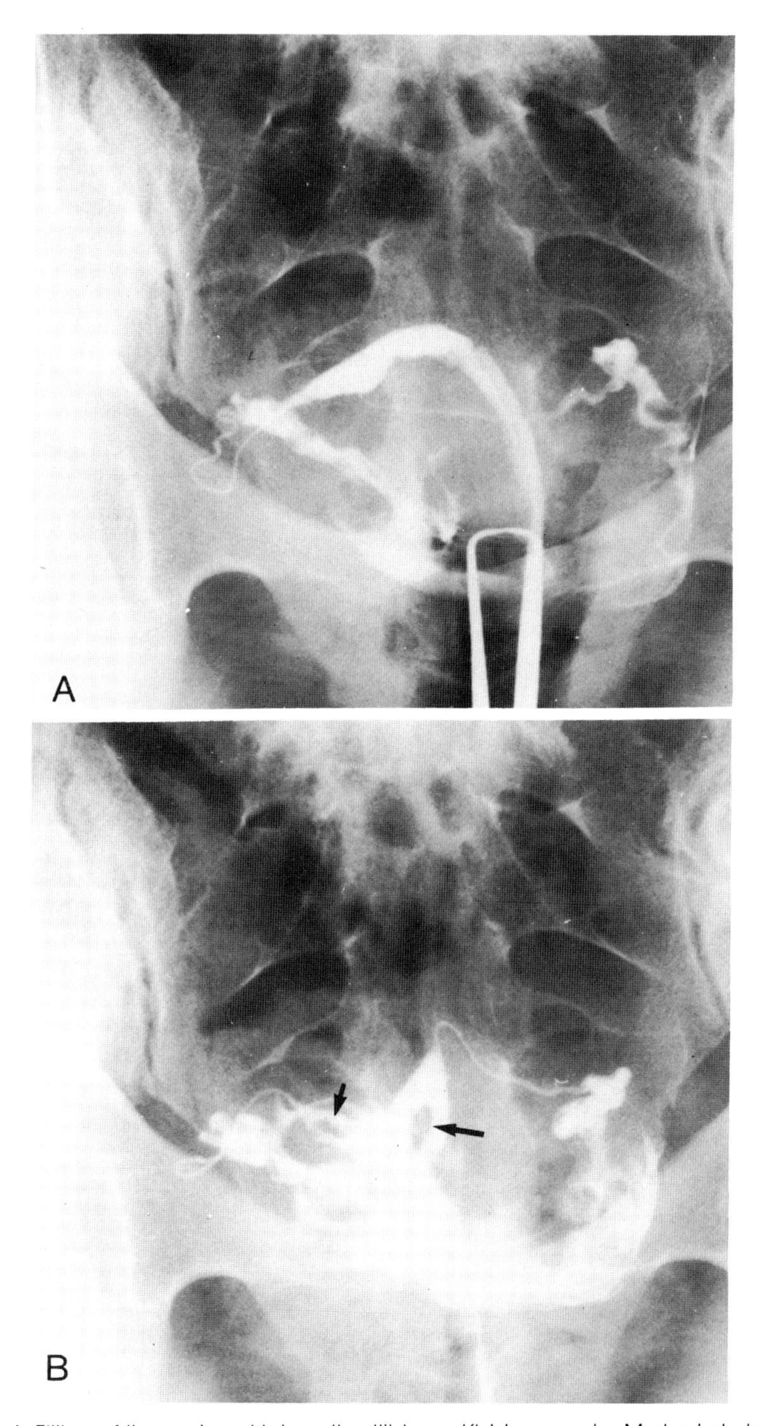

Figure 2.10. *A:* Filling of the endometrial cavity utilizing a Kidde cannula. Marked uterine flexion limits the ability to adequately visualize the uterine cavity. *B:* Same examination as in *A*. Traction applied via the tenaculum straightened the uterine cavity, permitting better visualization and allowing demonstration of two uterine synechiae (*arrows*).

rial injected through the catheter. The prone position may permit easier filling of uterus and fallopian tubes by utilizing their dependency, if the uterus is anteflexed (15). The supine position would be ideal for the women with a retroflexed uterus. The technique of hysterosalpingography proceeds routinely, and interval films are taken when appropriate. Following the procedure, the bulb is deflated, either by cutting the catheter or aspirating the water from the balloon, and the catheter is easily removed.

One disadvantage of the Foley catheter technique is the failure to visualize the endocervical canal and the lower uterine segment. A second disadvantage is the inability to move the uterus in the absence of a tenaculum. Adequate traction applied during filling of the uterus results in straightening of this organ and improved visualization of the uterine cavity (Fig. 2.10), which is impossible with the catheter technique because it invariably pulls the Foley bulb down the canal with considerable discomfort for the patient or totally dislodges the catheter. Filling defects are more readily detected when using the cannula technique and traction, and one can often avoid the need for extraneous oblique radiographic projections, thus reducing radiation exposure. Pushing the uterus up and then exerting traction may allow assessment of tubal mobility and the ability of the fimbria to move away from the pelvic side wall.

Complications of Hysterosalpingography

Some discomfort with hysterosalpingography is unavoidable. Use of a tenaculum adds one additional cramp when the tenaculum is placed on the anterior lip of the cervix. Introduction of anything into the uterine cavity causes uterine contraction and discomfort. The passage of contrast medium through the fallopian tube may be associated with some pain, which is maximized by any distention of the tube due to distal obstruction.

Mechanical complications of hysterosalpingography include uterine perforation or tubal rupture, both of which are very rare. Uterine perforation is unlikely to occur with the use of a polyethylene acorn with a flexible tip, but has been reported with metal instruments, particularly those of the Jarcho type. This cannula has a metal end with a rubber acorn fitted over the end and held in place with an adjustable-screw band. If the screw is loosened the band could be dislodged, allowing the metal cannula to slip farther through the rubber tip and extend into the uterine cavity. If this happened suddenly and with force, perforation could occur (Fig. 2.3).

Traumatic elevation of the endometrium by the cannula insertion may occur, usually without significant consequence (Figs. 2.11 and 2.12).

Overdistention of a hydrosalpinx can cause tubal rupture, and is accompanied by significant pain. A closed hydrosalpinx, opened by distention in this manner, can result in activation of pelvic inflammatory disease.

Complications of hemorrhage and shock are uncommonly encountered; no large vessels are likely to be encountered. Tearing of the tenaculum through cervical tissue is unlikely to result in hemorrhage, although the friable diethylstilbestrol-exposed cervix may bleed freely. Anaphylactic shock has rarely been reported; the performance of hysterosalpingography in patients thought to be allergic to iodine-containing contrast media or to other agents used in the procedure is discussed earlier in this chapter.

Lymphatic or venous intravasation of either water-soluble or oil-base contrast medium is a dramatic event. Lymphatic intravasation with water-base contrast results in a reticular pattern of small vessels in the broad ligament. With venous intravasation, the medium passes quickly through the uterine and/or ovarian veins to the lungs, delineating clearly the vascular architecture. If water-soluble contrast medium has been used, the medium is quickly dissipated, and no embolic symptoms or side effects have been reported. On the other hand, if embolization occurs with an oil-soluble contrast medium, the result may be

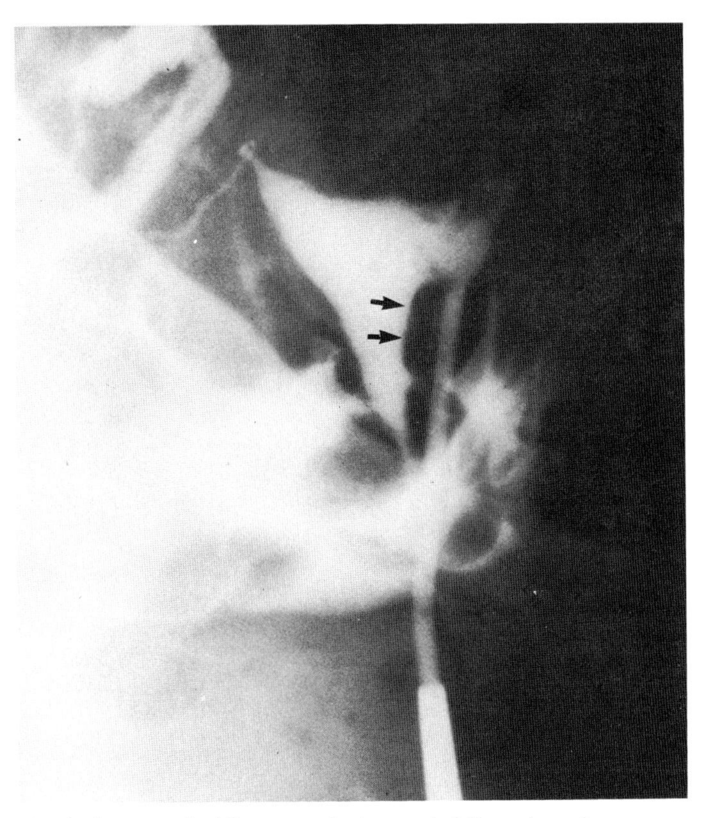

Figure 2.11. Subintimal placement of the cannula (*arrows*). Although such occurrence is unusual and generally without incident, significant bleeding may ensue.

detrimental to the patient. Intravasation has been reported to occur in between 0 and 6.3% of patients, and certain predisposing conditions may allow this to happen with increased likelihood. These include tubal disease and/or obstruction, recent uterine surgery, uterine malformation, synechiae, malplacement of the hysterosalpingogram cannula, and excessive injection pressure or quantity of the contrast medium. Most oil emboli are innocuous and should not be regarded as a major complication of hysterosalpingography (16). However, there have been 5 deaths attributed to emboli after hysterosalpingography, 4 of which were with oil-base media (17). Two of these deaths followed major operations performed after the hysterosalpingogram, and the radiographic procedure may only have been contributory. The fifth embolus-related death was in 1959, and occurred after the use of a water-soluble contrast medium containing carboxymethylcellulose. The 6 other deaths found in the world literature and attributed to hysterosalpingography have all been related to infection.

Intrauterine pregnancy, when recognized, is another major contraindication. Careful scheduling, prior to the anticipated date of ovulation, generally avoids this possibility. It is also helpful to verify the date of the last menstrual period and the cycle day with the patient, to look at her temperature chart to verify a low temperature, and to pay attention to the appearance of the cervical mucus. Hysterosalpingography has been performed inadvertently in the presence of an intrauterine pregnancy, but the occurrence is highly unlikely. Nonetheless, Wilson et al. (18) performed hysterosalpingograms during infertility investigations in 10 women who had apparently conceived, finding intrauterine filling defects in 2; normal term infants were delivered in all 10. The concerns in the event of such an occurrence include disruption of

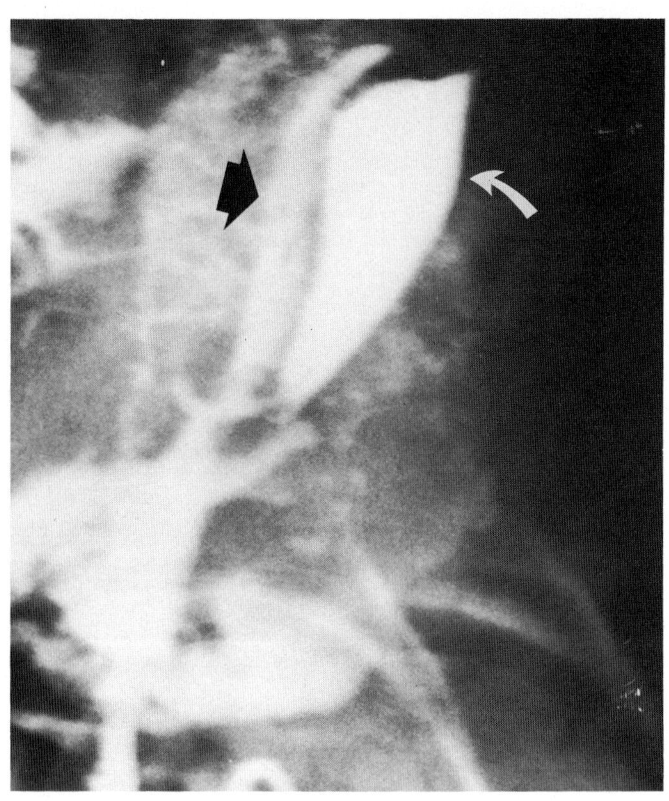

Figure 2.12. Intramural placement of the cannula has resulted in false passage and a second collection of contrast (*arrow*) posterior to the true uterine cavity (*curved arrow*).

the pregnancy, displacement of the fertilized ovum into the peritoneal cavity, and the mutative effect of the radiation on the developing fetus.

Genetic hazards associated with hysterosalpingography are related to the irradiation encountered. The amount of ovarian radiation from a hysterosalpingogram depends upon the technical equipment, the number of films obtained, the duration of fluoroscopy, the distance of the tube to the film, and the size of the patient. Using a two-film technique without fluoroscopy, Shirley (19) found that a posterior fornix dosimeter recorded a mean of 129 mrad, which increased to 1053 mrad when fluoroscopy was used. Using a different technique, Sheikh and Yussman (20) found the gonadal radiation dose to be between 75 and 550 mrad, depending almost solely on the duration of fluoroscopic time. No deleterious fetal effects from low doses of radiation received in pregnancy have been proven by epidemiologic investigations. Goldenberg and coworkers (21) evaluated the health of the infants of 26 women who had undergone hysterosalpingography during the cycle of conception; these children were all healthy and free of any congenital defects.

Another potential disadvantage of hysterosalpingography is the possibility of postprocedure infection. Reports suggest that the incidence of pelvic inflammatory disease following hysterosalpingography may be anywhere from 0.5 to 6%, depending upon the technique, population base, and criteria of diagnosis. It is likely that these infectious episodes are usually due to the reactivation of pre-existing disease rather than a de novo infection, and they are more frequently seen in patients demonstrating underlying abnormal tubal inflammatory disease. When such an appearance is encountered, prophylactic antibiotic administration is appropriate (see

Chapter 8). The presence of active pelvic infection is an absolute contraindication to hysterosalpingography; peritonitis following hysterosalpingography was reported to be responsible for 6 of the 11 deaths in the world literature (22, 23).

Chemical and toxic complications of hysterosalpingography include the formation of granulomas following oil-based contrast use, allergic reactions, and iatrogenic misdiagnosis of thyroid dysfunction due to the iodine contained within contrast media. Recently Beyth and coworkers (24) have recommended the use of oil-soluble contrast media followed by the instillation of a physiologic solution until no intraluminal contrast is seen, to avoid intratubal granuloma formation.

Pregnancy Rates Following Hysterosalpingography

Weir et al. (25) in 1957 compared the therapeutic effects of a carbon dioxide insufflation procedure (the Rubin test) and oil salpingography in patients with normal tubal patency. There was no difference pre- and postinsufflation in terms of cumulative conception rate, but the postsalpingogram rate of conception approached 30% after three ovulations, and 45% with eight cycles. Thus, as early as 1957, an increased incidence of conception occurring for the 4 and perhaps 8 months following salpingography with an oil-base medium was appreciated. In 1980, DeCherney and coworkers (26) established that, in the 4 months following hysterosalpingography, 13% of women in whom water-base and 29% of women in whom oil-base contrast had been used achieved pregnancy. More recently, Schwabe et al. (27) compared the two types of contrast media in women with unexplained infertility, and found significantly higher pregnancy rates after hysterosalpingography with oil (77.8%) than with aqueous (10%) contrast medium, results similar to those reported by Cooper et al. (28) using the water-base medium Sinografin. Oil-base contrast did not influence pregnancy rates in any other diagnostic group, or for the group of infertility patients as a whole. A potential benefit from oil-contrast medium must be balanced against the superior tubal mucosal detail and the rapid resorption from the peritoneal cavity provided by the aqueous medium. "Therapeutic" hysterosalpingography with oil contrast might be reserved for those patients with a diagnosis of unexplained infertility, and others with no contraindications to the use of oil-base contrast.

REFERENCES

1. Owens OM, Schiff I, Kaul AF, Cramer DC, Burt RAP: Reduction of pain following hysterosalpingogram by prior analgesic administration. *Fertil Steril* 43:146, 1985.
2. Cohen BM, Katz M: The significance of the convoluted oviduct in the infertile woman. *J Reprod Med* 21:31, 1978.
3. *Medical Radiation, A Guide to Good Practice*. American College of Radiology, Washington, DC, 1985.
4. Rindfleisch W: Darstellung des Cavum uteri. *Berl Klin Wochensher* No 17(Apr):780, 1910.
5. Palmer A: Ethiodol hysterosalpingography for the treatment of infertility. *Fertil Steril* 11:311, 1960.
6. Ekelund L, Karp W: Comparison between two radiographic contrast media for hysterosalpingography. *Acta Obstet Gynecol Scand* 60:393, 1981.
7. Winfield AC, Henderson-Slayden R, Wentz AC, Harding DR: Hysterosalpingography: Comparison of Conray 60 and Sinografin. *Am J Roentgenol* 138:599, 1982.
8. Moore DE: Pain associated with hysterosalpingography: Ethiodol versus Salpix media. *Fertil Steril* 38:629, 1982.
9. Shehadi WH: Clinical problems and toxicity of contrast agents. *Am J Roentgenol* 97:762, 1966.
10. Brasch RC: Allergic reactions to contrast media: Accumulated evidence. *Am J Roentgenol* 134:797, 1980.
11. Lasser EC, Walters AJ, Lang JH: An experimental basis for histamine release in contrast material reactions. *Radiology* 110:49, 1974.
12. Zweiman B, Mishkin MM, Hildreth EA: An approach to the performance of contrast studies in contrast material-reactive persons. *Ann Intern Med* 83:159, 1975.
13. Rapoport S, Bookstein JJ, Higgins CB, et al: Experience with metrizamide in patients with previous severe anaphylactic reactions to ionic contrast agents. *Radiology* 143:321, 1982.
14. Winfield AC, Pittaway D, Maxson W, Daniell J, Wentz AC: Apparent cornual occlusion in hysterosalpingography: Reversal by glucagon. *Am J Roentgenol* 139:529, 1982.
15. Spring DB, Wilson RE, Arronet GH: Foley catheter hysterosalpingography: A simplified technique for investigating infertility. *Radiology* 131:543, 1979.
16. Bateman BG, Nunley WC, Kitchin JD: Intravasation during hysterosalpingography using oil-base contrast media. *Fertil Steril* 34:439, 1980.

17. Soules MR, Spadoni LR: Oil versus aqueous media for hysterosalpingography: A continuing debate based on many opinions and few facts. *Fertil Steril* 38:1, 1982.
18. Wilson RV, Lee RA, Jensen PA: Inadvertent infertility investigations in pregnant women. *Fertil Steril* 17:126, 1966.
19. Shirley RL: Ovarian radiation dosage during hysterosalpingography. *Fertil Steril* 22:83, 1971.
20. Sheikh HH, Yussman MA: Radiation exposure of ovaries during hysterosalpingography. *Am J Obstet Gynecol* 124:307, 1976.
21. Goldenberg RL, White R, Magendantz HG: Pregnancy during the hysterogram cycle. *Fertil Steril* 27(11):1274, 1976.
22. Siegler AM: Dangers of hysterosalpingography. *Obstet Gynecol Surv* 22:284, 1967.
23. Bang J: Complications of hysterosalpingography. *Acta Obstet Gynecol Scand* 29:383, 1950.
24. Beyth Y, Navot D, Lax E: A simple improvement in the technique of hysterosalpingography achieving optimal imaging and avoiding possible complications. *Fertil Steril* 44(4):543, 1985.
25. Weir WC, Weir DR, Littell AS: A statistical comparison of the therapeutic value of carbon dioxide insufflation versus oil salpingography. *Am J Obstet Gynecol* 73:412, 1957.
26. DeCherney AH, Kort H, Barney JB, DeVore GR: Increased pregnancy rate with oil-soluble hysterosalpingography dye. *Fertil Steril* 33:407, 1980.
27. Schwabe MG, Shapiro SS, Haning RV Jr: Hysterosalpingography with oil contrast medium enhances fertility in patients with infertility of unknown etiology. *Fertil Steril* 40:604, 1983.
28. Cooper RA, Jabamoni R, Pieters CH: Fertility rate after hysterosalpingography with Sinografin. *Am J Roentgenol* 141:105, 1983.

3 The Normal Hysterosalpingogram

The uterus is a thick-walled, muscular organ, lying totally within the true pelvis, and measuring approximately 7 × 4 × 2.5 cm in its nongravid state (Fig. 3.1). The cavity of the uterus is quite small in comparison to the overall size of the organ, primarily because of the thickness of the myometrium. In its empty state, the uterine cavity is only a potential space with anterior and posterior walls in apposition, but it is distended by the introduction of 2.0–3.0 ml of contrast to create the radiographic appearance seen on a hysterosalpingogram (Fig. 3.2).

The uterine cavity can be conveniently divided into two regions. The inferior portion, the endocervical canal, extends from the cervix to the internal cervical os. The uterine body comprises the remainder of the cavity and is superior to the internal cervical os. The fundus represents that portion of the uterus above the level of the ostia of the fallopian tubes. In the adult woman of reproductive age, the length of the uterine body is longer than the endocervical canal, while the converse is true in the child. However, radiographic measurements of the two regions are inaccurate and may vary considerably with the degree of traction used during the examination (Fig. 3.3). If one

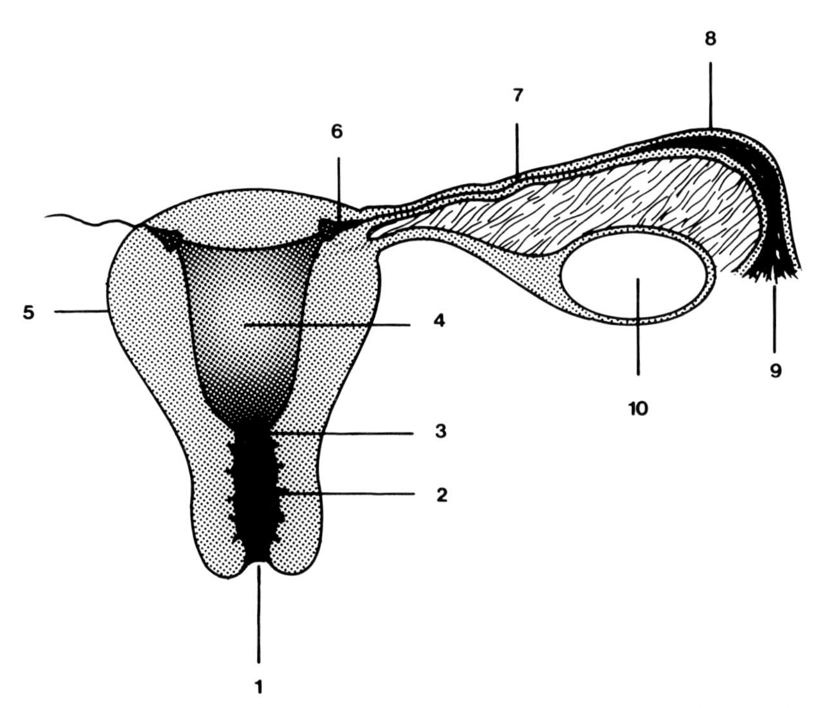

Figure 3.1. Normal uterus: (*1*) external cervical os; (*2*) cervical canal; (*3*) internal cervical os; (*4*) uterine cavity; (*5*) myometrium; (*6*) fallopian tube: interstitial segment; (*7*) fallopian tube: isthmus; (*8*) fallopian tube: ampulla; (*9*) fimbria; (*10*) ovary.

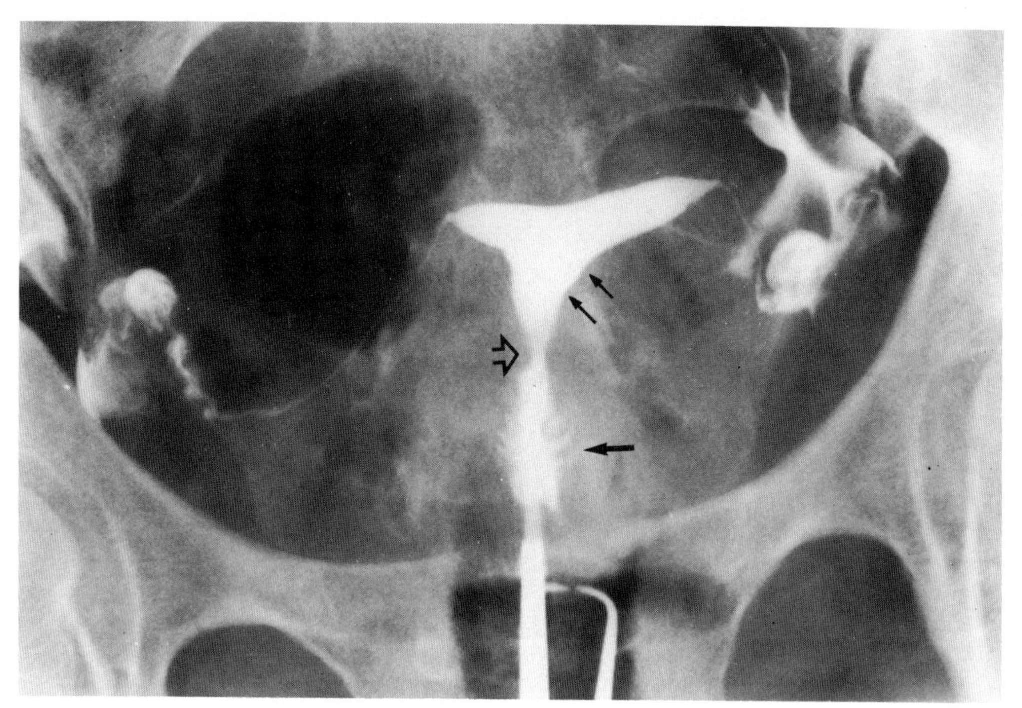

Figure 3.2. Hysterosalpingography of normal uterus. The internal cervical os (*open arrow*) is well defined. The cervical canal (*single arrow*) demonstrates coarse plicae palmatae. The uterine cavity (*double arrow*) is characterized by a relatively smooth contour.

uses an elongated cannula with its tip in the uterine cavity, or the Foley catheter technique with the balloon of the catheter in the uterine cavity, the endocervical canal may not be outlined with contrast in its entirety. For these reasons, measurements of the various components of the uterus and ratio calculations of uterine body to cervix are usually not of great diagnostic value.

The Endocervical Canal

The configuration of the endocervical canal, if visualized, is quite variable and strikingly different from the uterine cavity. Numerous mucosal folds, the plicae palmatae or arbor vitae, give an interdigitated frond-like appearance to the canal (Fig. 3.4). This appearance, most striking in the nulliparous uterus, is seen less commonly in multiparous women. The diameter of the endocervical canal varies considerably from patient to patient, and at times may appear significantly dilated in the normal uterus (Fig. 3.5). The course and direction of the endocervical canal is occasionally distorted and angulated and may in fact create technical problems in introduction of a cannula because of this irregularity. Flexible tips such as the Kidde cannula are of value in this regard and are introduced more easily if significant traction is applied to the cervix. A shortened cannula may be helpful. Dilated cervical glands may appear as diverticula-like projections of contrast in the endocervical canal (Fig. 3.6). Mesonephric remnants may produce linear filling defects that may be mistaken for endocervical synechiae or scarring (see Chapter 4). A well-defined and solitary diverticular projection in the lower uterine segment is frequently seen in patients having undergone previous lower segment cesarean section (Fig. 3.7). Mobile filling defects, artifactual in origin, due either to the introduction of air or displacement of cervical mucus, are occasionally seen (Fig. 3.8).

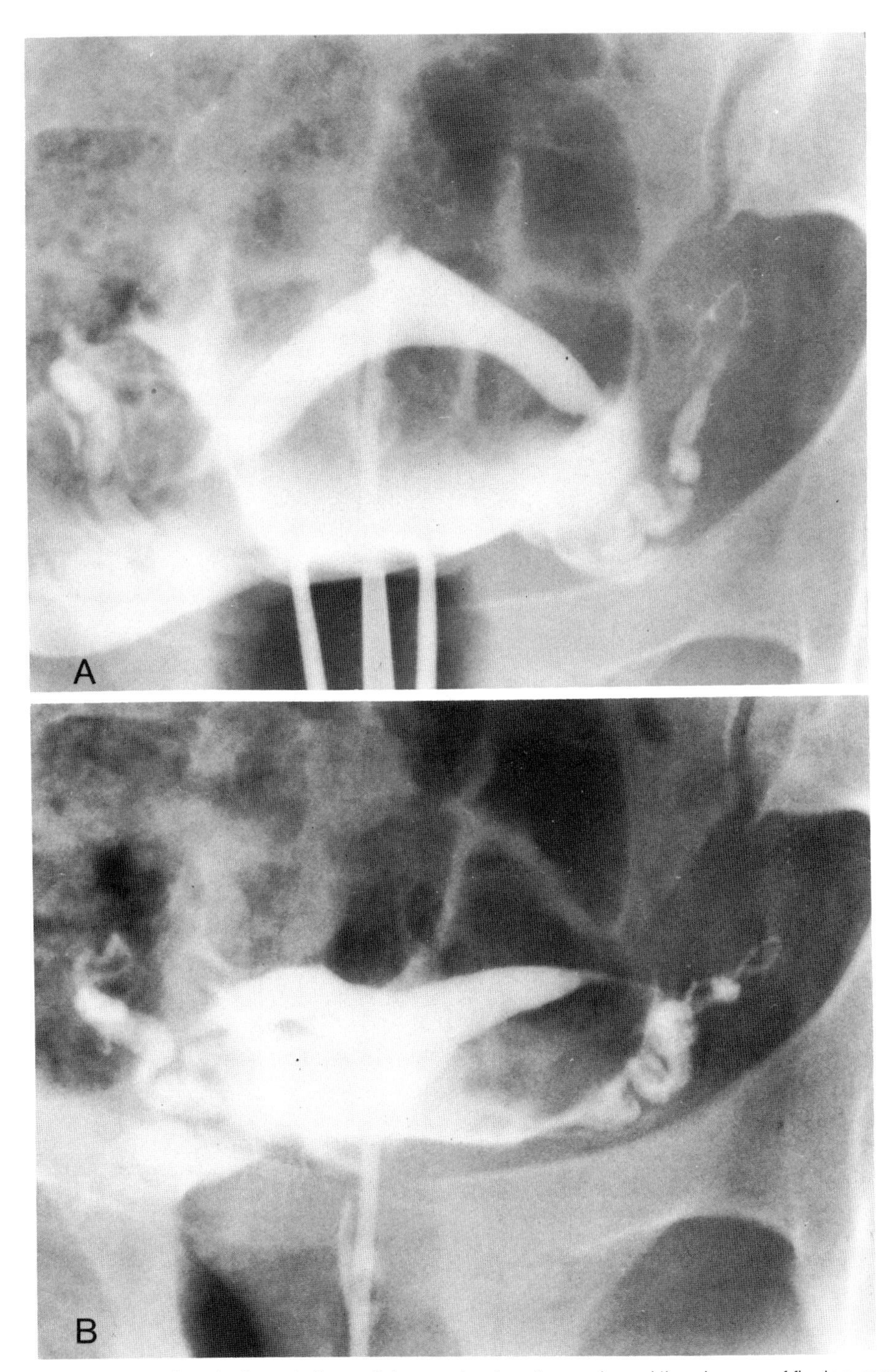

Figure 3.3. *A*: Markedly retroflexed uterus. *B*: Increasing traction reduced the degree of flexion, resulting in a differing appearance of the relationship of the uterine body and endocervical canal.

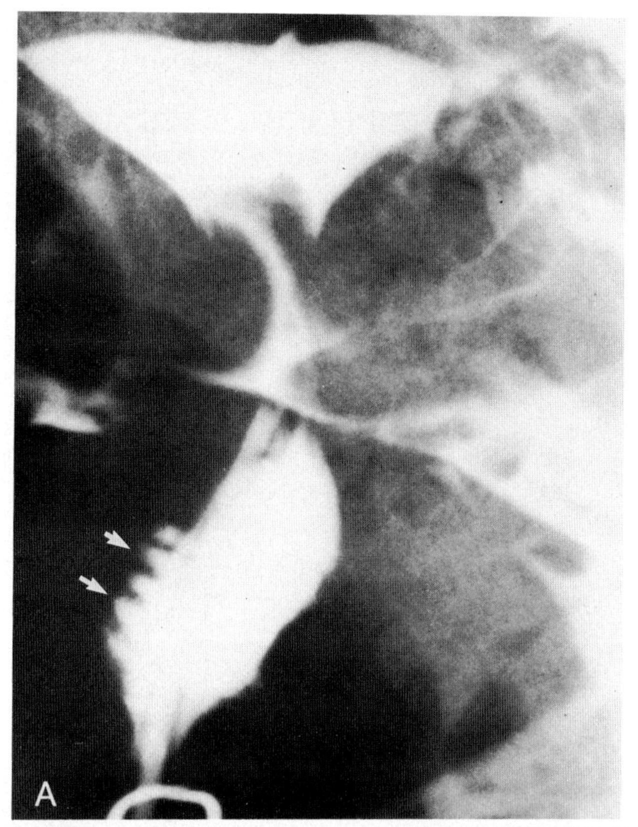

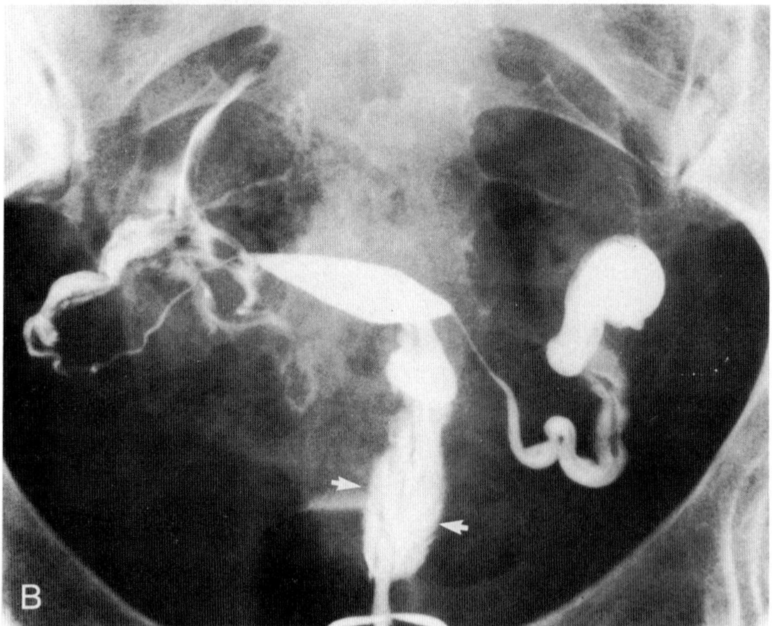

Figure 3.4. *A*: Prominent spiculated plicae palmatae are demonstrated in the cervical canal (*arrows*). Note the asymmetry. *B*: Plicae palmatae having a different pattern. A fine fern-like interdigitated mucosal pattern is noted throughout the cervical canal (*arrows*).

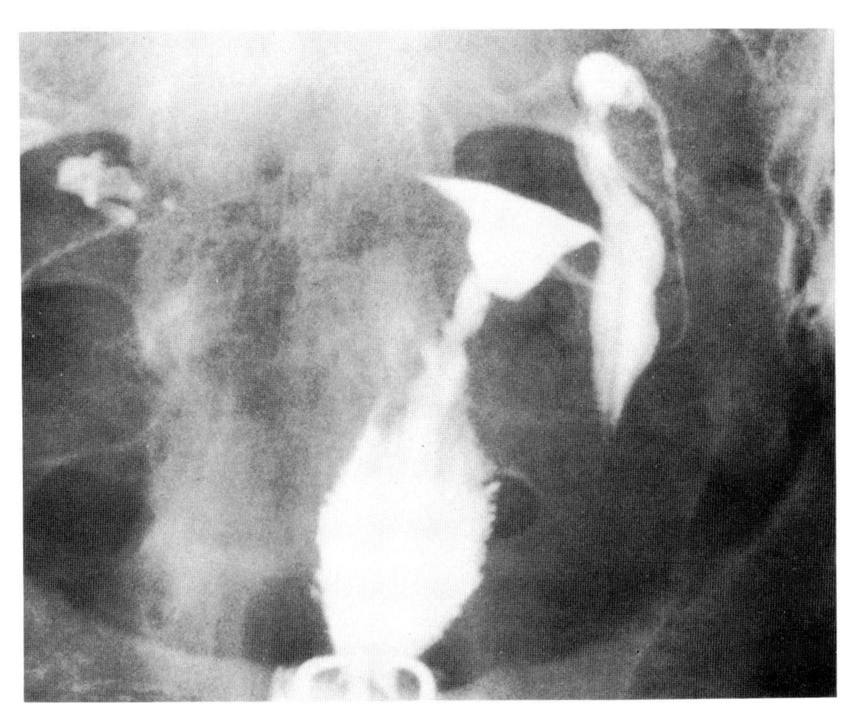

Figure 3.5. Normal uterus. Note the marked increased diameter of the endocervical canal, a normal variant. The cervical canal is longer than the uterine body in this nulliparous patient.

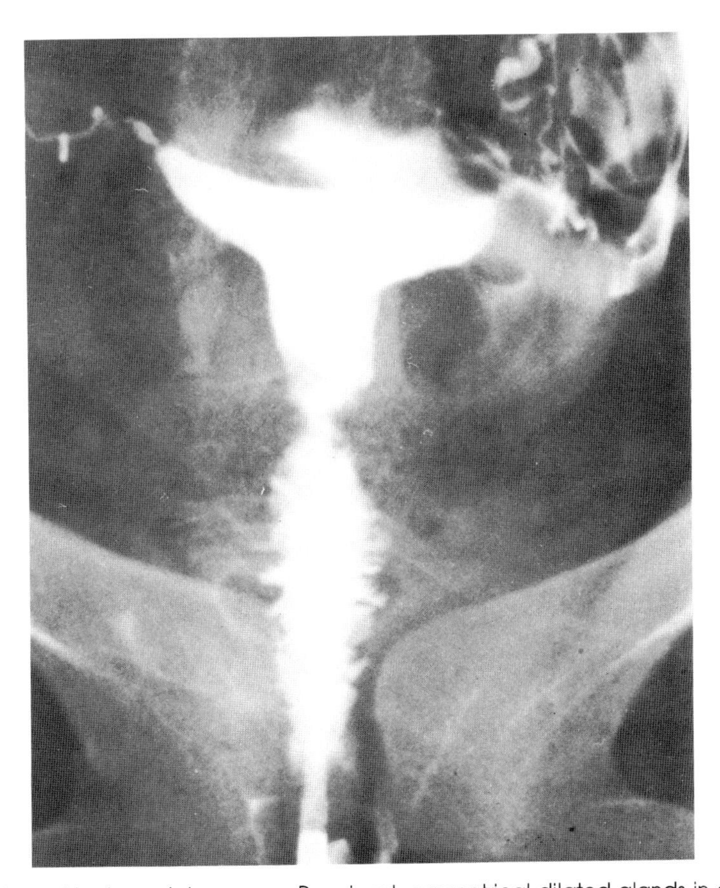

Figure 3.6. Normal hysterosalpingogram. Prominent, symmetrical dilated glands in cervical canal.

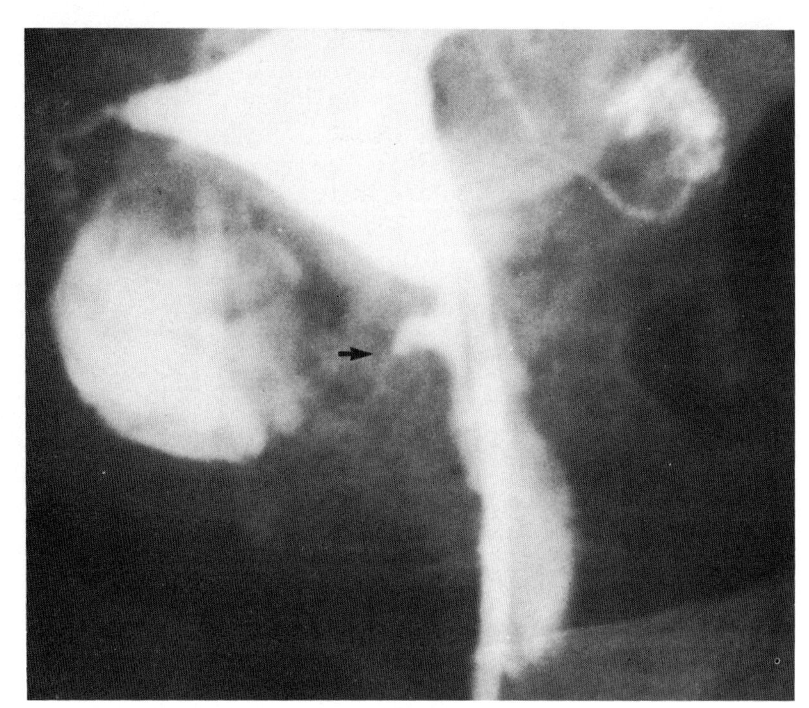

Figure 3.7. Typical pseudodiverticulum projecting from the lower uterine segment (*arrow*) in this patient with a history of previous delivery by cesarean section.

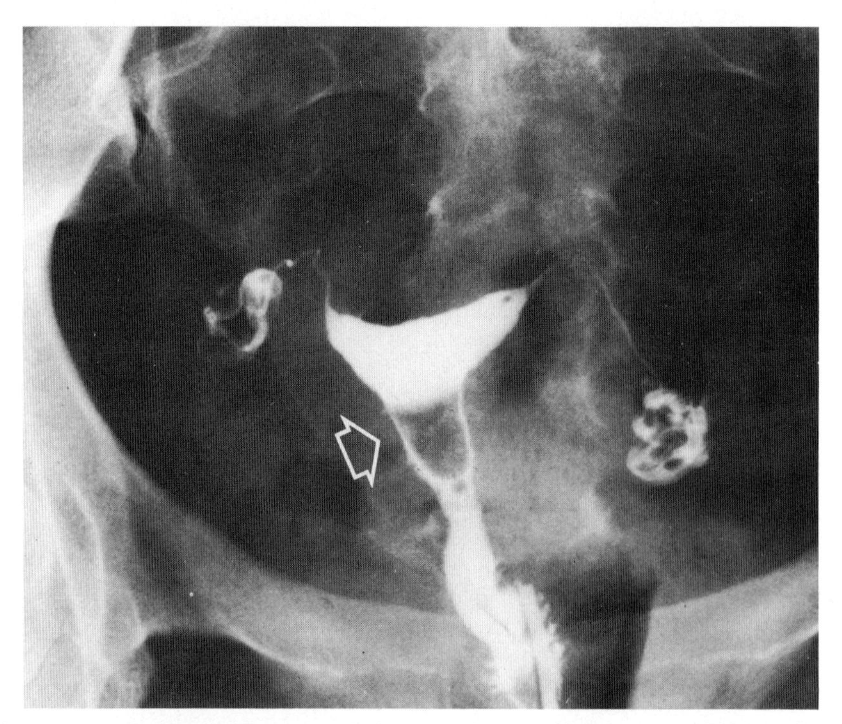

Figure 3.8. Large radiolucent filling defect (*open arrow*) caused by an air bubble introduced into the uterine cavity during hysterosalpingography.

The Internal Cervical Os

The region of the internal cervical os is also quite variable in its appearance. Some patients demonstrate a well-defined, markedly narrow internal os (Fig. 3.9). Others show virtually no definition of the internal os, having a gradual, funnel-shaped uterine cavity contour, which makes it impossible to identify the internal os with certainty (Fig. 3.10). Some authors have speculated that widening of the region of the os in this manner may be correlated with the clinical syndrome of "incompetent cervical os," which is responsible for recurrent second trimester fetal wastage. Such data are difficult to substantiate.

The diameter of the internal os may vary in the same patient during different phases of the menstrual cycle. The internal os is most narrow in the postovulatory phase of the cycle, and achieves its maximal width during menses and the follicular phase. The apparent length of the region of the internal os can also change during the course of a single hysterosalpingographic procedure (Fig. 3.11). This may be a reflection of asym-

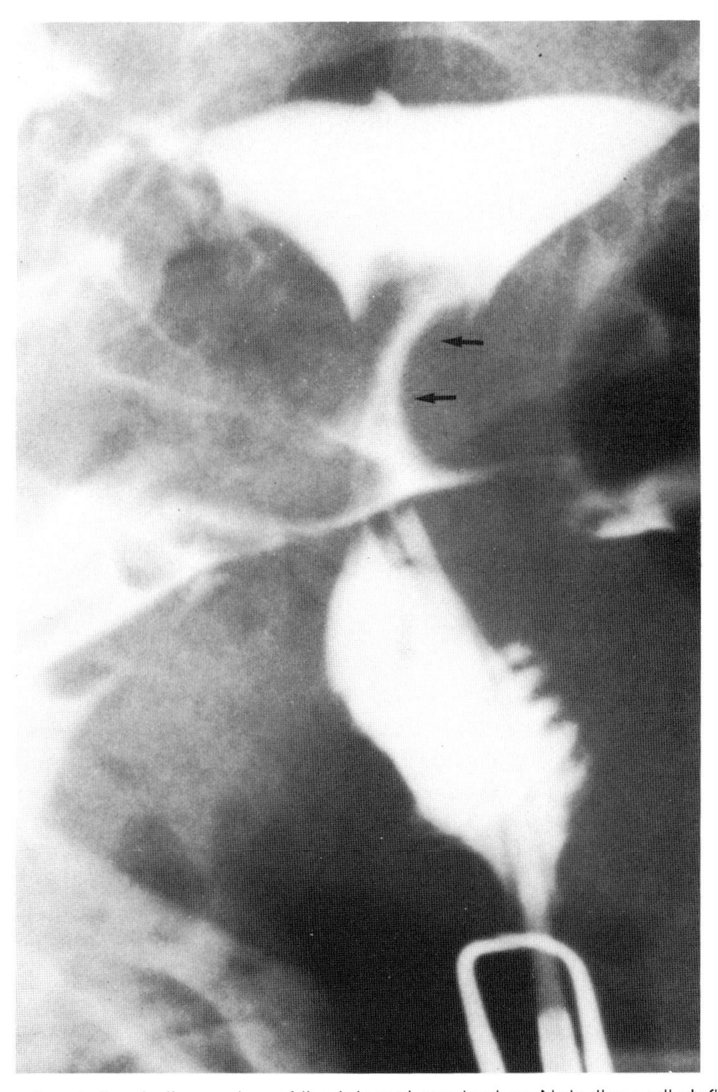

Figure 3.9. Normal variation in the region of the internal cervical os. Note the well-defined elongated narrow segment (*arrows*).

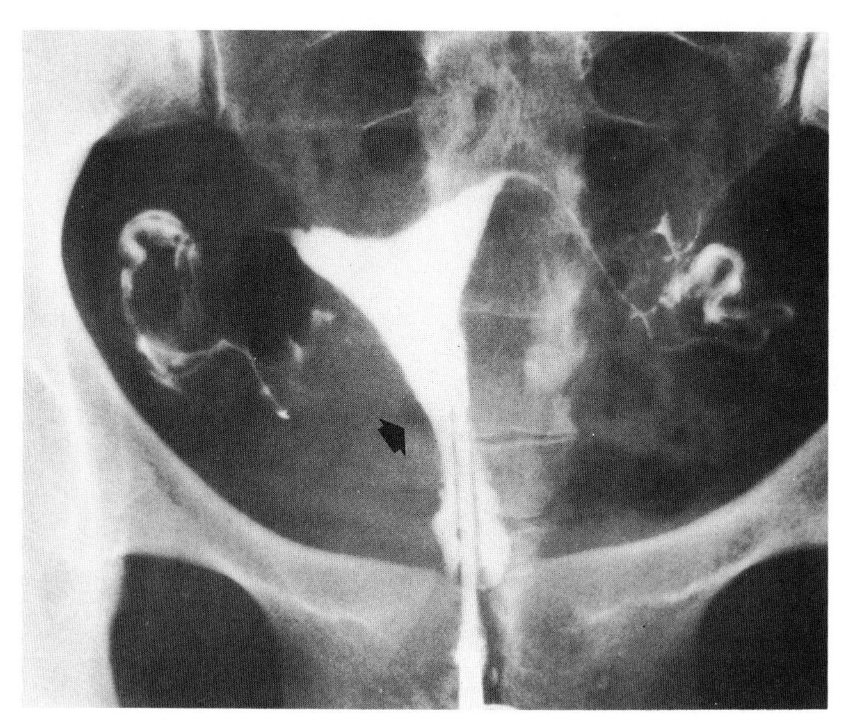

Figure 3.10. At times the internal cervical os cannot be defined. Note the funnel-shaped transition between the uterine body and the cervical canal (*arrow*).

metric contraction of the myometrium or of variations in the degree of distension of the cavity as the quantity of injected contrast medium changes. Such alterations in appearance are unrelated to identifiable pathology and seem to have no corollary at hysteroscopic examination.

The Uterine Cavity

The size and shape of the uterine body are quite variable. The alignment of the uterus, the traction applied during the procedure, the degree of flexion, and the position of the patient at the time of radiographic exposure may all alter the appearance of the shape of the uterine cavity. Such changes in the appearance are dependent upon the incident angle of the radiographic beam in reference to the alignment of the uterus.

Minimal variations of the normal radiographic appearance of the uterine cavity are frequently encountered. The abnormal configuration seen in patients with maternal diethylstilbestrol (DES) exposure (Fig. 3.12) demonstrates a T-shaped uterine cavity with a narrowed lower uterine segment and endocervical canal and may be confusing. The configuration of the fundus may be convex, straight, or concave, and is more fully discussed in Chapter 4 under müllerian fusion disorders (Fig. 3.13). The radiographic appearance of the contour of the fundus may appear to change with varying degrees of traction, depending upon the portion of the uterine surface tangential to the x-ray beam. These relatively minimal alterations in the contour of the fundus have been postulated to play a role in infertility and fetal wastage, and their proper identity and interpretation may be clinically significant.

The appearance of the walls of the uterine cavity margin is dependent upon the phase of the menstrual cycle at the time of the examination. Hysterosalpingography is optimally performed during the preovulatory phase of the menstrual cycle, when the endometrium is thickened and relatively smooth. Such timing is optimal and avoids

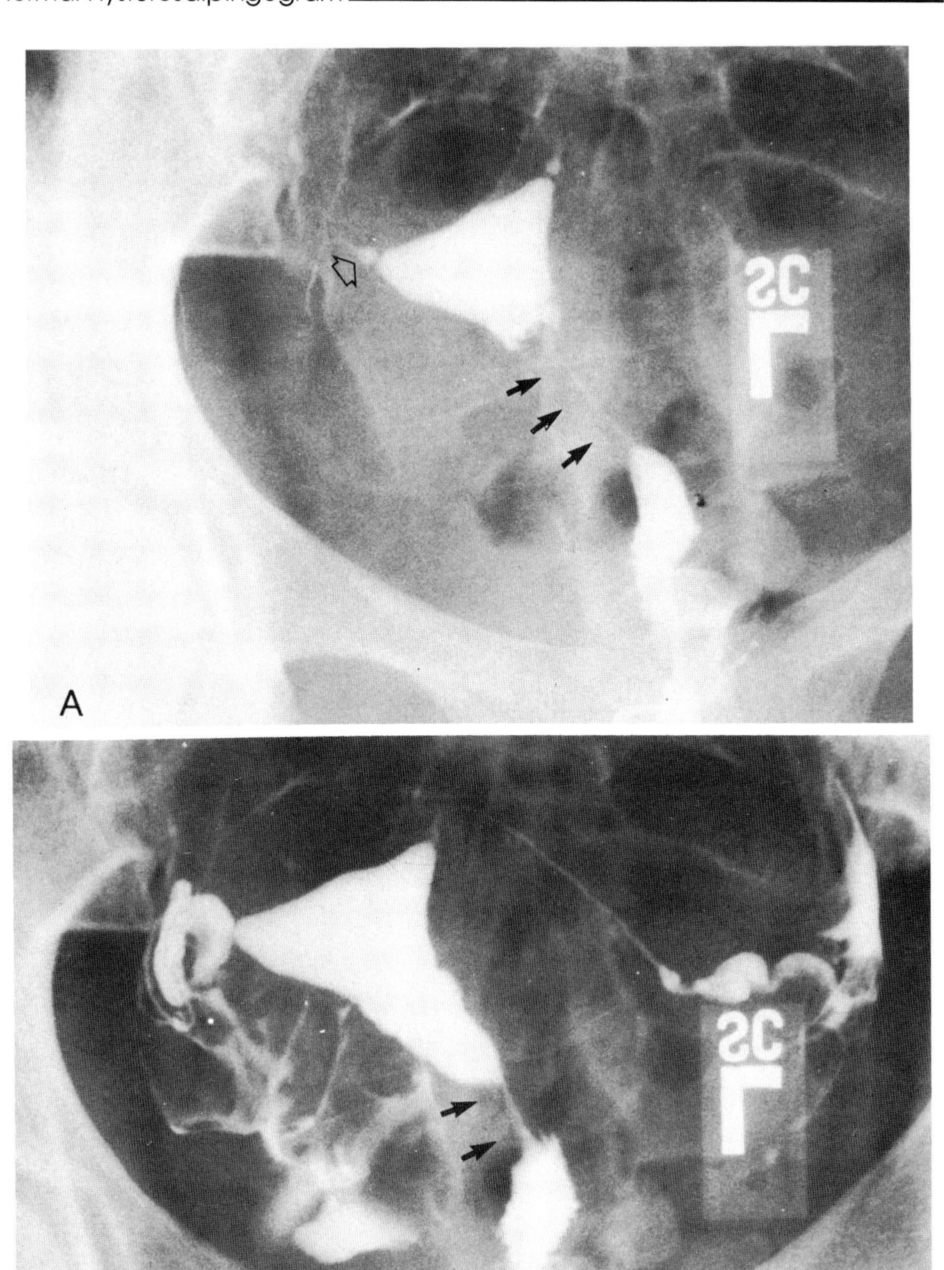

Figure 3.11. *A*: Early in the procedure, the transition zone between the uterine body and the cervical canal is markedly elongated and quite narrow (*arrows*). Note also the normal, very well defined tubal bulge at the interstitial segment of the fallopian tube (*open arrow*). *B*: With increasing amount of contrast, the transition zone changes contour considerably and is much shorter and slightly wider (*arrows*).

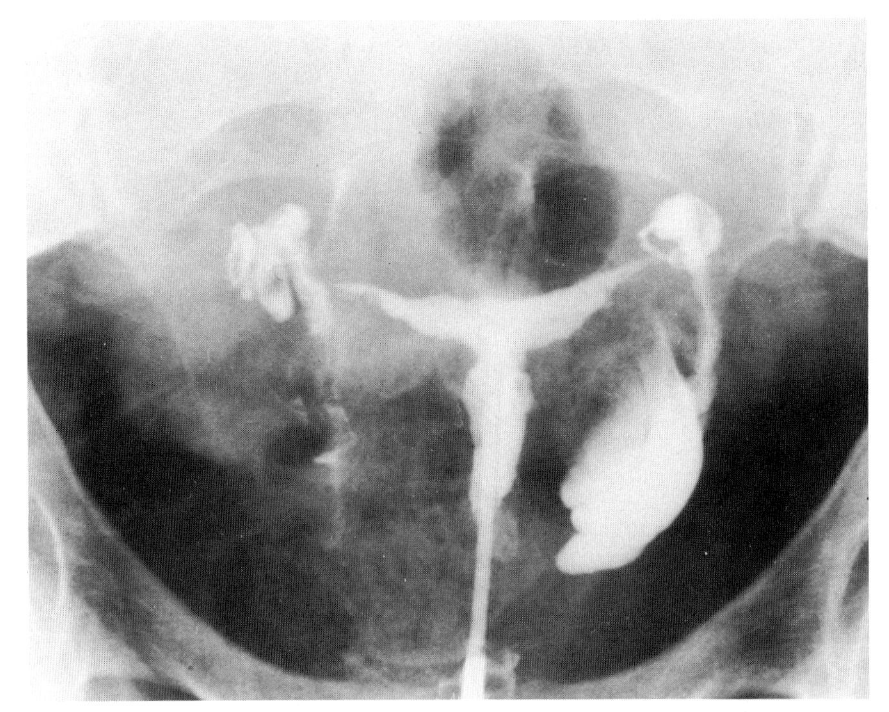

Figure 3.12. A T-shaped uterine cavity secondary to maternal diethylstilbestrol exposure.

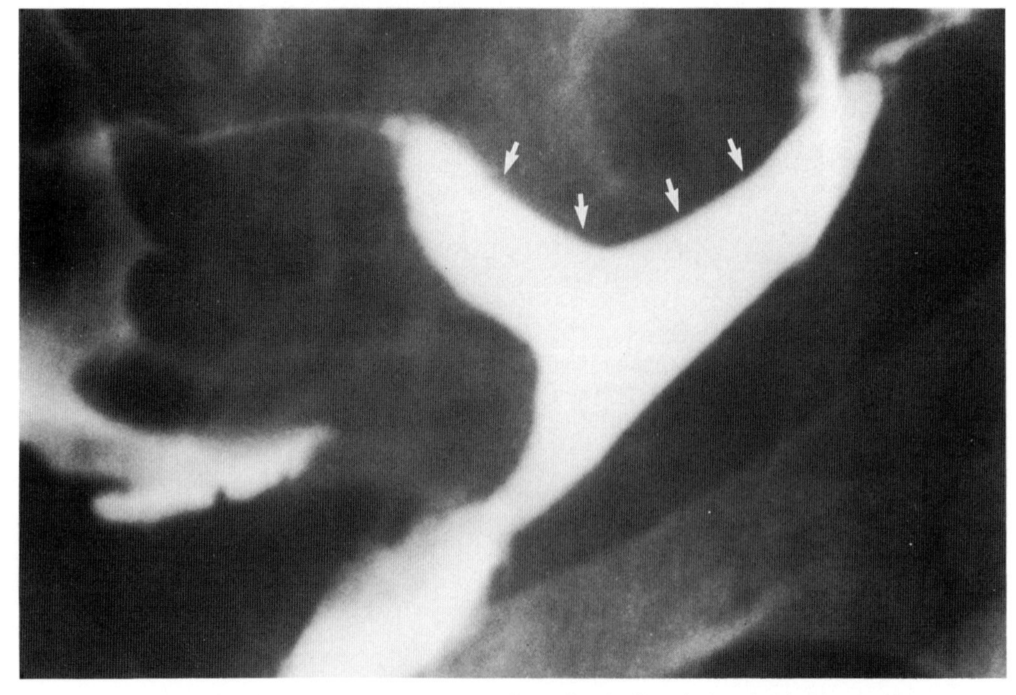

Figure 3.13. Concave or arcuate configuration to the uterine fundus (*arrows*).

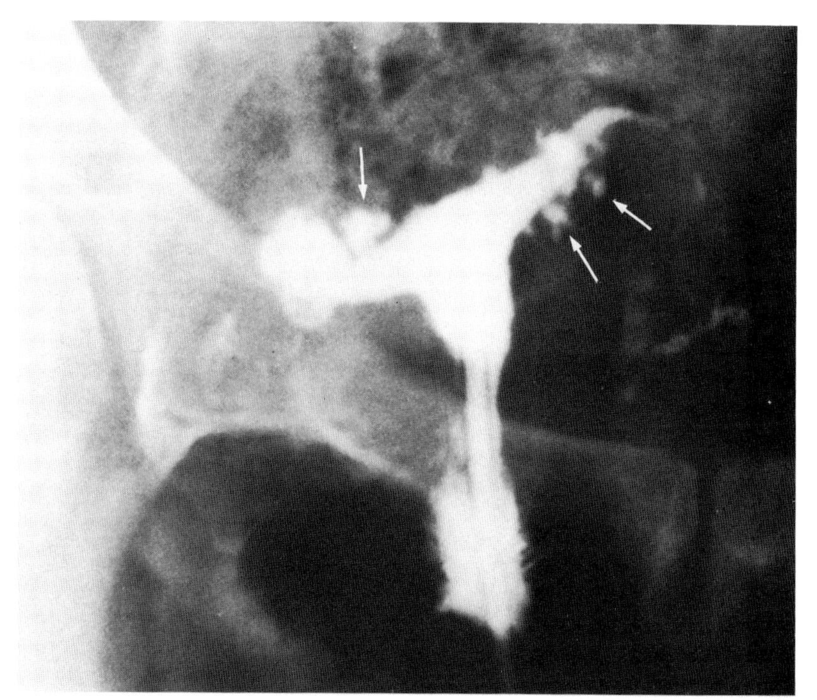

Figure 3.14. Adenomyosis. Note the coarse and irregular collections of contrast invaginating the myometrium (*arrows*).

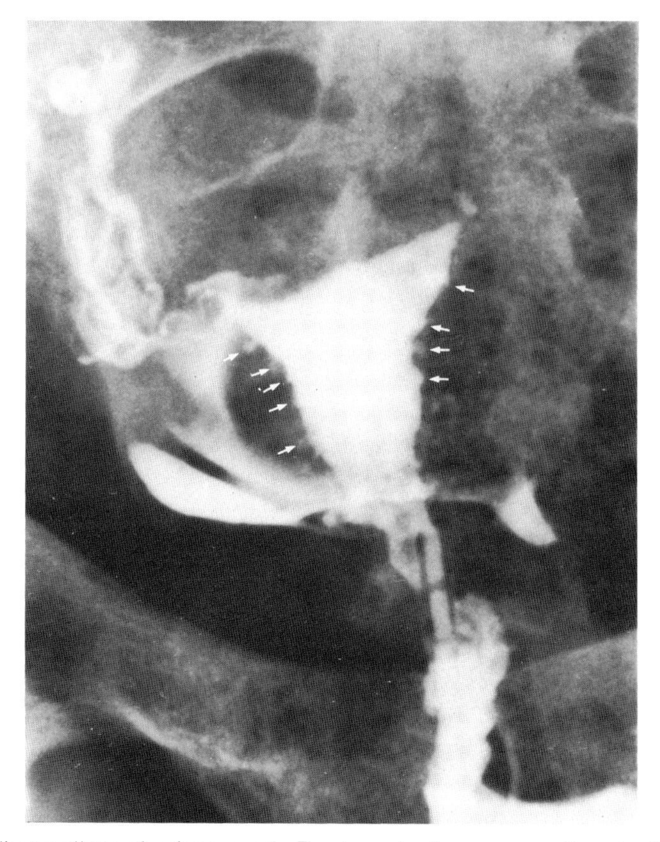

Figure 3.15. Another pattern of adenomyosis. The invaginations are small and relatively symmetrical (*arrows*).

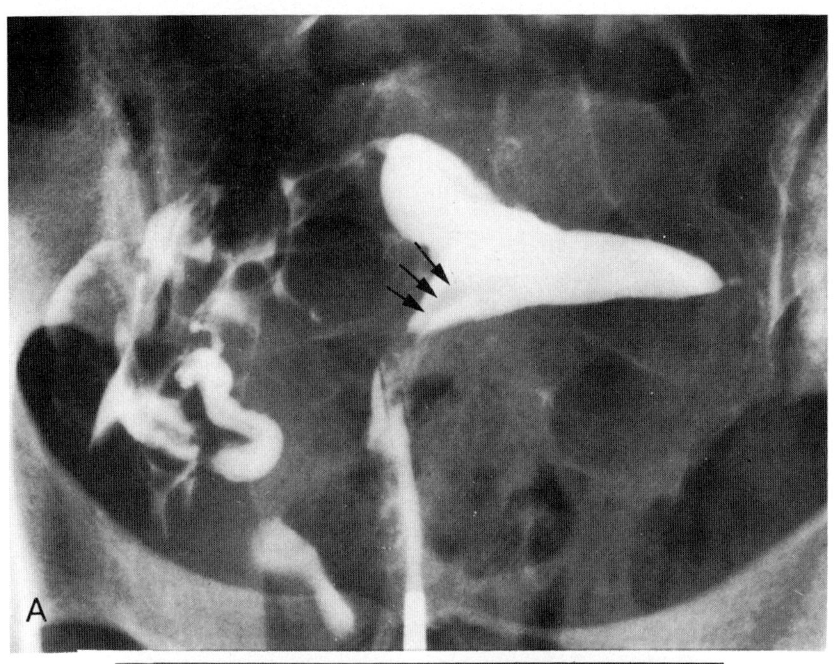

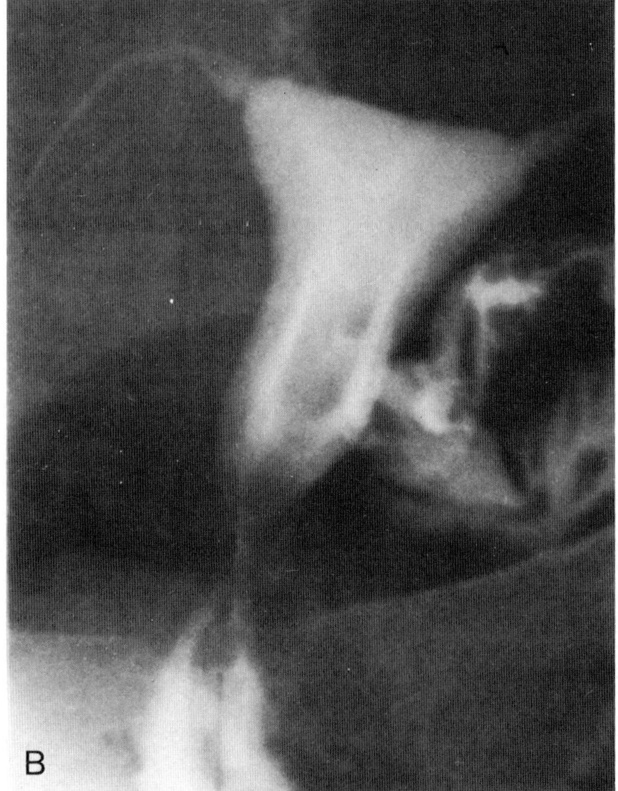

Figure 3.16. Ridging, a normal variation. *A*: The ridged configuration is sometimes thin and delicate (*arrows*). *B*: At other times the ridged contours are rather broad and prominent. These have no apparent corollary at hysteroscopic examination.

performance of the examination after ovulation, when conception may have occurred. However, if examined during the postovulatory secretory phase, the endometrium usually shows a more irregular contour. Such coarsely irregular margins are also noted in patients with adenomyosis. This latter entity, reflecting a benign proliferation of endometrium into the uterine musculature, results in an easily recognized radiographic pattern. The phenomenon may be local or generalized, but is typified by small collections of contrast material invaginating the myometrium and suggesting a series of palisading diverticula likened to "lollipops" (Figs. 3.14 and 3.15). Another unusual but innocuous normal variation of endometrial pattern is a ridged contour, consisting of longitudinal folds involving the uterine body (Fig. 3.16). These parallel the long axis of the uterus, have no obvious corollary at hysteroscopy, and seem to have no clinical significance.

Extravasation of contrast material into the wall of the myometrium is occasionally seen (Fig. 3.17). The most common causative factor is excessive pressure within the uterine cavity during introduction of the contrast. This may be seen with tubal occlusion, when intracavitary pressure is markedly elevated, but may merely be a reflection of the force of injection of contrast. If extensive synechiae are present and partially obliterate the uterine cavity, the high pressure needed to distend the potential space of the uterine cavity may also result in extravasation. Performing a hysterosalpingogram shortly after endometrial instrumentation (biopsy, dilatation and curettage) is also causative. Such extravasation can be localized or generalized (Fig. 3.18), and, when excessive, is associated with pelvic vascular filling. The contrast deposits in the myometrium disappear after several minutes and seem to cause no ill effects. The vascular channels, when outlined by contrast, are

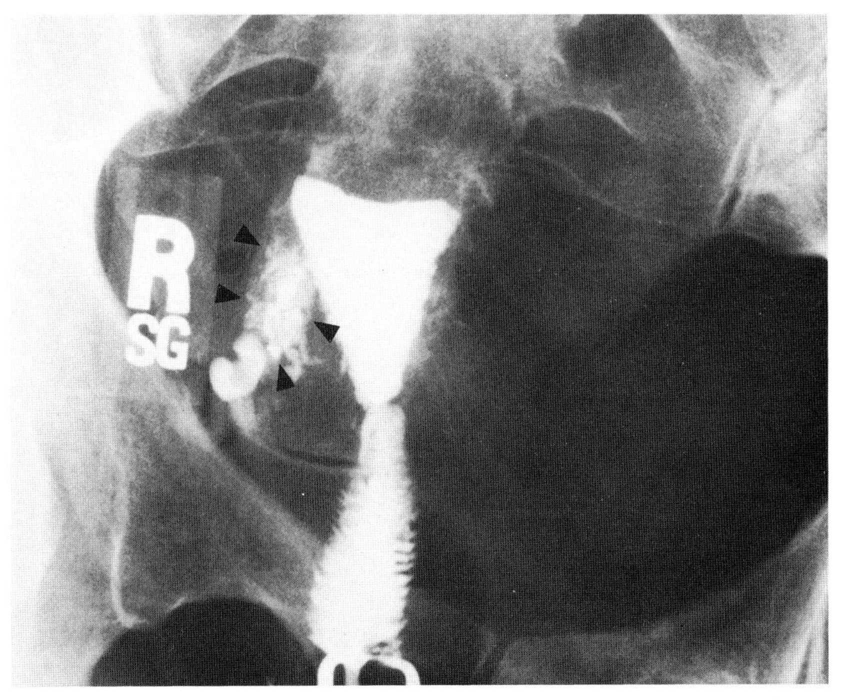

Figure 3.17. Rather large but localized area of extravasation of contrast medium into the myometrium (*arrowheads*). In this patient, the excessive pressure that resulted in the extravasation is secondary to bilateral tubal occlusion. Note also the normal variation in the appearance of the mucosal contour of the cervical canal.

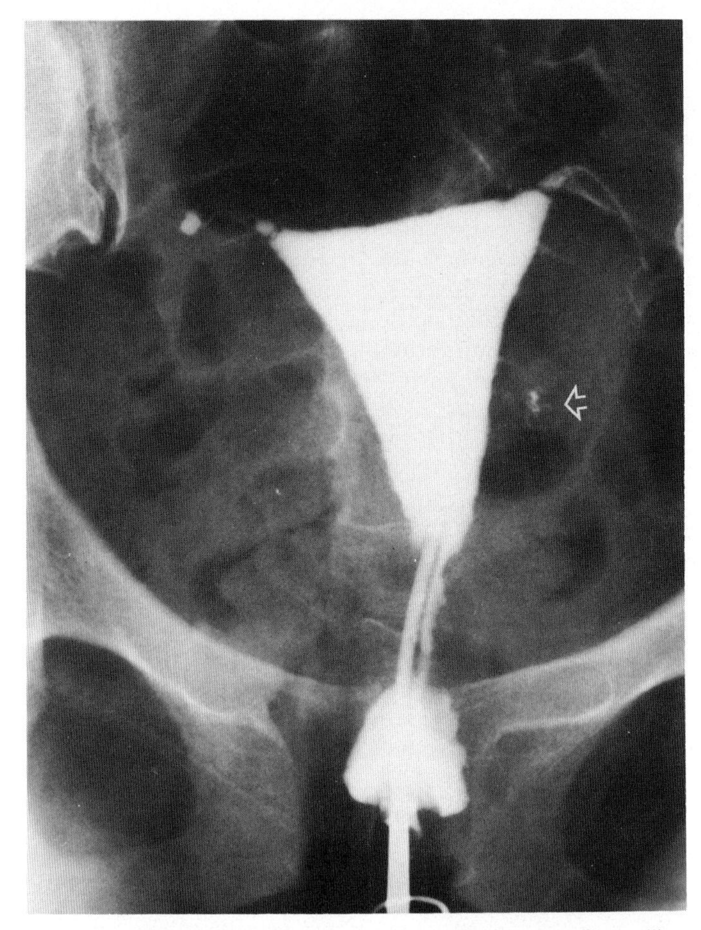

Figure 3.18. Very localized intravasation of contrast (*open arrow*) in a patient with a previous bilateral tubal ligation.

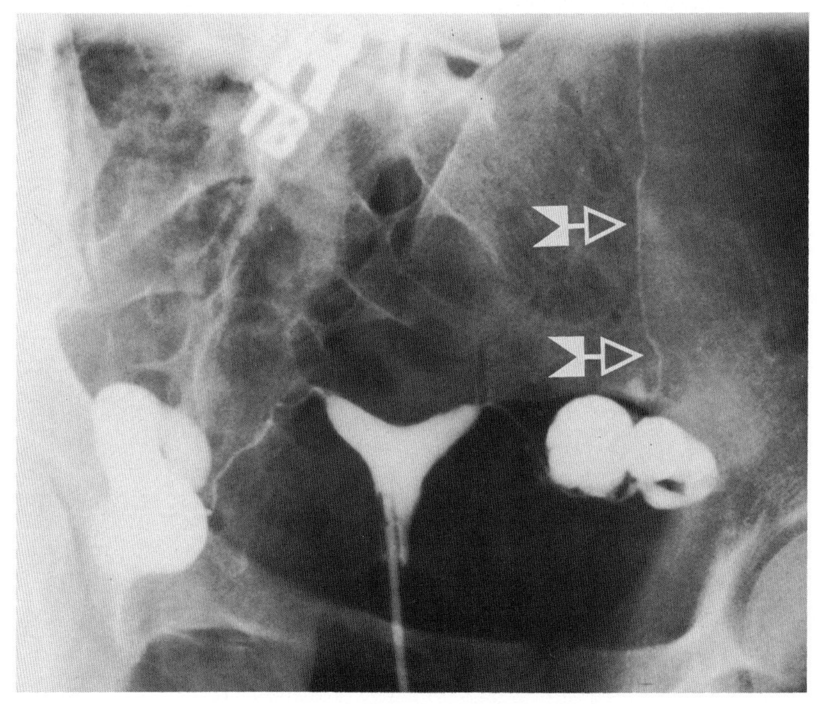

Figure 3.19. Intravasation of contrast into lymphatic channels (*open arrows*). Note the bilateral hydrosalpinges in this patient with stigmata of chronic pelvic inflammatory disease.

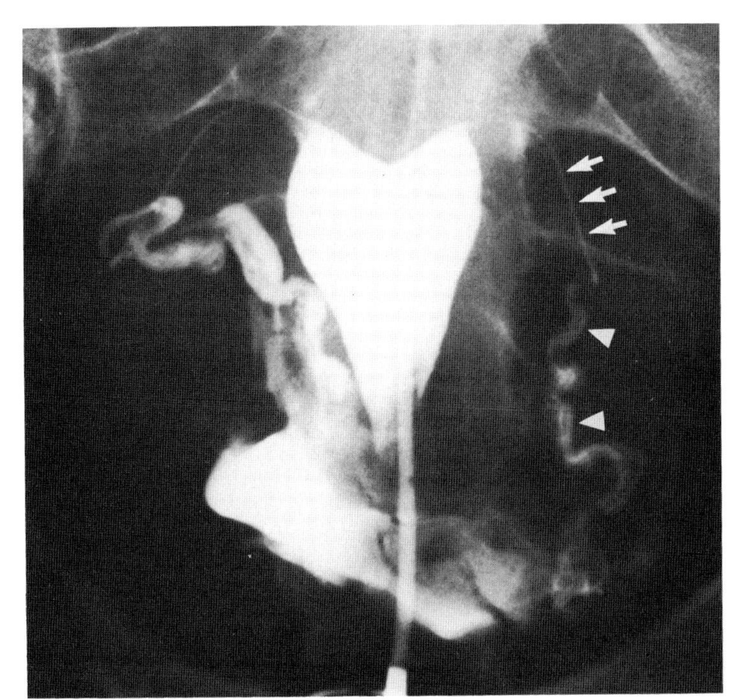

Figure 3.20. The isthmic (*small arrows*) and ampullary (*arrowheads*) portions of the fallopian tube are clearly different in appearance. Note the rugal fold in the ampullary segment of the tube.

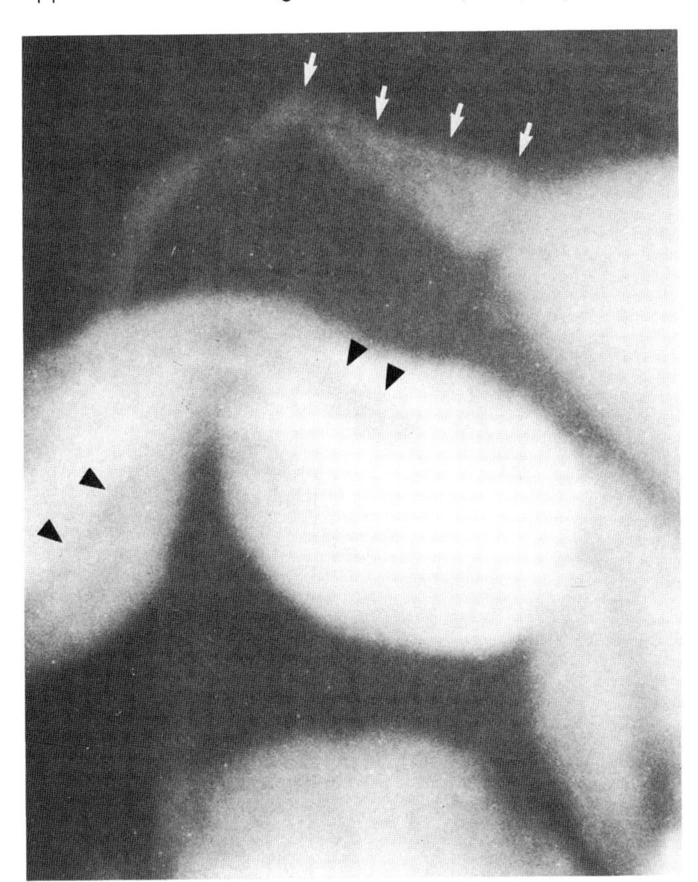

Figure 3.21. The interstitial segment of the fallopian tube is frequently conical or triangular in shape (*arrows*). Note the rugal folds in the ampullary portion of the tube (*arrowheads*).

transitory in their appearance, clearing in seconds as a reflection of normal blood flow. Contrast in thin delicate lymphatic channels may also be observed; lymphatics are differentiated from blood vessels by their thinner caliber and slower emptying (Fig. 3.19).

The Fallopian Tubes

The fallopian tubes are paired structures of 10–12 cm in length, arising from the uterus at the fundal cornua, and extending laterally. Three segments, the interstitial portion, the isthmus, and the ampulla, are defined and apparently have differing functional and anatomic distinctions (Figs. 3.1 and 3.20). The interstitial segment is 1–2 cm in length and lies completely within the myometrium. A triangular or funnel-shaped zone of dilatation, the so-called tubal bulge, is commonly observed (Figs. 3.21 and 3.22). The junction of the interstitial and isthmic segments of the fallopian tubes is anatomically demarcated where the tube exits from the myometrium, a point not identifiable on a hysterosalpingogram. Obstruction at this point is not uncommon. Such occlusion may be anatomic or mechanical in etiology, or due to "cornual spasm" (more properly, myometrial contraction). This resistance to filling may often be overcome by gentle, persistent pressure during contrast induction. If the obstruction to flow persists, the administration of antispasmodic agents may be attempted to relieve the apparent spasm. Varying degrees of success have been reported with amyl nitrate, atropine, nitroglycerin, and dihydroergotamine. Glucagon has been the

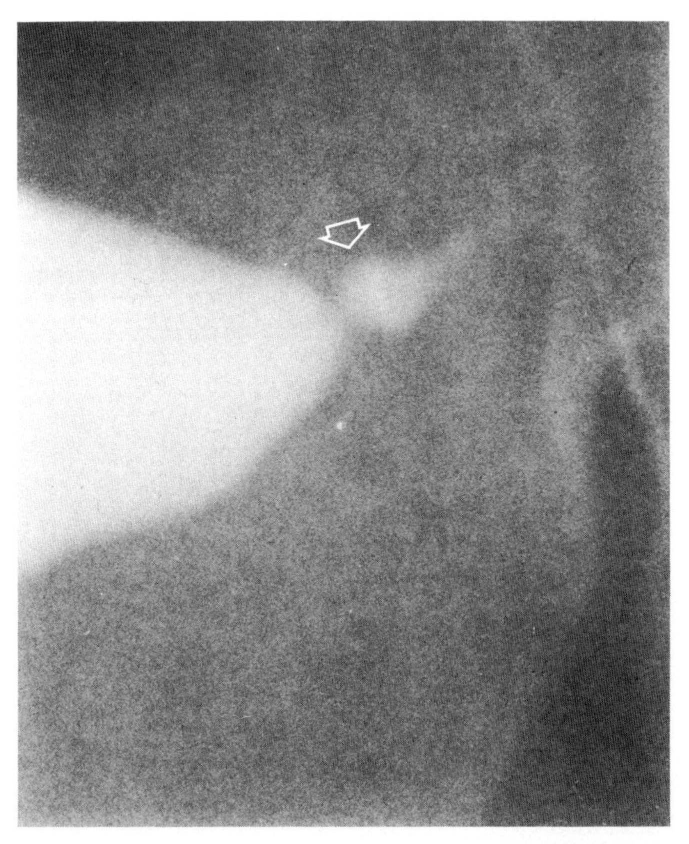

Figure 3.22. The interstitial segment of the tube at times is diamond-shaped in configuration (*open arrow*). This is a normal variation and is of no clinical significance.

most satisfactory agent in our hands, although the success rate for reversing obstruction is still only 25% (Chapter 2).

The isthmus of the fallopian tube is tortuous, delicate, and narrow; measurements of the length of this segment are difficult on radiography because of the tortuous nature of the tube and the irregular direction of its course. The diameter of the isthmus rarely exceeds 2 mm and may be so thin as to be virtually imperceptible radiographically.

The ampulla is the widest and longest segment, with a gradual increase in diameter to an average of 5 mm. Inherent magnification may make this appear even larger on a hysterosalpingogram. Rugal folds are frequently noted and may be recognized in both the normal and pathologic tube (Figs. 3.23 and 3.24). Their presence cannot be construed as excluding inflammatory change. Numerous descriptions of peristaltic activity in the fallopian tubes are recorded, but are not usually observed during the course of the examination. Conjecture as to a possible physiologic role for the isthmic-ampullary junction is beyond the scope of this text, but this zone may well have a specific function during the process of fertilization of the oocyte.

Demonstration of the mobility of the isthmic and ampullary portions of the tubes may be accomplished by varying the degree of traction on the cervix during the examination. Tubes demonstrating mobility are suggested to be free of peritubal adhesive change, but the inability to detect such mobility may simply reflect insufficient movement of the uterus.

Patency of the tube, a vital determination during hysterosalpingography, is usually readily recognizable by spill of contrast from the fimbriated end of the ampulla into the peritoneal cavity. The contrast usually puddles along a peritoneal reflection or, occasionally, interjects between bowel loops. Persistence of contrast in an immediate peritubal location may signify inflammatory adnexal disease (see Chapter 8). The free flow of contrast deep into the pelvis and along peritoneal reflections is the common criterion of patency.

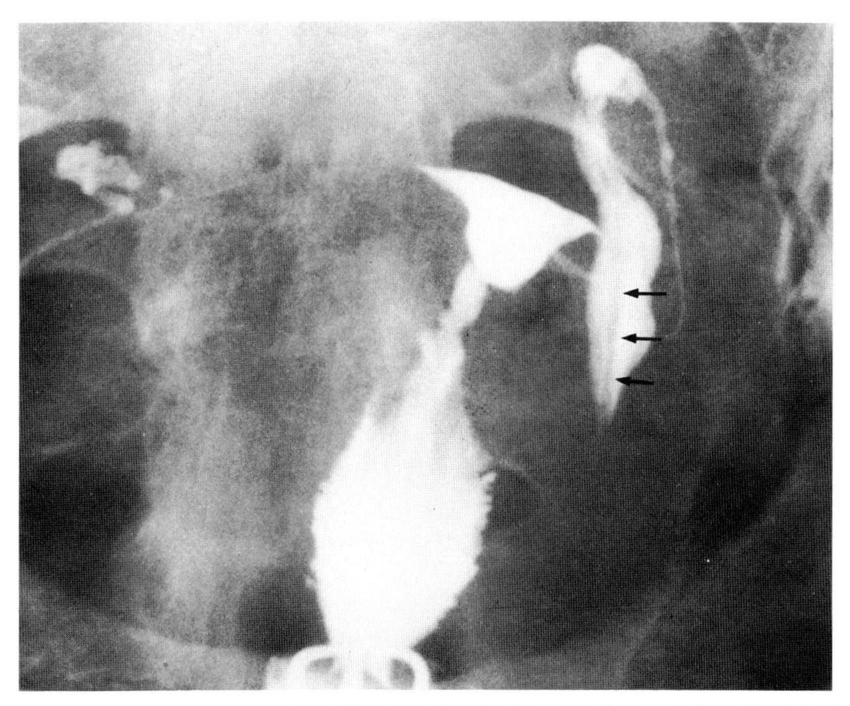

Figure 3.23. Ampullary segment of the fallopian tube is significantly wider than the isthmic portion. Rugal folds (*arrows*) are striking. This appearance is well within the limits of normal.

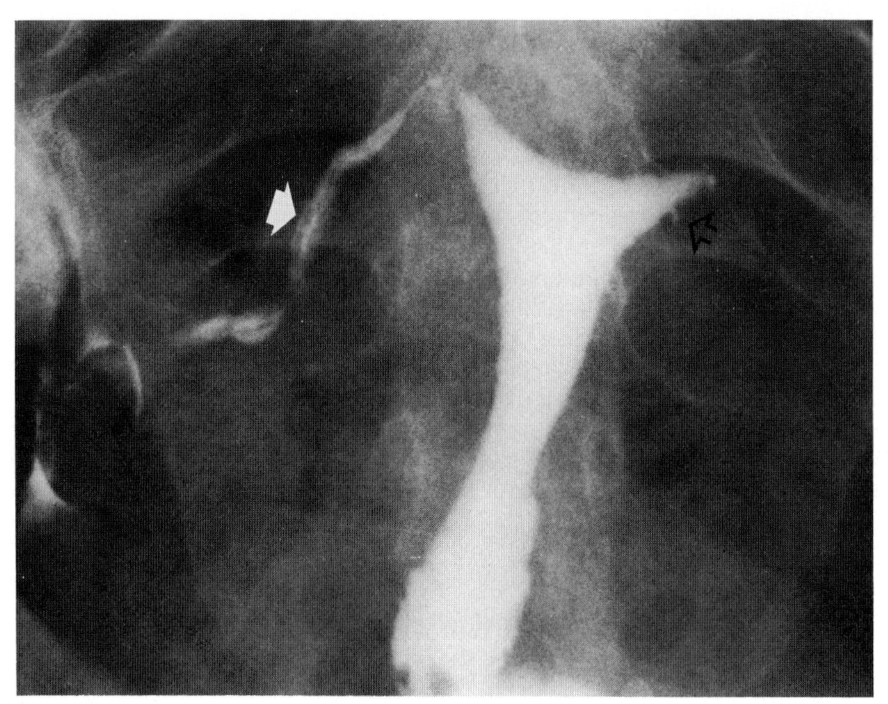

Figure 3.24. Prominent rugal folds in ampullary segment of right fallopian tube (*arrow*). The left tube is occluded. Minimal adenomyosis (*open arrow*) is present.

Summary

The multiple appearances that may be discovered during hysterosalpingography are extensive. It is critical that these variations be recognized for what they are and not misinterpreted as abnormal structural findings. Meticulous technique and careful observations are necessary to accomplish this end.

REFERENCES

Abyholm NHT, Borgersen A: Hysterosalpingography in the evaluation of infertility. *Acta Radiol Diagn* 24:253,1983.

Sanfilippo JS, Yussman MA, Smith O: Hysterosalpingography in the evaluation of infertility: A six year review. *Fertil Steril* 30:636, 1978.

Siegler AM: Hysterosalpingography. *Mod Trends* 40:2, 1983.

Stern WZ: Gynecologic radiology. *JAMA* 238:1763, 1977.

4 Congenital Anomalies of the Uterus and Fallopian Tubes

Congenital malformations of the müllerian paramesonephric system are expressed in multiple ways, including complete absence of a structure, duplication, and atresia. Patients with improper development of the müllerian ducts may present with a variety of symptoms. Gynecologic complaints include amenorrhea, dysmenorrhea, dyspareunia, infertility, fetal wastage, and poor reproductive performance. Menstrual outflow obstruction may lead to endometriosis, hematometra, and hematocolpos. Müllerian malformations may be completely asymptomatic and never detected, or they may present as surgical emergencies such as a gestation in a rudimentary uterine horn, rupture of a noncommunicating rudimentary horn, or an obstructed labor. A thorough knowledge of the embryologic development of the müllerian system is necessary for accurate diagnosis of a particular case, and for planning potential surgical therapy for such diverse conditions as vaginal agenesis, imperforate hymen, uterine septa, or other anomalies.

Embryologic Development of the Uterus

The müllerian, or paramesonephric, duct system begins to develop in both sexes by 40 days of gestation, appearing as a funnel-shaped opening of celomic epithelium, originating near the base of the dorsal mesentery in close association with the urinary tract. The mesonephric, or wolffian, duct system forms first, and becomes the male reproductive tract. Normal development of the male system requires both the presence of a testis and the production of testosterone. In the male after 60 days of gestation, the embryonic testis begins to produce a müllerian inhibiting factor that causes the müllerian ducts to regress and prevents development of the female system. In the absence of the müllerian inhibiting factor the ducts continue to proliferate unimpeded, developing into paired, undifferentiated tubes that later fuse to become the primordia of the uterine corpus, cervix, and a portion of the vagina. In the male, failure of inhibition of the müllerian duct or failure of masculinization of the internal genitalia may allow development of the female ductal system or the formation of an ambiguous, incompletely masculinized system.

In the female, the paramesonephric or müllerian ducts develop on the lateral side of the mesonephric ridge, and extend caudally parallel to the mesonephric ducts. The mesonephric duct serves as a guide for the growing müllerian duct, and any interruption of the downward growth of the mesonephric duct will result in arrested growth of the müllerian duct beyond this point. Canalization of the solid müllerian duct proceeds simultaneously with its downward growth. In the female, the wolffian ducts, for lack of stimulus, eventually persist only as microscopic islands or short segments of ductal epithelium buried in the anterolateral vaginal wall (Gartner's duct), cervix, broad ligament (epoophoron), and paraovarian (paroophoron) tissues. At the caudal end, the paired müllerian ducts cross medially and fuse by the eighth week with the urogenital sinus to form a solid mass, the müllerian tubercle. The process of fusion of the two hollow müllerian ducts into the uterus starts at the caudal tip, the müllerian tubercle, and proceeds cranially up to the junction of the future round ligaments. Resorption of the fused median septum next occurs, and this may start at any level and proceed in either or both directions. This fusion of the previously paired müllerian ducts and subsequent resorption of the septum leaves a single uterovaginal canal that differentiates further into a recognizable vagina, cervix, uterine corpus, and fallopian tubes. Toward the end of the third month of embryonic life, distinct muscular and connective tissue layers can be seen in the uterus, and by the end of the sixth month the endometrium appears.

Tissue of both müllerian and urogenital sinus origin normally participates in the development of the vagina (1,2). The müllerian tubercle, where the hymen will ultimately develop, is the site of contact between the solid tip of the fused müllerian ducts and the dorsal wall of the urogenital sinus. The mesonephric ducts, which are in the process of degeneration, enter the urogenital sinus immediately lateral to the tubercle. A solid mass of proliferating sinus cells, called the sinovaginal bulb, develops between the openings of the mesonephric ducts and the müllerian tubercle, and forms a cord from the dorsal urogenital sinus wall to the solid tips of the müllerian ducts, the vaginal plate. The vaginal plate later canalizes, starting at the hymenal ring and proceeding cranially, to end at the cervix; the process is completed between the 20th and 22nd weeks.

There is still major controversy regarding the origin of the stratified squamous epithelium of the vagina. The three possible sources are the müllerian ducts, the wolffian ducts, and the urogenital sinus. Most think that the squamous epithelium of the urogenital sinus extends cranially to the level of the future external cervical os, and displaces the müllerian columnar cells (3), but others have contended that the entire vaginal epithelium is of mesonephric origin (4). The major point to be understood is that müllerian anomalies affect more than just the corpus and cervix of the uterus, and can involve the upper portion of the vagina as well. The major defects can be categorized as occurring because of failure of one or more aspects of the normal embryologic development. For example, failure of initial fusion and descent of the paired caudal ends of the müllerian ducts could result in complete atresia of uterus and upper vagina. An abnormality of vertical fusion would occur with a fault in the junction between the down-growing müllerian tubercle and the up-growing derivative from the urogenital sinus; this could cause an obstructive transverse vaginal septum. A defect in lateral fusion of the two müllerian ducts could result in persistence of the two developing cords, with partial or complete duplication; an abnormality of resorption of the septum could explain some forms of communicating uteri. The sequence in which development occurs makes cetain anatomic problems impossible and others more likely to occur.

Incidence of Congenital Uterovaginal Anomalies

The incidence of congenital defects of the reproductive tract varies with the population studied and the zeal and thoroughness of the investigating clinician. Many patients with uterovaginal anomalies are never detected because they never have obstetric or gynecologic difficulties. Several retrospective surveys have estimated the prevalence of incomplete müllerian fusion as approximately 0.1–0.5% (5,6), although Tulandi and coworkers (7) reported an incidence in excess of 1 in 100. If manual exploration of the uterine cavity is performed at delivery, a higher prevalence of congenitally anomalous uteri will be reported, in the range of 2–3% (8,9). Family clusters with incomplete müllerian fusion have been reported (10–

12), but patients identified as having müllerian defects have a low frequency of affected relatives (13), which is more consistent with a polygenic/multifactorial etiology.

The most common malformations include septate, bicornuate, and didelphic uteri. The unicornuate uterus is very rare (14,15). When uterine anomalies are grouped according to degree of failure of normal uterine development, those with the greatest degree of abnormality seem to have the worst fetal survival. Although most investigators (16–18) report poor obstetric performance with septate uteri, Heinonen and coworkers (19) reported that uteri with complete septa have the best fetal survival rate, around 86%, the complete bicornuate about 50%, and the unicornuate about 40%. Worthen and Gonzalez (20) diagnosed and followed by ultrasound alone 7 patients with various degrees of uterine septum who had 17 pregnancies, 13 of which were viable. Others describe a very poor reproductive outcome in patients with a unicornuate uterus; intrauterine growth retardation has been reported in pregnancies with this uterine abnormality (15), and the frequency of breech presentation is high. Women with unicornuate uteri also have the highest (15%) incidence of primary infertility (21), although experience with the reproductive performance of this very rare uterine anomaly is limited, and nonuterine causes of infertility must be investigated.

Concomitant malformations of uterus and urologic tract might be expected, and up to 20% of women with reproductive tract anomalies have abnormalities of the urinary tract (22,23). In patients with unilateral renal agenesis, over 50% will be found to have genital anomalies (24). Gurin and Leiter (25) make the important point that an intravenous pyelogram (IVP) is frequently forgotten, because only 12 of 55 patients with abnormal müllerian development had had an IVP, and inversely, only 2 of 21 women with renal anomalies had had a gynecologic evaluation. Thus, the association of müllerian duct and renal anomalies is clear, but the prevalence of the association has been inadequately evaluated (Fig. 4.1).

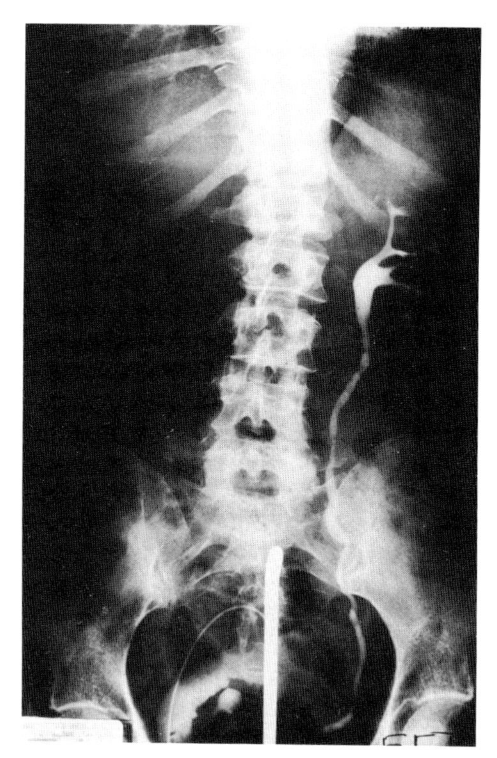

Figure 4.1 Renal agenesis, a common although inconstant abnormality associated with genital tract anomalies.

Congenital Malformations

Either initial failure of fusion of the müllerian duct or later failure of the resorption of the septum results in a continuum of malformations. Fusion begins caudally and extends cranially; if there is incomplete fusion, it will most likely be seen in the upper system, causing duplication of the upper uterus; what cannot occur is a fused single structure in the cephalad area with duplication of the cervix. Failure of lateral fusion may result in partial or complete duplication, which may be symmetrical or asymmetrical, and obstructive or nonobstructive.

Symmetric malformations include those in which external division of the uterus leads to two separate uterine cavities, typified by the bicornuate uterus, and those in which external separation is lacking, for example, the arcuate uterine pattern. Externally, the arcuate uterus may have only a midline groove and a somewhat flattened and transversely elongated fundus; internally, a midline septum may subdivide the cavity either partially (uterus subseptus) or totally (uterus septus), as exemplified in Class III, IV, and V anomalies (see below).

Most *asymmetric malformations* involve uterine duplication in which one uterine horn is fully developed and represented as a semiuterus and the other exhibits rudimentary development or is totally absent. The rudimentary horn is usually attached to the normal contralateral uterine horn by a thin fibrous band. There may be considerable disproportion between the two sides, and, if the endometrium is functional, significant symptomatology due to the obstructed (absent) outflow tract. If the rudimentary horn ends blindly, progressive accumulation of menstrual flow and hematometra gradually leads to increasing distension of the cavity. When rudimentary horns are unconnected and the endometrium is nonfunctional, there will be no clinically detectable symptoms. If the rudimentary horn connects with the normal cavity through a small channel, implantation of a viable conceptus may simulate ectopic pregnancy.

Classification of Müllerian Anomalies

Buttram and Gibbons (18) proposed a classification of müllerian anomalies (Table 4.1) and analyzed 144 cases with respect to pregnancy rates and reproductive success (26).

Class I: Müllerian agenesis or hypoplasia. Since the urogenital sinus contributes to formation of the vagina, vaginal agenesis is not strictly a "müllerian" problem; nevertheless, vaginal agenesis has been included in the Class I anomalies. Vaginal agenesis occurs in about 1 of 5000 phenotypic females and it is difficult to clinically distinguish vaginal agenesis from transverse vaginal septum. If uterus and cervix are intact, and only the vagina has not formed, a surgical approach can be attempted. Anastomosis has been reported with restoration of menses, but there are no documented pregnancies. Those cases in which pregnancy has been reported are probably due to a septum and not to vaginal agenesis. In cervical agenesis, conception is highly unlikely because the fistulous tracts created may have contributed to the occurrence of ascending infection. Fundal agenesis has rarely been observed, and bilateral tubal

Table 4.1.
Müllerian Anomalies: Classification of Buttram and Gibbons

Class I:	Müllerian agenesis or hypoplasia
	A. Vaginal agenesis
	B. Cervical agenesis
	C. Fundal agenesis
	D. Tubal agenesis
	E. Combined
Class II:	Unicornuate uterus
	A. with rudimentary horn
	1. with endometrial cavity
	2. without endometrial cavity
	a) with communication with opposite horn
	b) without communication with opposite horn
	B. without rudimentary horn
Class III:	Uterus didelphus
Class IV:	Bicornuate uterus
	A. complete
	B. partial
	C. arcuate
Class V:	Septate uterus
	A. complete septum
	B. incomplete septum
Class VI:	"DES-exposed" uterus

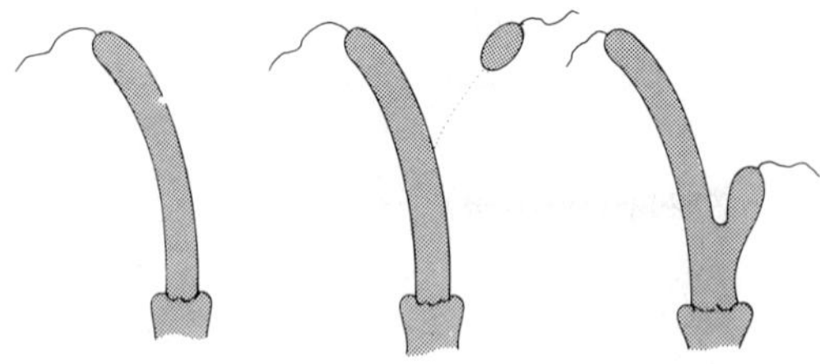

Figure 4.2. Class II anomaly. A unicornuate cavity may reflect any of the depicted configurations. The true unicornuate uterus (*left*) is most rare. Rudimentary horns of the underdeveloped müllerian duct may be either noncommunicating (*center*) or communicating (*right*).

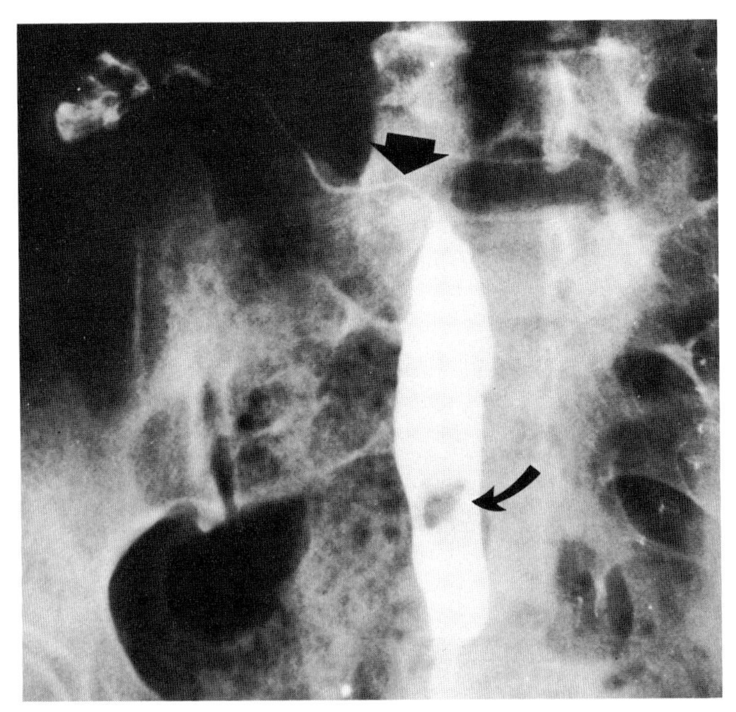

Figure 4.3. Unicornuate uterus. Single fallopian tube extending from cornual tip (*arrow*). Note incidental polyp in endometrial cavity (*curved arrow*).

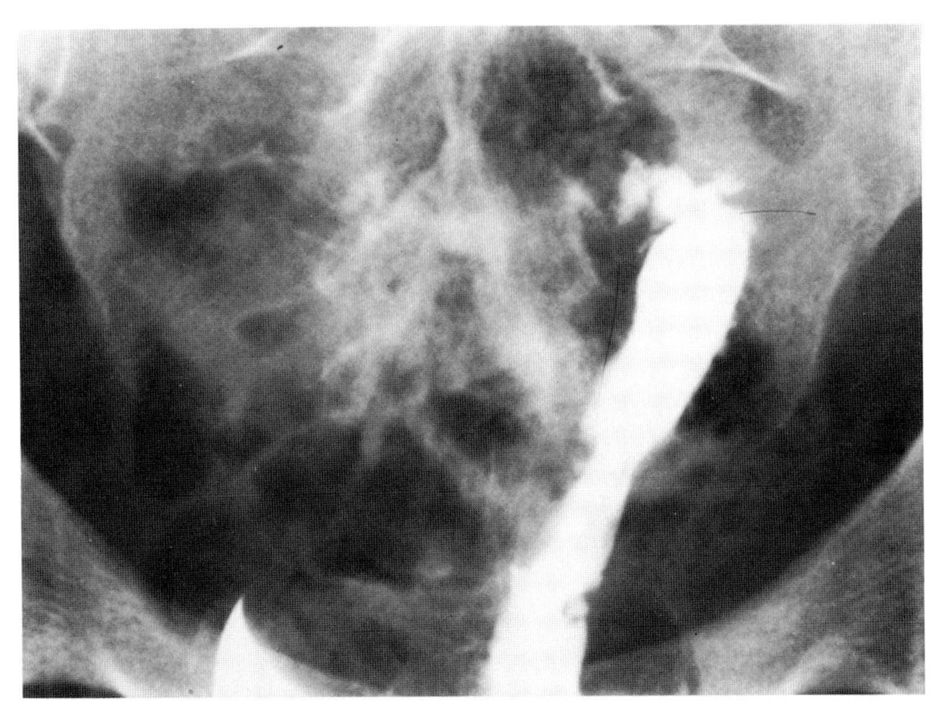

Figure 4.4. Extensive fundal synechia (see Chapter 7) can result in configuration similar to unicornuate uterus.

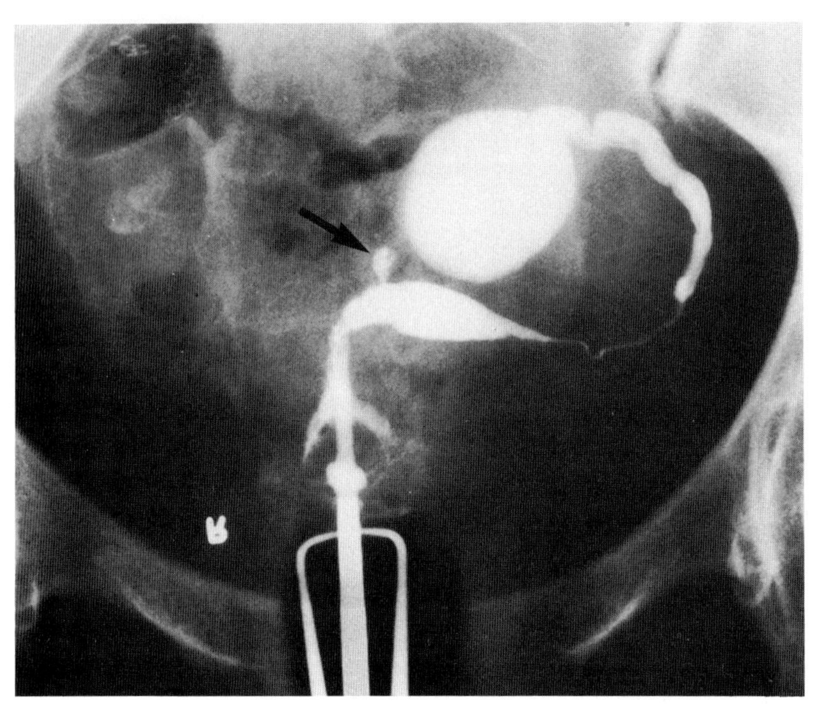

Figure 4.5. Duplication anomaly. Unicornuate uterus with a small communicating rudimentary horn (*arrow*). Large hydrosalpinx on left.

agenesis is associated with an increased incidence of ectopic pregnancy. Overall, a combined uterovaginal agenesis, the so-called Mayer-Rokitansky-Kuster-Hauser syndrome, is the most common class I anomaly. Because of the nature of the deformity, hysterosalpingography plays no significant role in the diagnosis of Class I anomalies.

Class II: Unicornuate uterus. A unicornuate configuration may result from one of several possible anomalies. Rarely, this appearance is due to agenesis of one of the müllerian pair, a true unicornuate uterus. More commonly, however, the unicornuate chamber is associated with a rudimentary horn (representing a remnant of the contralateral müllerian system) that may be either communicating or noncommunicating (Figs. 4.2, 4.3, 4.4, and 4.5).

Class III: Uterus didelphys. The true didelphic uterus, reflecting complete failure of müllerian fusion, has two separate uterine cavities with a vertical septum in the proximal portion of the vagina. It is not unusual, however, for partial fusion and resorption to cause the vaginal septum to disappear (Figs. 4.6 and 4.7). As mentioned, urinary tract anomalies are common (Figs. 4.1 and 4.8).

Class IV: Bicornuate uterus. A bicornuate uterus results from incomplete müllerian pair fusion. The degree of separation of the paired lumens may vary from a minimal arcuate contour at the fundus to a deep cleft with division to the level of the lower uterine segment. There will be only one cervix (Figs. 4.9, 4.10, 4.11, and 4.12).

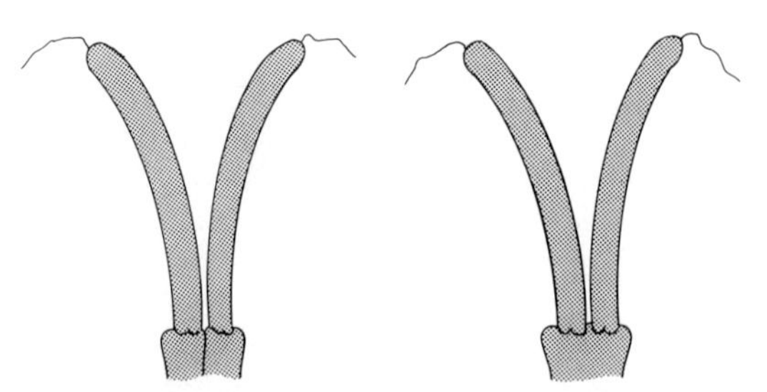

Figure 4.6. Class III anomaly. The didelphys uterus consists of two entirely separate chambers. The vagina may be partitioned in its upper portion by a vertical septum (*left*) or may be unipartite (*right*).

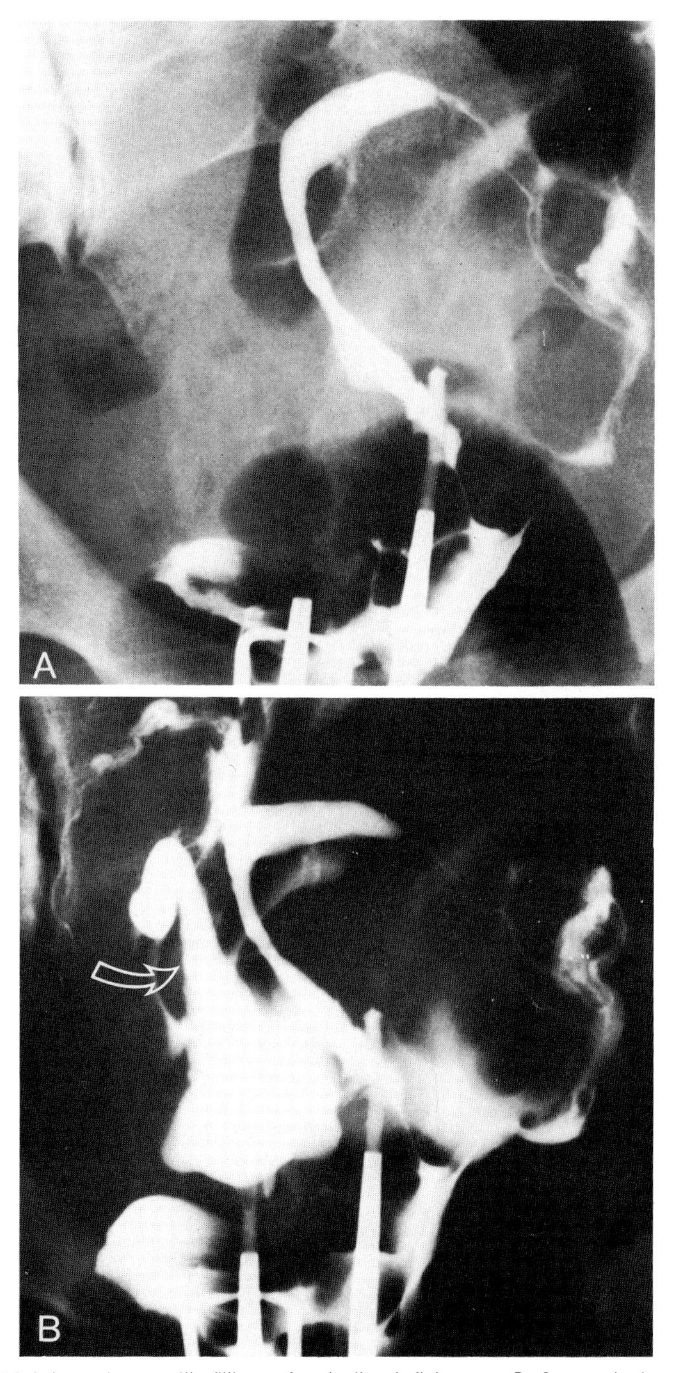

Figure 4.7. *A:* Didelphys uterus with filling of only the left lumen. *B:* Second chamber is now filled (*arrow*), using a second cannula.

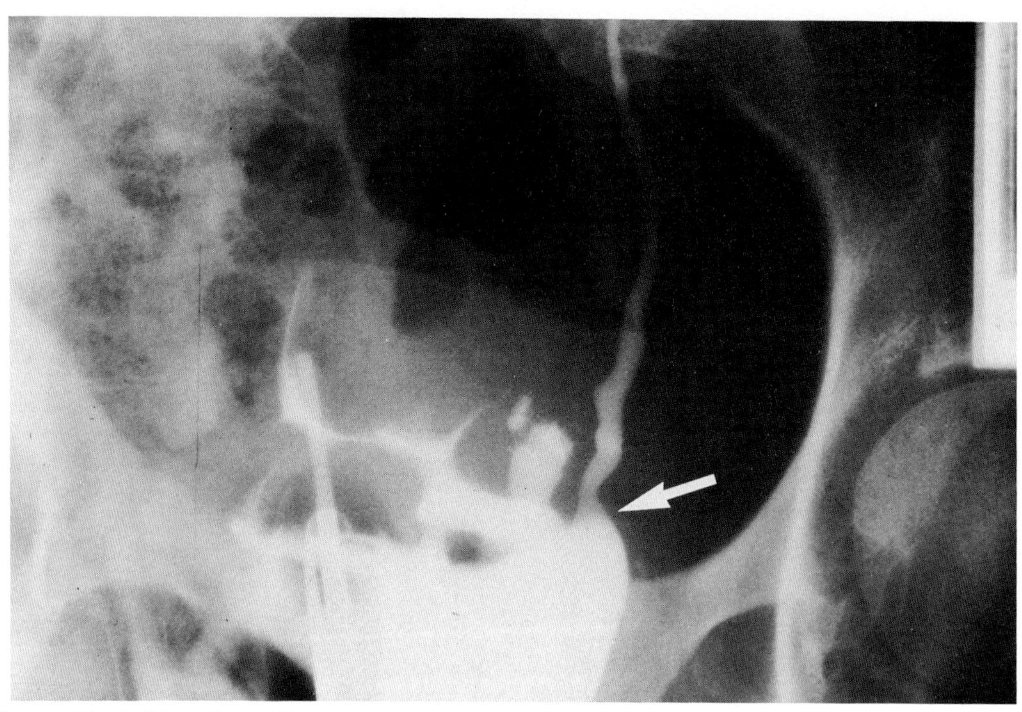

Figure 4.8. In this patient with a didelphys uterus, one also notes ectopic insertion of the left ureter into the vagina (*arrow*).

Class V: Septate uterus. Fusion of the müllerian pair with failure of resorption of all or part of the intervening septum results in a configuration that may mimic the appearance of the bicornuate uterus on a hysterosalpingogram (Fig. 4.13). Differentiation of septate and bicornuate uteri requires laparoscopy or occasionally ultrasound examination in order to appreciate the presence or absence of myometrial tissue between the cornua (Fig. 4.14). Although it is noted that the cornua of the septate uterus frequently create a more acute angle than seen with a bicornuate uterus, the degree of overlap on a hysterosalpingogram is uncertain (Figs. 4.13, 4.15, 4.16, and 4.17).

Class VI: Diethylstilbestrol (DES)-exposed uterus. This uterine malformation has internal luminal changes, such as a T shape, constricting band affecting the uterine corpus, or a widened lower two-thirds of the uterus. These anomalies are almost entirely drug induced, and are further discussed in Chapter 5.

In addition, Toaff and colleagues (27) have reviewed and classified nine types of communicating uteri. These are a distinct class of uterine malformations identified by the presence of a communication between two otherwise separate uterocervical cavities. The different types of communicating uteri are induced by a teratogenic process that is active at different stages of the embryologic development. An early process would in-

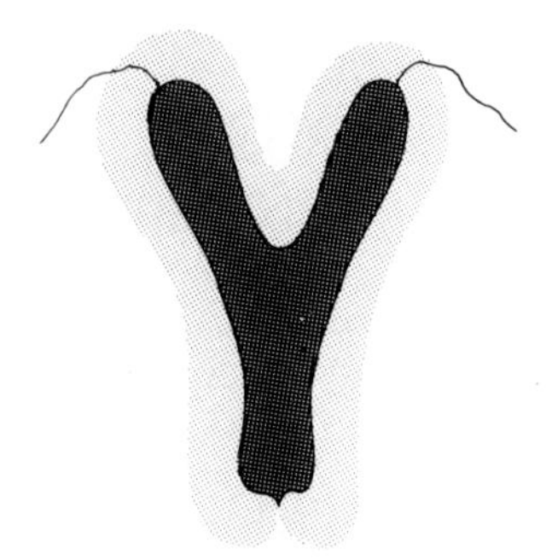

Figure 4.9. Class IV anomaly: the bicornuate uterus.

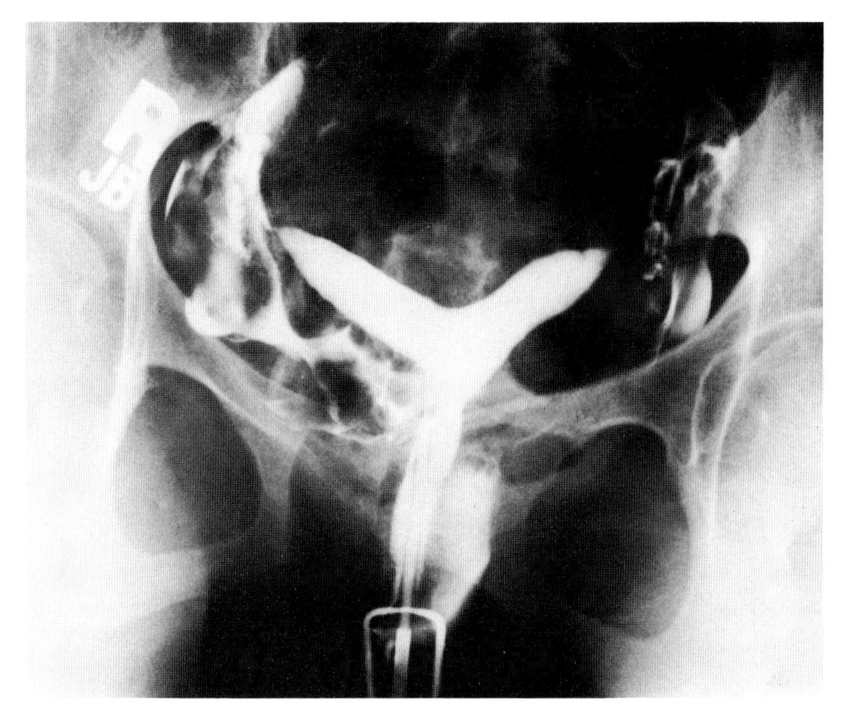

Figure 4.10. Typical appearnce of bicornuate uterus.

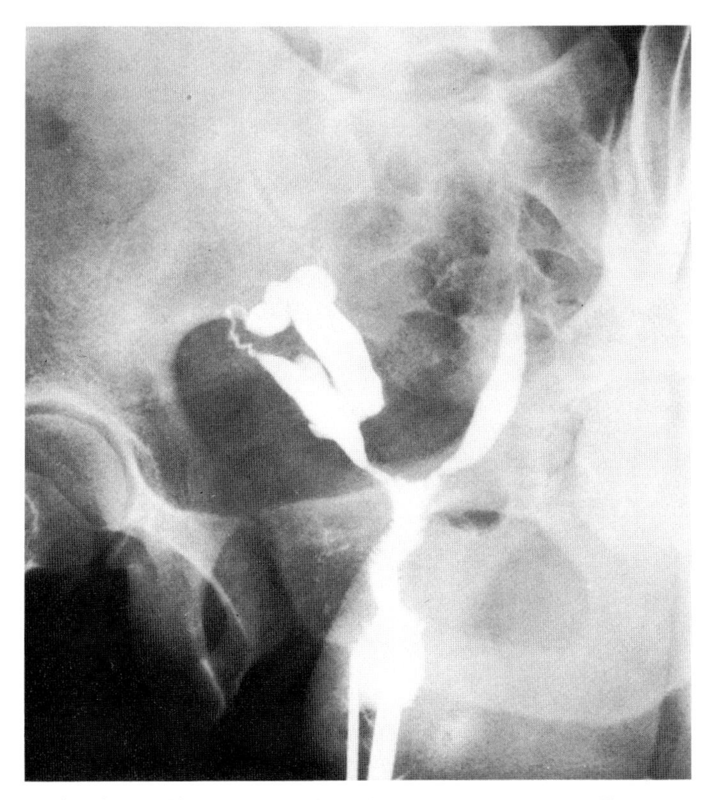

Figure 4.11. Bicornuate uterus with a very prominent lower uterine segment. The left tube is incompletely filled. The right tube is obstructed and dilated.

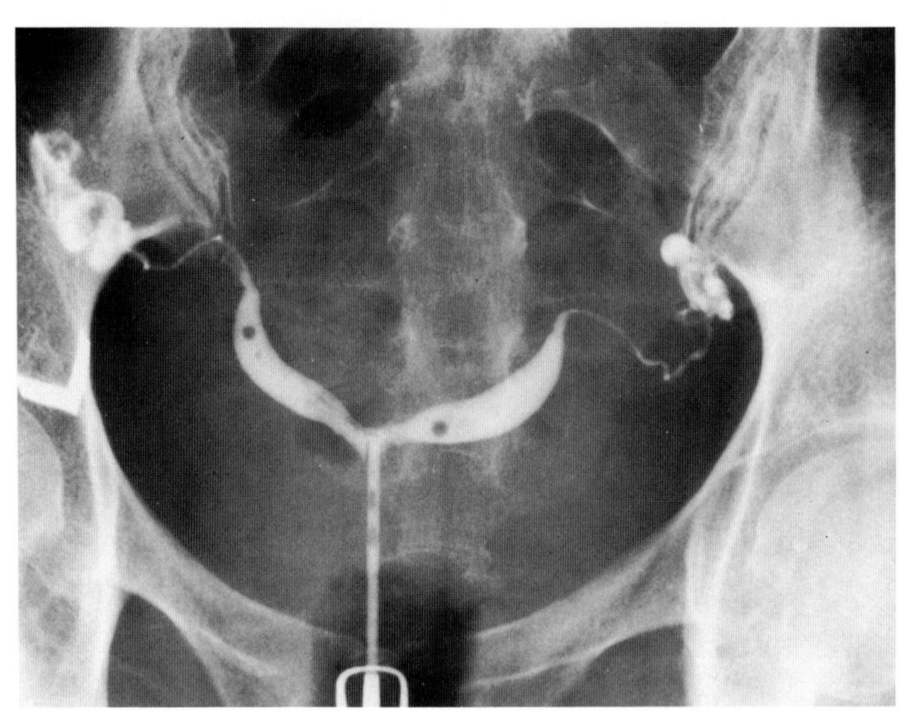

Figure 4.12. Bicornuate uterus. Radiolucencies within the chambers represent air bubbles introduced during the procedure.

volve the development of both urinary and genital ducts and may result in the more severe anomalies of ipsilateral renal agenesis and ductal atresia. A later process could result in a failure of fusion or of complete resorption of the septum, resulting in atypically located communications, often at the level of the uterine cavity (Figs. 4.18, 4.19, and 4.20).

Residual segments of the wolffian ducts may persist in the anterolateral vagina as well, adjacent to the cervix, or in the paraovarian tissues (paroophoron). Referred to as Gartner's ducts, these wolffian remnants may communicate with the vagina or endometrial cavity. These duct lumens may fill with contrast during hysterosalpingography and are identified as linear channels of contrast immediately lateral to the lateral

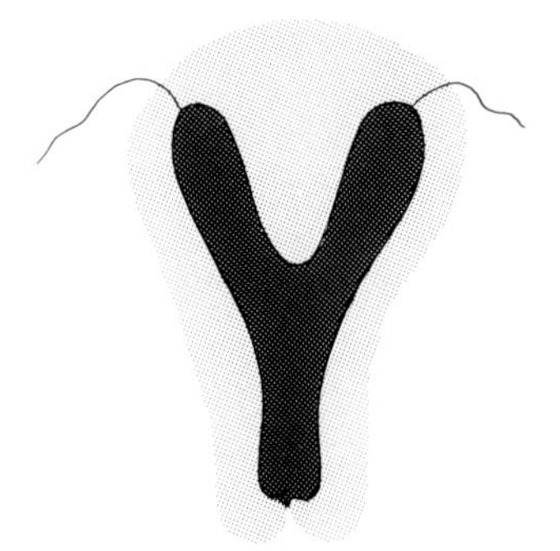

Figure 4.13. Class V anomaly: the septate uterus. The müllerian tubes have fused but there is failure of resorption of a varying amount of the septum separating the chambers.

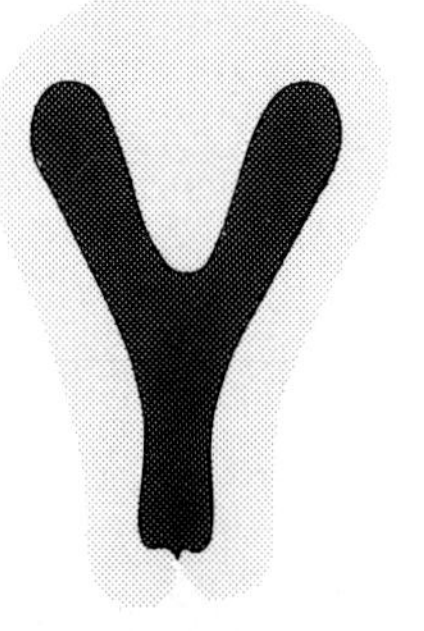

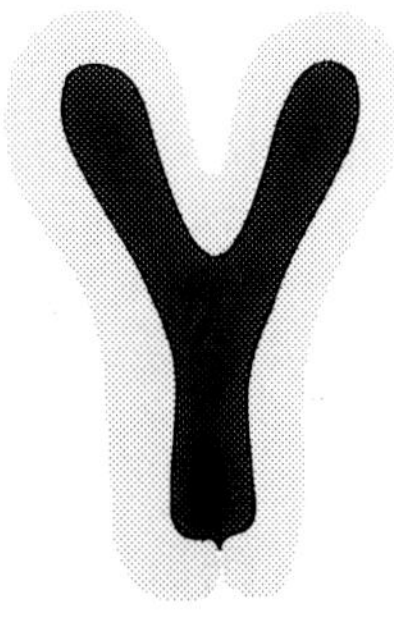

Figure 4.14. Hysterosalpingographic appearance of septate (*left*) and bicornuate (*right*) uterus may not be differentiated. External evaluation of the uterine fundus is necessary for precise diagnosis.

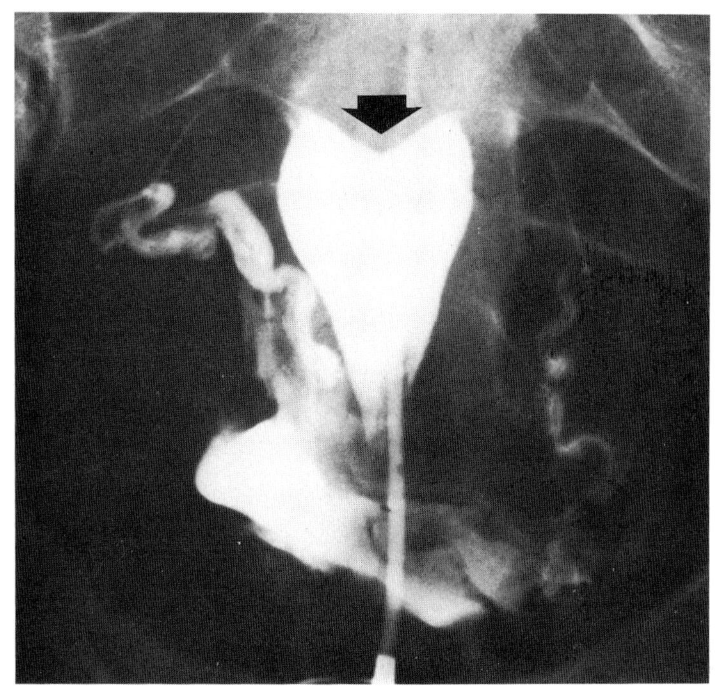

Figure 4.15. Septate uterus. Small septum characterized by cleft-like defect in the fundus (*arrow*).

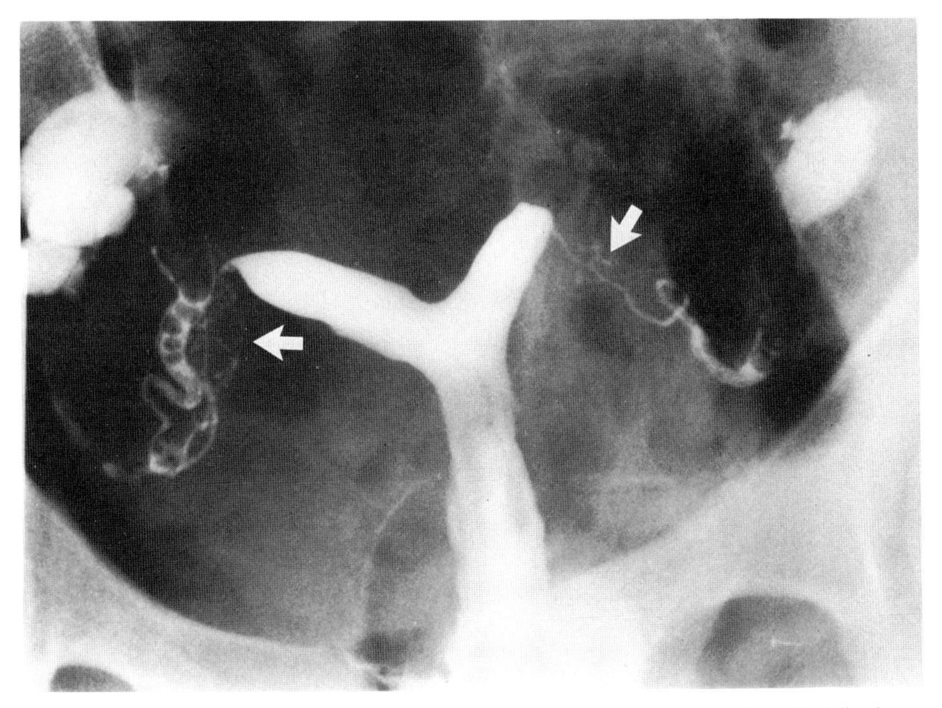

Figure 4.16. Septate uterus in woman with bilateral hydrosalpinges and salpingitis isthmica nodosa (*arrows*).

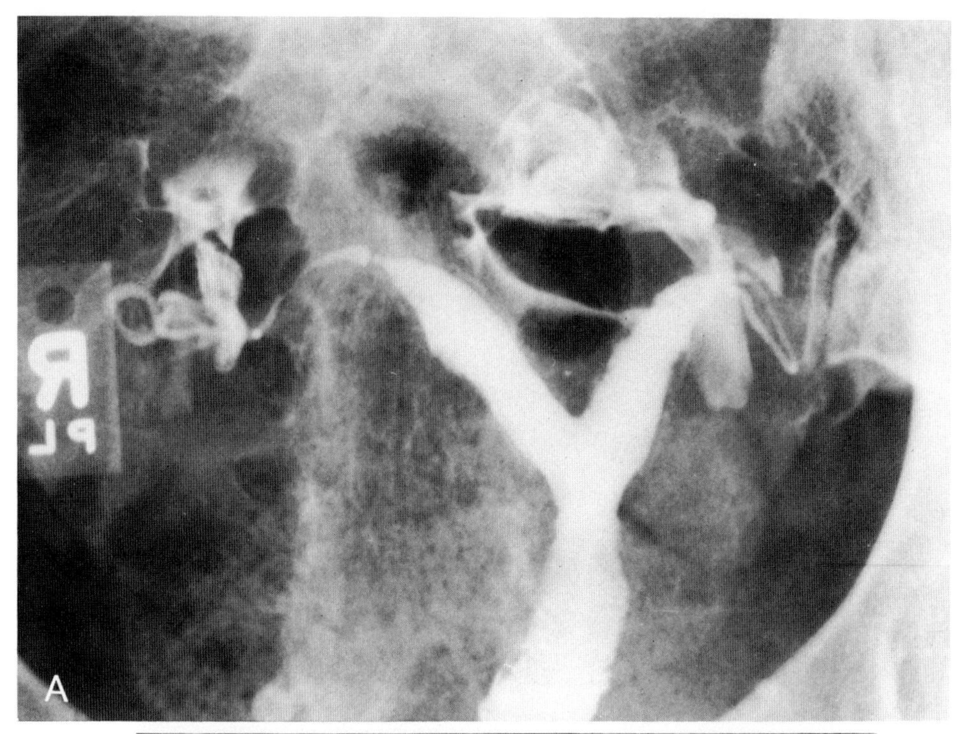

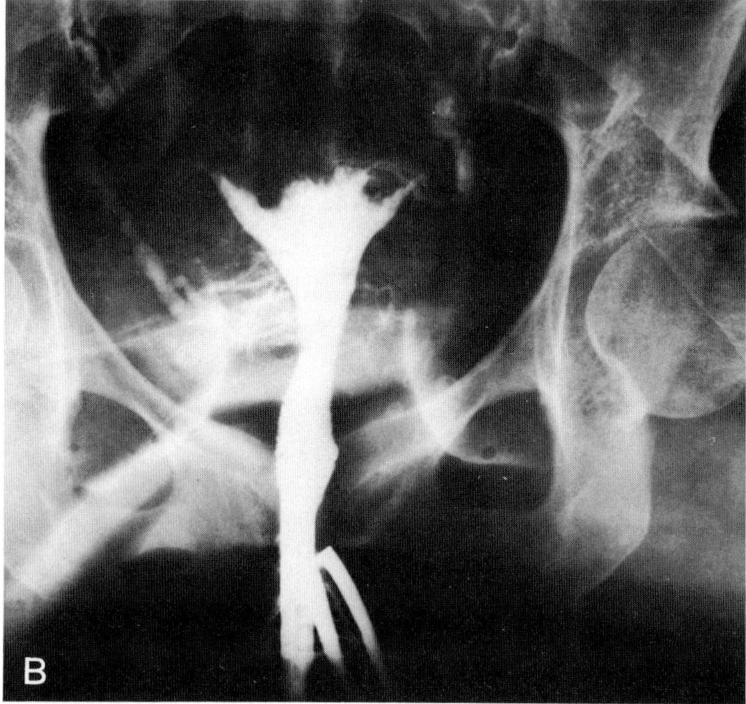

Figure 4.17. *A:* Septum formation. *B:* Postoperative hysterosalpingogram after metroplasty to correct defect (see Chapter 9).

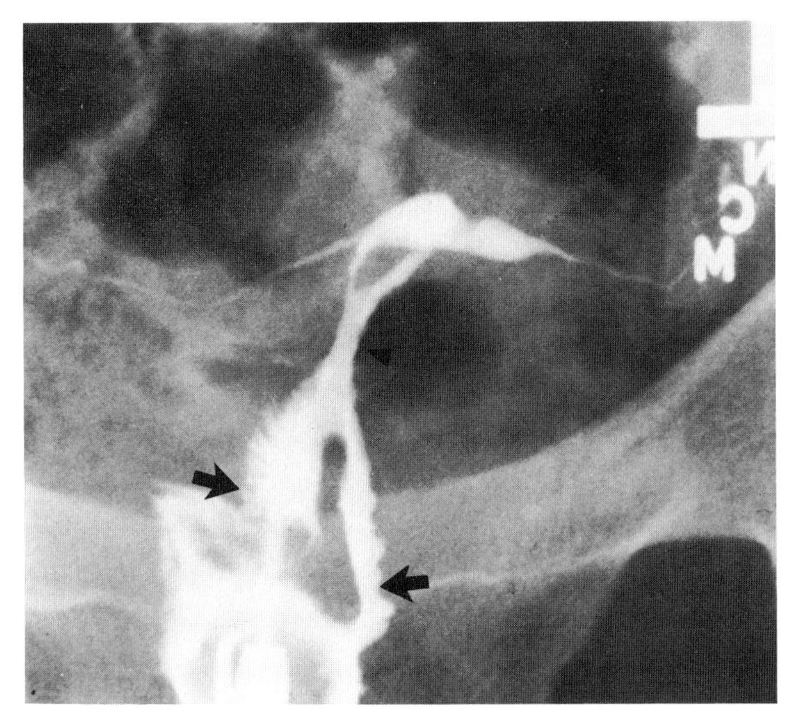

Figure 4.18. Communicating uterus. Two cervices (*arrows*) and luminal separation of the fundus (*small arrows*) with a zone of communication centrally (*arrowhead*).

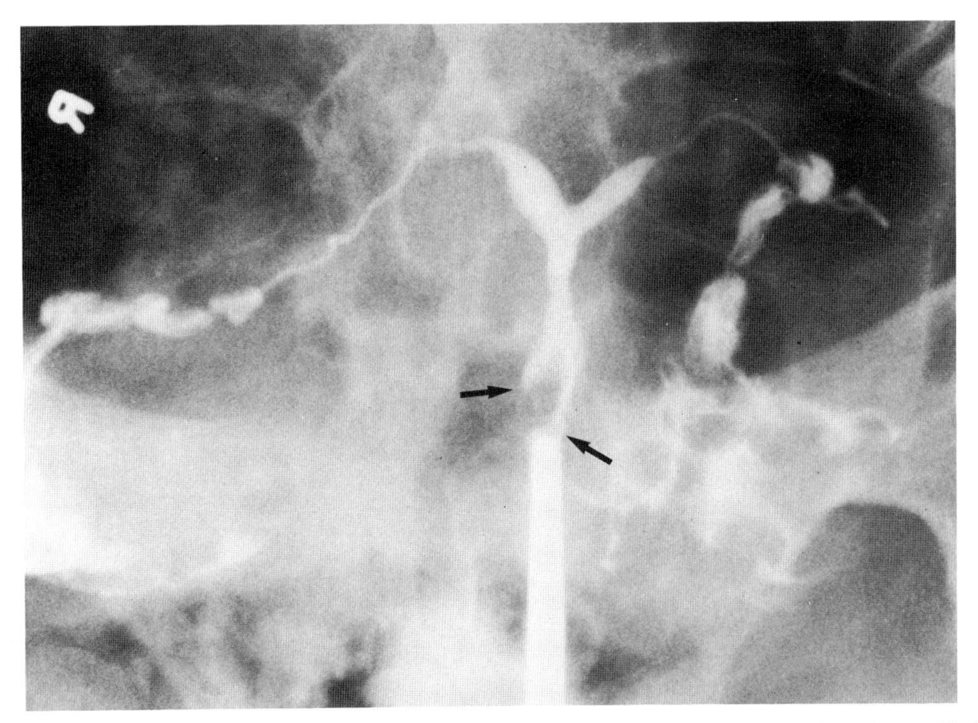

Figure 4.19. Unusual communication. Clinically, this patient presented as a didelphys uterus with two cervical ora (*arrows*). The major portion of the uterine lumen is united with a bicornuate configuration at the fundus.

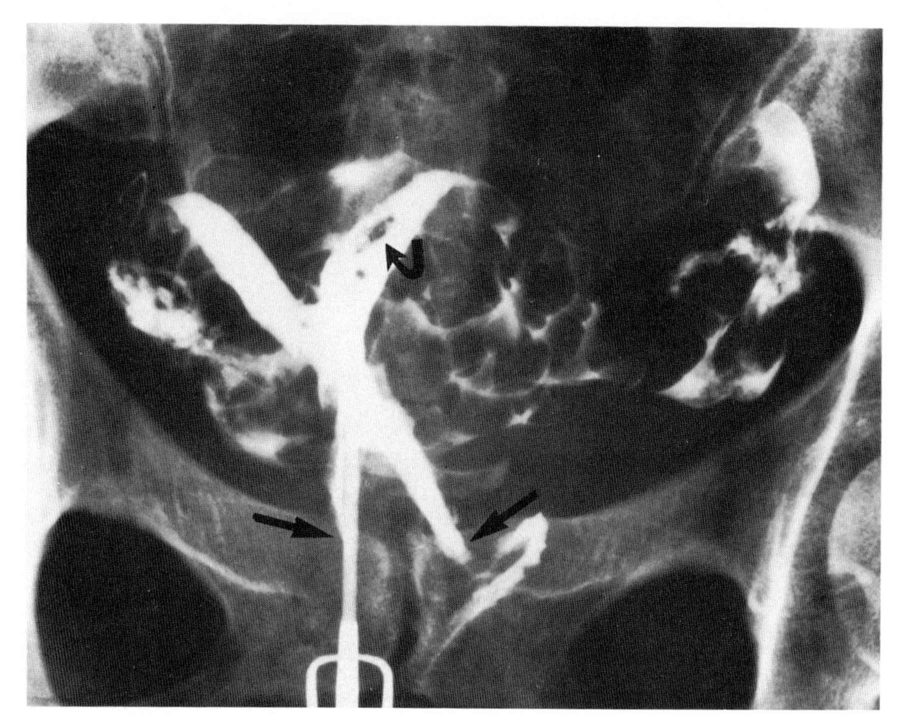

Figure 4.20. Uterus didelphys presentation with two cervices (*arrows*). The bodies communicate and there are two distinct horns at the fundus. Laparoscopic observation demonstrated that the fundal configuration was septate rather than bicornuate. Incidental note is made of the synechia in the left uterine horn (*curved arrow*).

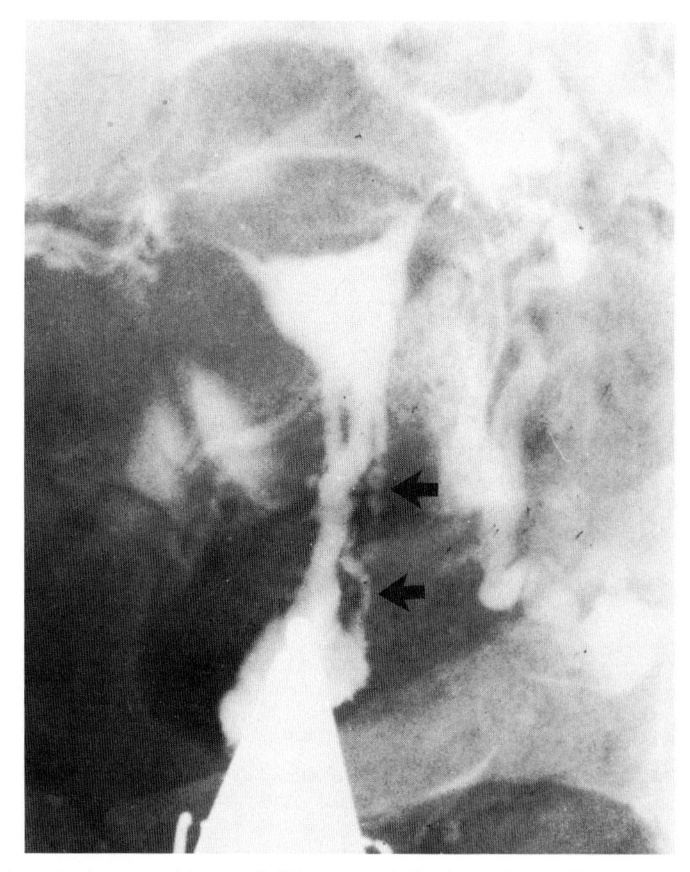

Figure 4.21. Gartner duct remnant (*arrows*). This anomaly is characterized by a line of contrast parallel to the uterine cavity. (Courtesy of L. Ekelund, Lund, Sweden.)

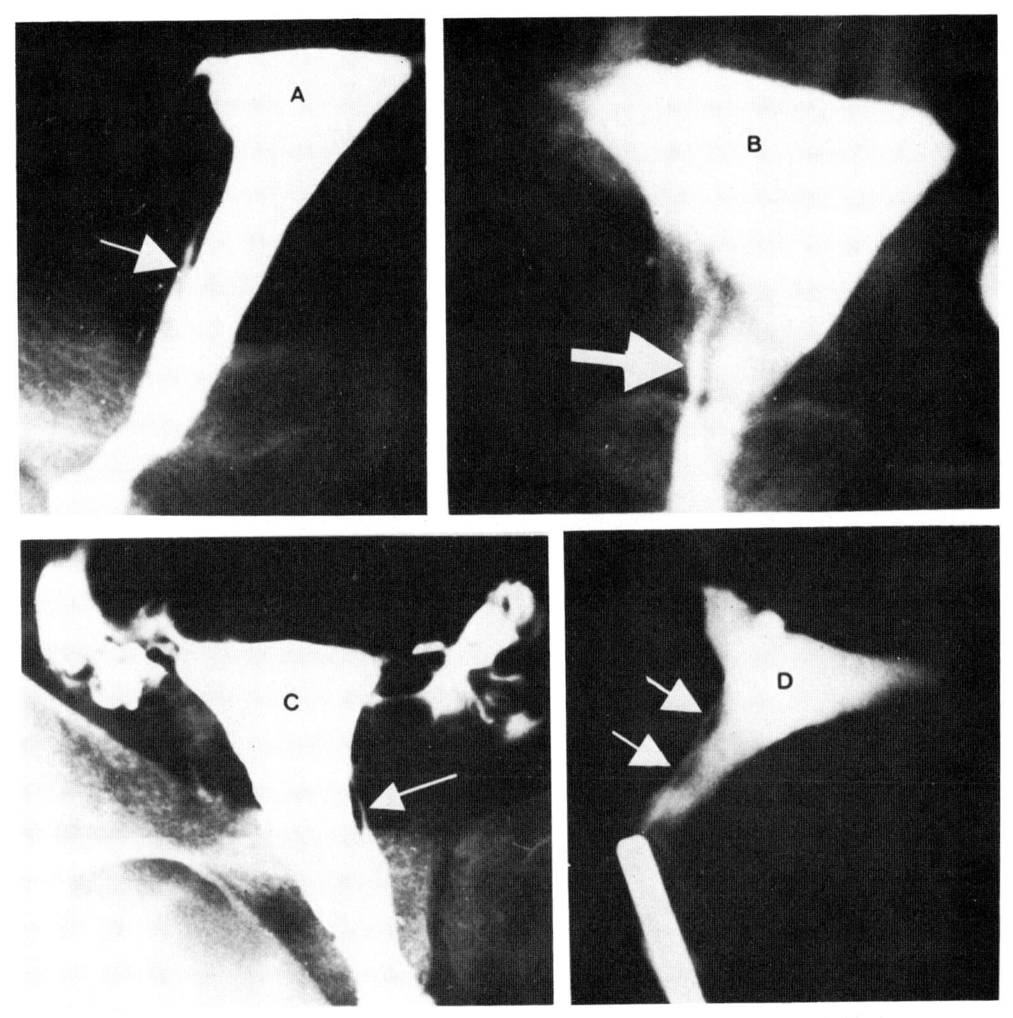

Figure 4.22. Four examples of Gartner duct remnants characterizing their variability in appearance. (Reprinted by permission from Katz Z, Bernstein D, Lancet M: A possible causal relationship between mesonephric remnants and infertility of uterine origin. *Int J Fertil* 27:125, 1982.)

margins of the endometrial cavity (Figs. 4.21 and 4.22) (8).

Reproductive Potential

Pregnancy outcome for patients with congenital anomalies of the uterus obviously depends upon the degree of anomaly, and to a certain extent upon the operative procedure employed. Clearly, patients fitting the Buttram and Gibbons Class I have a very low potential for fertility. Uterus didelphys, and a bicornuate uterine cavity, have little effect on fertility potential. Nonetheless, unification procedures were attempted in the past (Fig. 4.23). Significant septal deformities are more frequently associated with fetal wastage than infertility. On the other hand, patients with a septate uterus undergoing wedge metroplasty (29, 30), Tompkins metroplasty (31), or other operative procedures (32) have an expectation of greater than 75% chance of live birth. Although postoperative hysterosalpingography may show a markedly contracted or irregular uterine cavity, pregnancy rates and full-term delivery rates are excellent, and complications of pregnancy

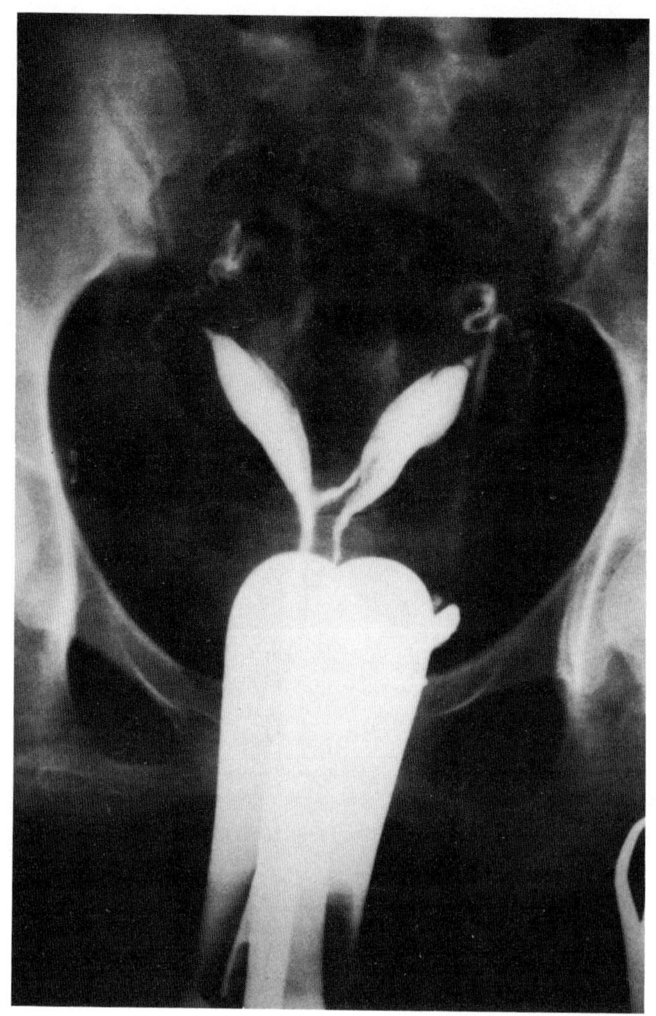

Figure 4.23. Unification procedure surgically connecting the two chambers of a didelphys uterus. This procedure has no place in the modern management of such anomalies.

not markedly greater than in the infertile population in general.

Diagnostic Techniques in Patients with Congenital Abnormalities

Hysterosalpingography in the patient with genital tract anomalies may require modifications of technique and meticulous care to assess the variation adequately. The limitations of differentiating septate and bicornuate uteri have been mentioned previously. The appearance of a unicornuate uterus demands particular care. Such a configuration may indeed represent a true unicornuate uterus, but is much more likely to signify the presence of either an associated rudimentary horn or, even more commonly, a didelphic uterus of which only one side is filled with contrast (Fig. 4.24). Diligent search for a second cervical os is required, particularly since a vaginal septum may obscure this os from obvious view. Removal or replacement of the vaginal speculum may be needed.

In most of these situations, there is not a great deal of maneuvering room available for placement of instrumentation. Knowl-

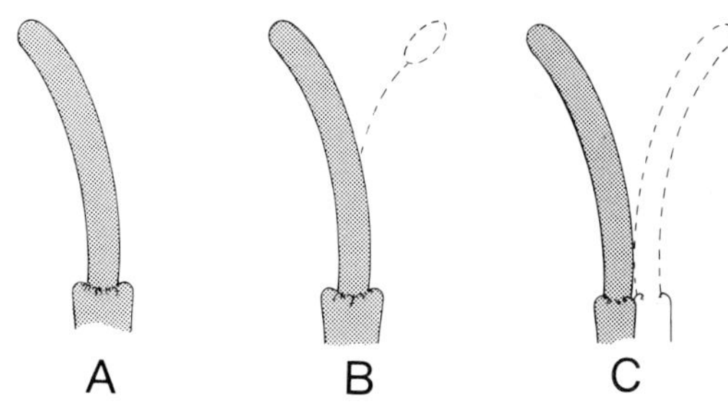

Figure 4.24. The radiographic appearance of a unicornuate uterus may have any of the following three explanations: (*A*) true unicornuate uterus; (*B*) a unicornuate uterus with a noncommunicating or "blind" horn; and (*C*) filling of only one side of a uterus didelphys. The third possibility is the most common.

edge of previous findings is valuable to the radiographer prior to initiation of the examination. Frequently, a double cervix will appear with the external cervical openings facing the lateral vaginal walls. Under these circumstances, a flexible means of instrumentation, perhaps with the use of a flexible polyethylene acorn or a small Foley catheter, will be likely to succeed. Under certain circumstances, when the vaginal septum does not extend all the way to the cervix, it may be difficult to ascertain which cervix is which. Placing methylene blue or indigo carmine dye on the anterior lip of one cervix will allow identification of a second cervix as the vaginal septum is pressed to one or another side. The use of the Kidde cannula in conjunction with the Foley catheter is frequently helpful, because of the need of one rigid and one more flexible instrument.

Once the instrumentation has been selected, it is useful to place both within the separate cervices, and to make absolutely sure that occlusion of the external os is accomplished on both sides. Then one may elect to introduce contrast into one side alone and to watch its progression under fluoroscopy. Having demonstrated the contours and anatomy of one side, the next side then can be injected with contrast and further information about the anatomic details delineated. It is not wise to introduce contrast simultaneously, because information is lost about midline septa within the uterine cavity, and sometimes about tubal patency.

On occasion, a rudimentary horn may open into the endocervical canal. The radiographer should be aware of this possibility, because the use of a short cannula is recommended in such a case unless it is impossible to get a decent fit with such rigid apparatus. Similarly, a short cannula is helpful in the presence of a deep septum to allow filling of both cornual areas.

Clearly, the use of ultrasonography, laparoscopy, hysteroscopy, and IVP is valuable to complete delineation of anatomic abnormalities in patients with congenital defects. The major point is that flexibility must be maintained, and plans made for delineation of the defects before the patient appears in the radiologist's office. Amassing the necessary equipment in duplicate is of importance to establishing the most information with the least time and inconvenience.

REFERENCES

1. Prins RP, Morrow CP, Townsend DE, Disaia PJ: Vaginal embryogenesis, estrogens, and adenosis. *Obstet Gynecol* 48:246, 1976.
2. Ulfelder H, Robboy SJ: The embryologic development of the human vagina. *Am J Obstet Gynecol* 126:769, 1976.
3. Koff AK: Development of the vagina in the human foetus. *Contrib Embryol Carnegie Inst* 24:59, 1933.
4. Forsberg JG: Cervicovaginal epithelium: Its origin and development. *Am J Obstet Gynecol* 115:1025, 1973.
5. Craig CFT: Congenital abnormalities of the uterus and foetal wastage. *S Afr Med J* 48:2000, 1973.
6. Semmes JP: Congenital defects of the reproductive tract: Clinical implications. *Contemp Obstet Gynecol* 5:95, 1975.
7. Tulandi T, Arronet GH, McInnes RA: Arcuate and bicornuate uterine anomalies and infertility. *Fertil Steril* 34:362, 1980.
8. Greiss FC Jr, Mauzy CH: Genital anomalies in women. An evaluation of diagnosis, incidence and obstetric performance. *Am J Obstet Gynecol* 82:330, 1961.
9. Green LK, Harris RE: Uterine anomalies. Frequency of diagnosis and associated obstetric complications. *Obstet Gynecol* 47:427, 1976.
10. Polishuk WZ, Ron MA: Familial bicornuate and double uterus. *Am J Obstet Gynecol* 119:982, 1974.
11. Carson SA, Simpson JL, Malinak LR, Elias S, Gerbie AB, Buttram VC Jr, Sarto GE: Heritable aspects of uterine anomalies. II. Genetic analysis of Mullerian aplasia. *Fertil Steril* 40:86, 1983.
12. Verp MS, Simpson JL, Elias S, Carson SA, Sarto

GE, Feingold M: Heritable aspects of uterine anomalies. I. Three familial aggregates with Mullerian fusion anomalies. *Fertil Steril* 40:80, 1983.
13. Elias S, Simpson JL, Carson SA, Malinak LR, Buttram VC Jr: Genetics studies in incomplete Mullerian fusion. *Obstet Gynecol* 63:276, 1984.
14. Williamson JG: True unicornuate uterus. A report of two pregnancies. *Int J Gynaecol Obstet* 11:233, 1973.
15. Andrews MC, Jones HW Jr: Impaired reproductive performance of the unicornuate uterus: Intrauterine growth retardation, fertility, and recurrent abortion in five cases. *Am J Obstet Gynecol* 144:173, 1982.
16. Jones HW Jr, Wheeless CR: Salvage of the reproductive potential of women with anomalous development of the Mullerian ducts: 1868–1968–2068. *Am J Obstet Gynecol* 104:348, 1969.
17. Zourlas PA: Surgical treatment of malformations of the uterus. *Surg Gynecol Obstet* 141:57, 1975.
18. Buttram VC Jr, Gibbons WE: Mullerian anomalies: A proposed classification (an analysis of 144 cases). *Fertil Steril* 32:40, 1979.
19. Heinonen PK, Saarikoski S, Pystynen P: Reproductive performance of women with uterine anomalies. *Acta Obstet Gynecol Scand* 61:157, 1982.
20. Worthen NJ, Gonzalez F: Septate uterus: Sonographic diagnosis and obstetric complications. *Obstet Gynecol* 64(Suppl):34S, 1984.
21. Heinonen PK, Pystynen PP: Primary infertility and uterine anomalies. *Fertil Steril* 40:311, 1983.
22. Woolf RB, Allen WM: Concomitant malformations of the reproductive and urinary tracts. *Obstet Gynecol* 2:236, 1953.
23. Muller P: Association of genital and urinary malformations in women. *Gynecology* 165:285, 1968.
24. Wiersma AF, Peterson LF, Justema EJ: Uterine anomalies associated with unilateral renal agenesis. *Obstet Gynecol* 47:654, 1976.
25. Gurin J, Leiter E: Associated anomalies of Mullerian and Wolffian duct structures. *S Med J* 74:805, 1981.
26. Buttram VC Jr: Mullerian anomalies and their management. *Fertil Steril* 40:159, 1983.
27. Toaff ME, Lev-Toaff AS, Toaff R: Communicating uteri: Review and classification with introduction of two previously unreported types. *Fertil Steril* 41:661, 1984.
28. Katz Z, Bernstein D, Lancet M: Possible causal relationship between mesonephric remnants and infertility of uterine origin. *Int J Fertil* 27:125, 1982.
29. Rock JA, Jones HW Jr: The clinical management of the double uterus. *Fertil Steril* 28:798, 1977.
30. Muasher SJ, Acosta AA, Garcia JE, Rosenwaks Z, Jones HW Jr: Wedge metroplasty for the septate uterus: An update. *Fertil Steril* 42:515, 1984.
31. McShane PM, Reilly RJ, Schiff I: Pregnancy outcomes following Tompkins metroplasty. *Fertil Steril* 40:190, 1983.
32. Musich JR, Behrman SJ: Obstetric outcome before and after metroplasty in women with uterine anomalies. *Obstet Gynecol* 52:63, 1978.

5 Diethylstilbestrol Exposure In Utero

Both the specialist in infertility and the radiologist will encounter at hysterosalpingography the anomalies of the uterine cervix, corpus, and tubes associated with diethylstilbestrol (DES) exposure in utero. This synthetic estrogen was administered to pregnant women for almost 25 years, beginning about 1948, for the indications of recurrent miscarriage, hypertension, diabetes mellitus, threatened abortion, previous stillbirth, and/or premature labor. Peak usage was in the early 1950s, and declined abruptly after 1971, when an association was observed between the administration of DES during pregnancy and the subsequent development years later of a clear cell adenocarcinoma of the vagina in the female offspring exposed in utero (1–3). Various structural and cytologic abnormalities of the reproductive system have since been described and characterized (4–9). A brief description of normal vaginal embryogenesis is necessary to understand the development of these abnormalities (10–11).

Embryologic Development of the Human Vagina

Tissue of both the müllerian ducts and the urogenital sinus are required for normal vaginal and cervical development. At 4 weeks of embryonic life, the paired müllerian (paramesonephric) duct system forms an invagination of celomic epithelium in the urogenital fold just lateral to the cranial end of the mesonephric ducts. As development proceeds, the paired müllerian ducts meet at the midline, and then grow caudally to reach the urogenital sinus or future hymen at 7 weeks of gestational age. The paired ducts then fuse. The septum formed by the fusion disappears both caudally and cranially, resulting in the formation of a single cavity, the uterovaginal canal, lined by columnar epithelium. Beginning at about the 10th week of gestation, squamous epithelium derived from the urogenital sinus invades the distal end of the uterovaginal canal from below and extends cranially, uniformly displacing the columnar müllerian epithelium to the level of the future external cervical os. The vagina, which was initially lined by simple columnar epithelium, now has acquired a stratified squamous epithelium, a process that is completed by the 18th week of gestation.

Robboy (12, 13) has theorized and evidence is increasingly available that mesenchyme in the developing cervix has inductive functions with respect to overlying epithelial cells. Normal müllerian tube stroma, probably by producing a "stromal inductive factor," is capable of inducing the development of a columnar epithelium that later differentiates into the adult-type tuboendometrial lining. The squamous epithelium of urogenital sinus origin, which lines the vagina and future exocervix, is impervious to any inductive stimulus to differentiate into glandular cells and therefore remains a squamous layer of cells.

In the 14th week postovulation the cervical wall thickens, and the vaginal fornices form from the vaginal plate a week later. The muscular wall of the uterus is completely developed by 26 weeks postovulation. In the DES-exposed fetus, the estrogen causes derangement of the formation of the two mesenchymal layers. DES administered during the time of differentiation of the cervix and vagina will induce deformities (see Figs. 5.1, 5.2, and 5.3). The layers fail to segregate, becoming clinically manifest in the young adult as stromal hyperplasia (ridges), hypoplasia (hypoplastic fornices), or an abnormally contoured uterine cavity (revealed by an abnormal hysterosalpingogram). The superficial stroma in the region of the future cervix splays centrifugally, causing an unusually wide zone of mucinous epithelium that surrounds the anatomic portio vaginalis and upper vagina. The DES-exposed vaginal mesenchyme impedes further upward growth of squamous cells of urogenital sinus origin, which results in the induction of the native (residual) embryonic müllerian epithelium to differentiate into adult-type tuboendometrial epithelium (adenosis). Thus, the normal sequence of developmental events is deranged, resulting in characteristic physical, anatomic, and histologic changes. DES exposure in very early pregnancy and after 22 weeks gestational age is unlikely to cause structural abnormalities.

Clinical Findings

Examination of the vagina and cervix is frequently diagnostic of DES exposure. The squamocolumnar junction is displaced caudad from its ordinary site just within the external cervical os to varying levels of the vaginal canal (Fig. 5.1). Columnar epithelium on the outer cervix and upper vagina has the appearance of a wet, mucinous, inflamed surface, which exudes mucoid material and bleeds easily (Fig. 5.2).

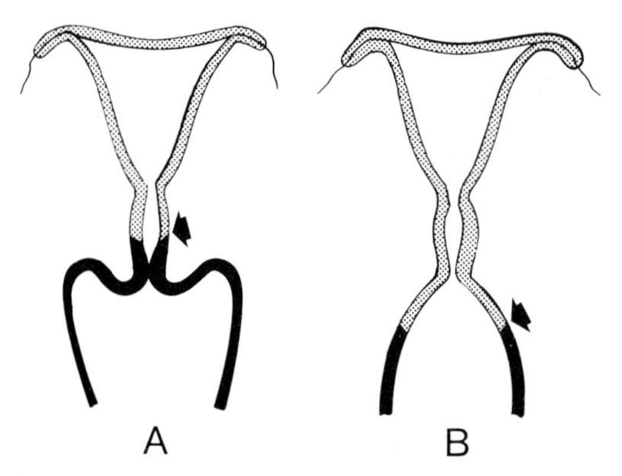

Figure 5.1. *A*: Diagram demonstrating the squamocolumnar junction (*arrow*) in a normal position above the external cervical os. *B*: Squamocolumnar junction (*arrow*) displaced caudally and located on the cervix or within the vagina in a patient exposed in utero to DES.

Structural abnormalities of the cervix and vagina are noted in such women, with varying frequency and severity. Cervical hypoplasia, cervical pseudopolyps, and a transverse or circumferential ridge, the so-called cock's comb or cervical hood, have been described (Fig. 5.3). Marked cervical stenosis may occur, and a pinpoint cervical os may be difficult to cannulate with the Kidde cannula or other hysterosalpingographic equipment. The os may be completely obscured by a cervical hood that folds over like a shelf, and the whole area may be so friable that persistent probing to find the os may result in significant bleeding. Islands of reddish tissue, which are identified as vaginal adenosis, may be observed on the vaginal walls in some women.

Abnormalities of the cavity of the uterus or the fallopian tubes are found in two-thirds of DES-exposed women (Fig. 5.4) (14). Up to 50% of exposed women may have gross structural abnormalities of the cervix and vagina, including ridges, hoods, and cock's combs. Of those with uterine anomalies, many will have associated cervical and vaginal changes. Over three-quarters of women with cervical anomalies will be found at hysterosalpingography to have uterine defects. On the other hand, over 50% of exposed women *without* cervical anomalies had uterine abnormalities, verifying the need for hysterosalpingography in all DES-exposed women, not just those with cervical abnormalities. The occurrence of clear cell adenocarcinoma of the vagina has been noted in this population of women, and in fact directed much attention to the study of this abnormality. Fortunately, the incidence of this otherwise rare vaginal neoplasm is relatively low, 0.14 to 1.4 per 1000 exposed (2), and it remains an unusual tumor. However, it is estimated that 100% of exposed women have vaginal adenosis.

Surprisingly, the anatomic findings in DES-exposed women may change with time, and are not necessarily permanent (15). Of women initially described as having a cervicovaginal hood, about one-half had a de-

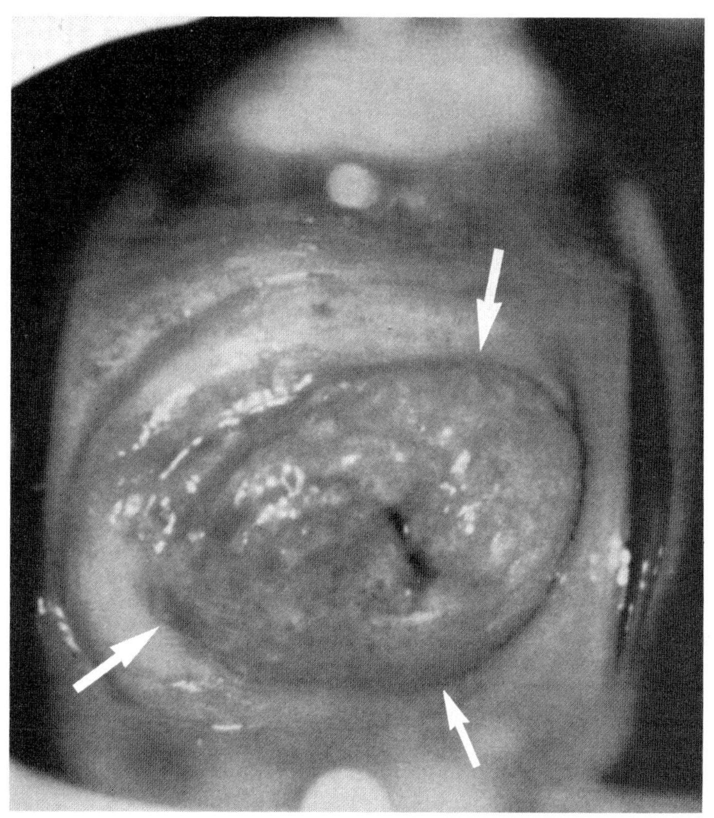

Figure 5.2. The displacement of the normal stratified squamous epithelium of the cervix by the columnar epithelium is shown. The squamocolumnar junction (*arrows*) is displaced out onto the cervix, for a so-called cervical collar. Although the external cervical os is easily visible, this area is friable and will bleed easily. (Courtesy of H.W. Jones, III, M.D., Nashville, Tennessee.)

crease in its size and one-quarter had complete disappearance of the hood over 5 years of observation. The longer the follow-up period, the greater the extent of resolution of these lesions.

Findings on Hysterosalpingography

The classic hysterosalpingographic finding is that of a T-shaped uterus, which is characterized by a small endometrial cavity, shortened upper uterine segment, and narrow endocervical canal (Fig. 5.5) (Table 5.1). In addition, the endometrial cavity may be lumpy and irregular, rather than smooth and symmetrical (Figs. 5.6 through 5.11). The external configuration of the uterus at laparoscopy may remain entirely within normal limits, or there may be varying degrees of severity of malformation. The fallopian tubes may also exhibit abnormalities, but these are only observed externally at laparoscopy. A unique tubal morphologic feature consisting of a foreshortened, convoluted tube with a "withered" fimbria and pinpoint fimbrial opening has been described (8), but the fimbrial abnormality and tubal shortening cannot be distinguished at hysterosalpingography.

In terms of size, the cavity of the uterus is usually compromised, and the area of the endometrial cavity, length of the upper uterine segment, and diameter of the endocervical canal are all significantly smaller in the DES-exposed group (5).

The urinary tract appears to have been spared abnormalities of development, and DES-exposed women with upper genital

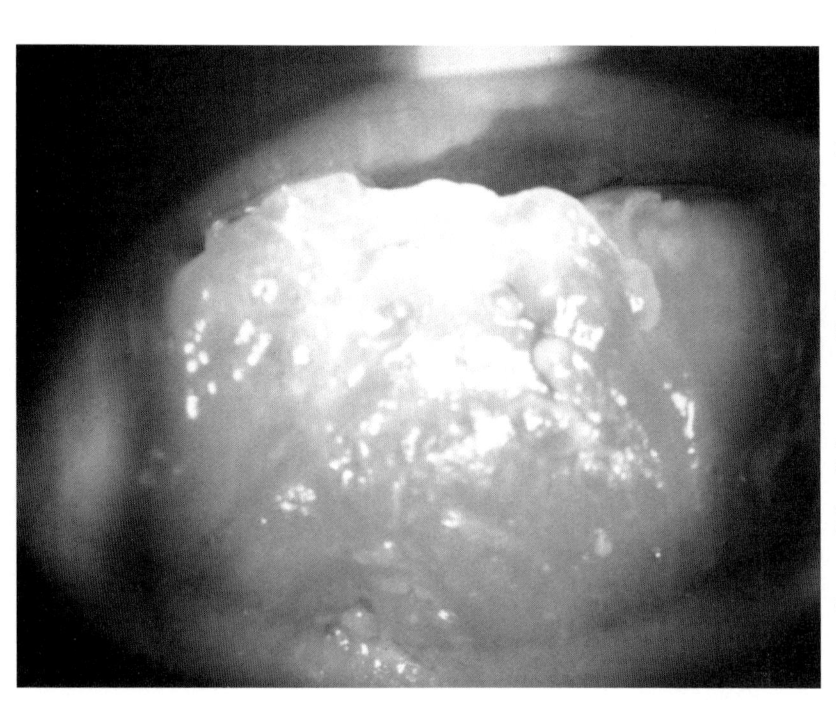

Figure 5.3. Columnar epithelium arranged in a cock's comb pattern, obscuring the external cervical os. This epithelium is extremely friable and bleeds readily on touch.

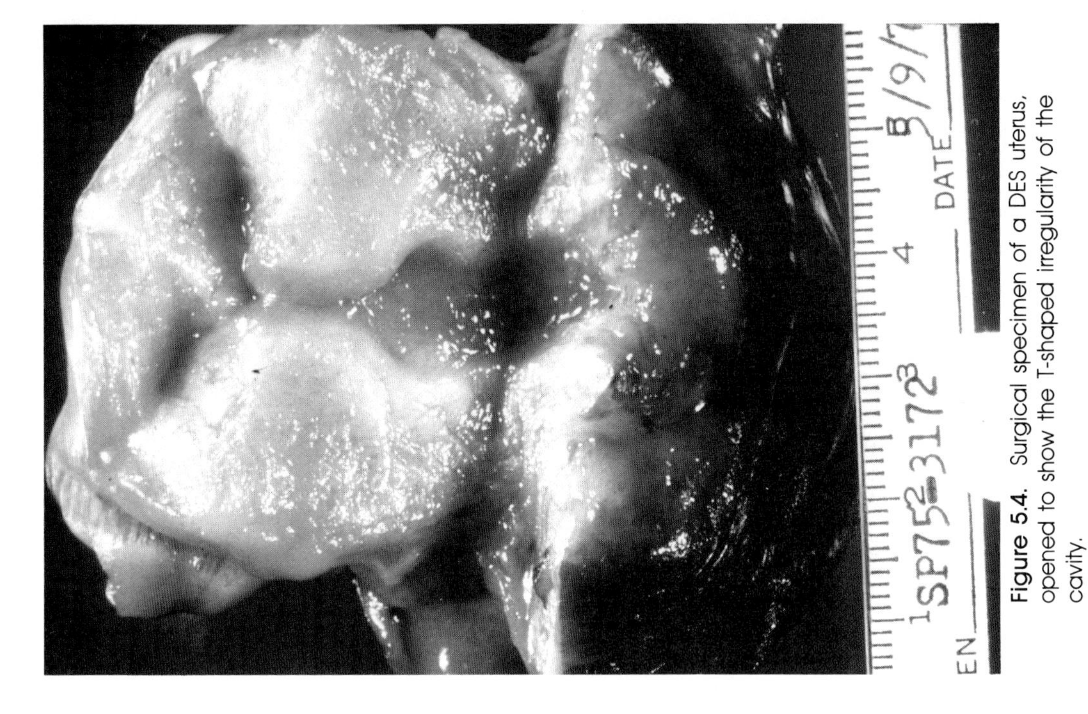

Figure 5.4. Surgical specimen of a DES uterus, opened to show the T-shaped irregularity of the cavity.

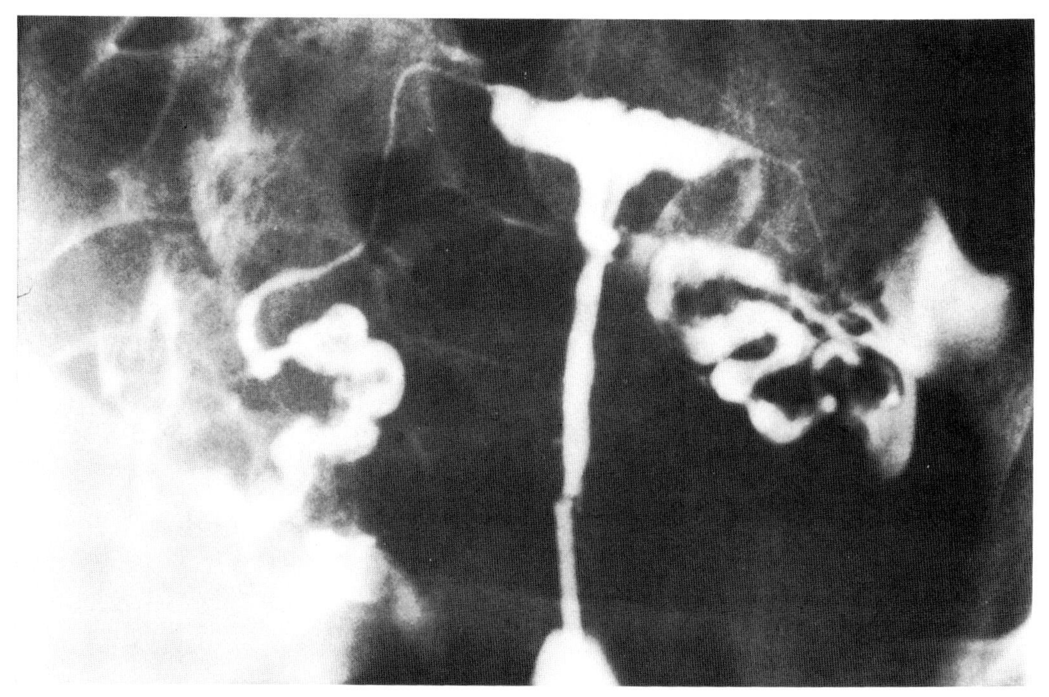

Figure 5.5. Characteristic hysterosalpingographic appearance of uterus in a woman exposed to DES in utero. Note the T-shaped configuration, the very long endocervical canal, and the lumpy and irregular contour to the endometrial cavity.

tract abnormalities on hysterosalpingograms have no increase in urinary tract abnormalities seen on intravenous pyelograms compared to a control population.

Fertility of DES-exposed Women

Conflicting reports have been published concerning the fertility of DES-exposed women (16–19). Cousins et al. (20) did not observe any differences between the incidence of pregnancy, mean number of pregnancies, or frequencies of infertility problems, ectopic pregnancy, or spontaneous abortion in 71 DES-exposed women compared to 69 nonexposed women (Fig. 5.12); the majority of other investigators dispute these findings in the early pregnant group (14, 16–19, 21–28). On the other hand, there is general agreement that late pregnancy complications, premature delivery, and perinatal death are all increased in the group with obvious changes. Herbst and associates (19), studying sexually active women who did not practice contraception, found an 86% conception rate in controls versus 67% in DES daughters. Herbst also found that twice as many exposed as unexposed women had tried unsuccessfully to become pregnant for at least 1 year. Mangan et al. (27) reported that DES daughters achieved

Table 5.1
Hysterosalpingographic Changes in DES-exposed Women

1. A T-shaped uterus with bulbous cornual extensions arising from the upper end of the uterine cavity.
2. Lumpy, irregular appearance of the walls of the uterine cavity.
3. Hypoplastic uterus.
4. Relatively long endocervical canal.
5. Narrowed lower two-thirds of the uterine cavity.
6. Band-like constrictions causing narrowing of the interstitial tubal segments.

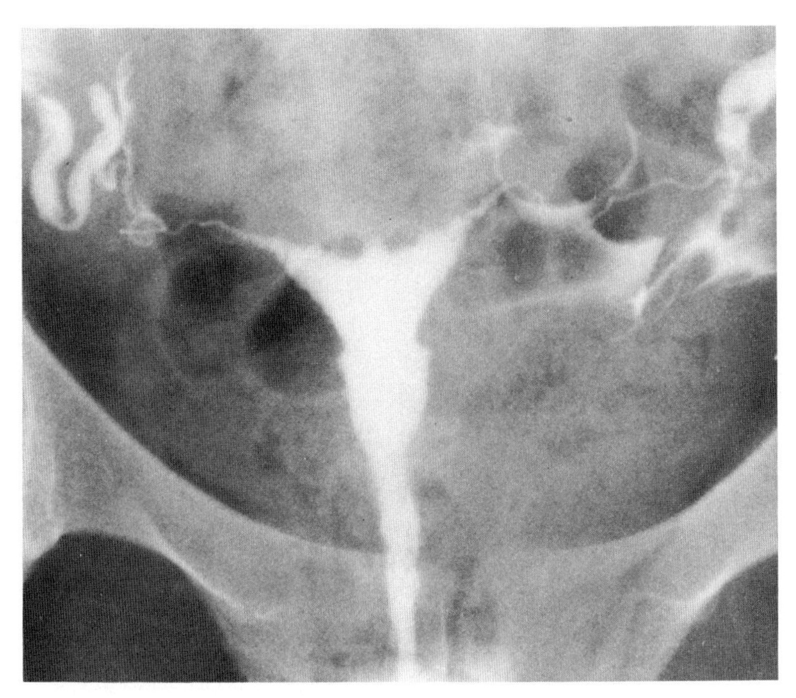

Figure 5.6. Typical T-shaped uterus with characteristic lumpy configuration. There is considerable symmetry of contour, not an unusual characteristic.

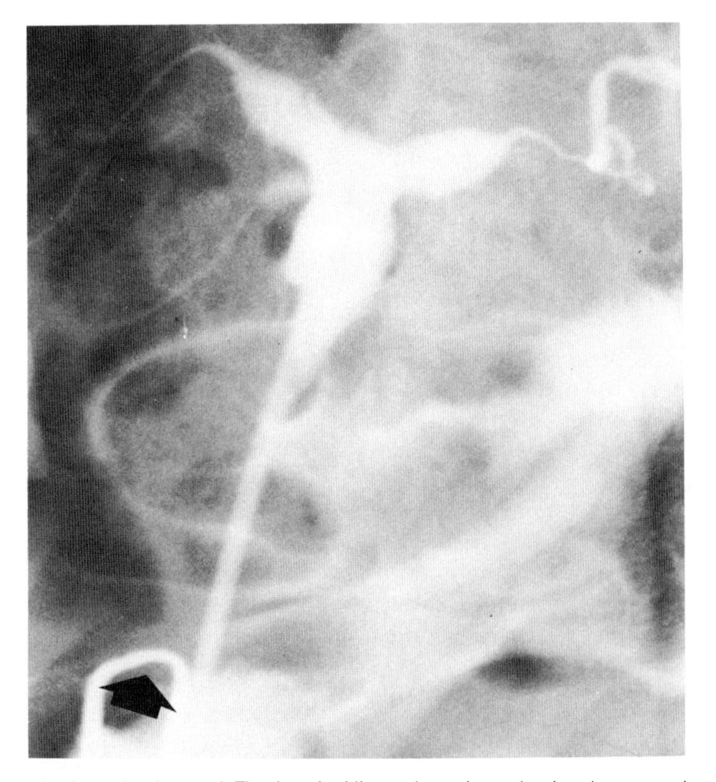

Figure 5.7. Elongated cervical canal. The level of the external cervical os is recognized by the position of the tenaculum (*arrows*). Note lumpiness of the endometrial cavity.

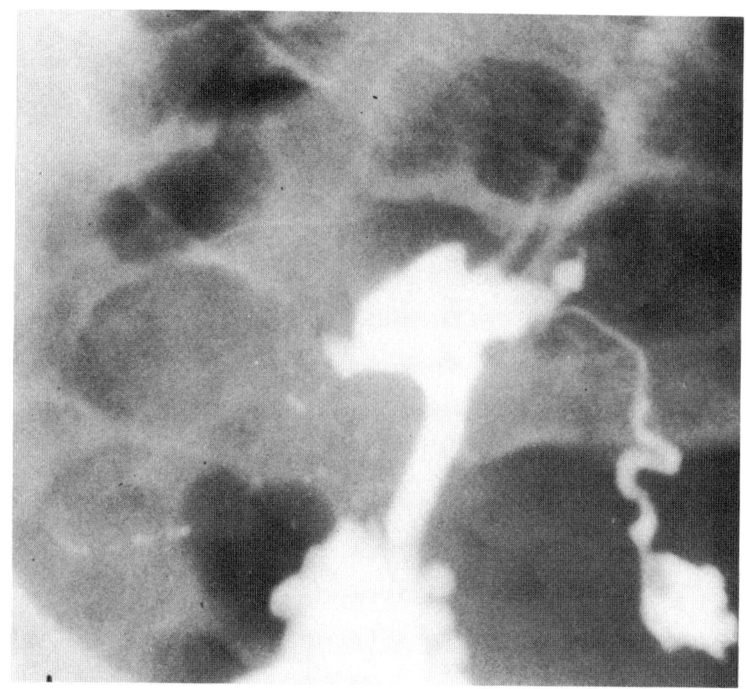

Figure 5.8. Use of oil-soluble rather than water-soluble contrast alters slightly the sharpness of contour of the endometrial cavity. Otherwise the characteristic appearance is unchanged.

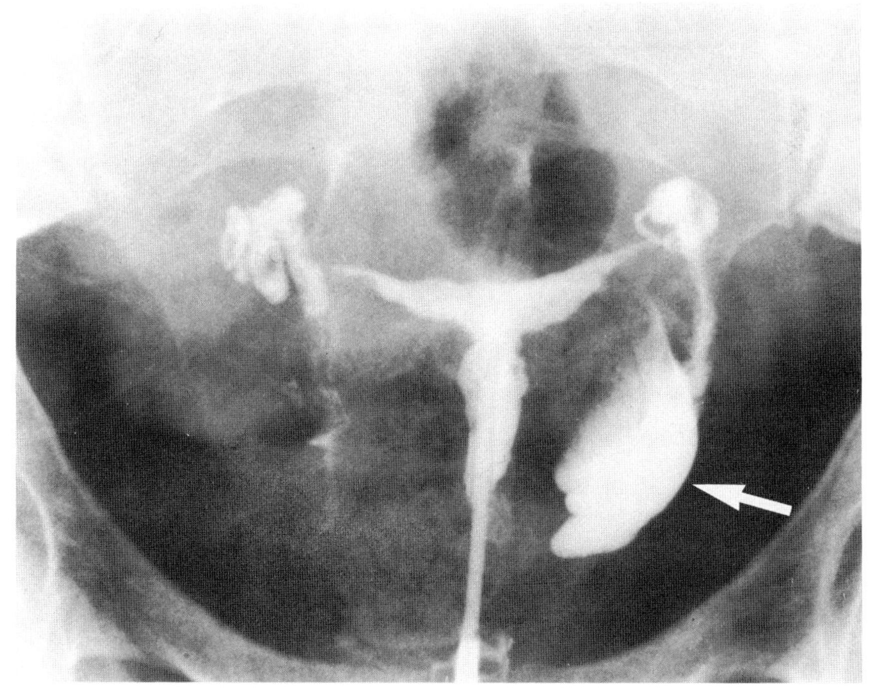

Figure 5.9. T-shaped uterine cavity. Configuration typical for intrauterine DES exposure. Unrelated coincidental hydrosalpinx on left (*arrow*).

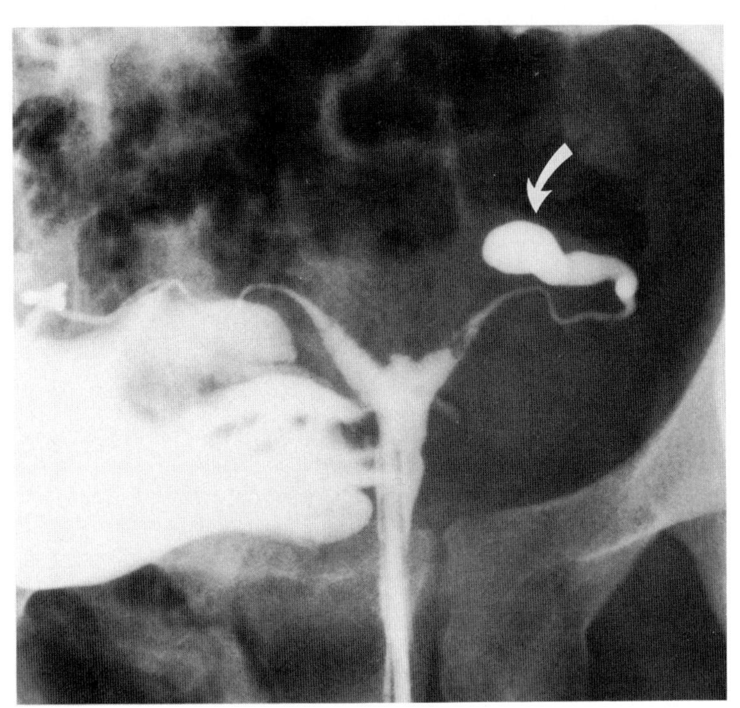

Figure 5.10. DES exposure characterized by the long endocervical canal and a lumpy configuration to the endometrial cavity. Bicornuate configuration, an anomalous variant unrelated to the diethylstilbestrol exposure, alters the characteristic T-shaped pattern. Coincidental hydrosalpinx on left (*arrow*).

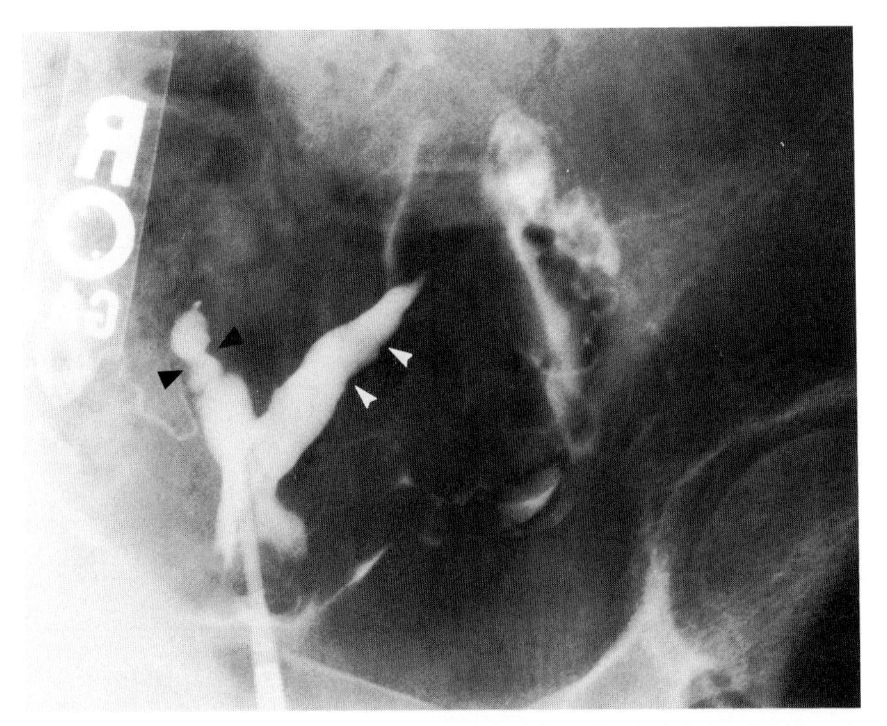

Figure 5.11. Documented history of maternal exposure to diethylstilbestrol during first trimester of pregnancy. Only stigma of DES exposure seen on hysterosalpingogram is the very lumpy configuration of the endometrial cavity (*arrowheads*). The bicornuate configuration is an unrelated müllerian variation.

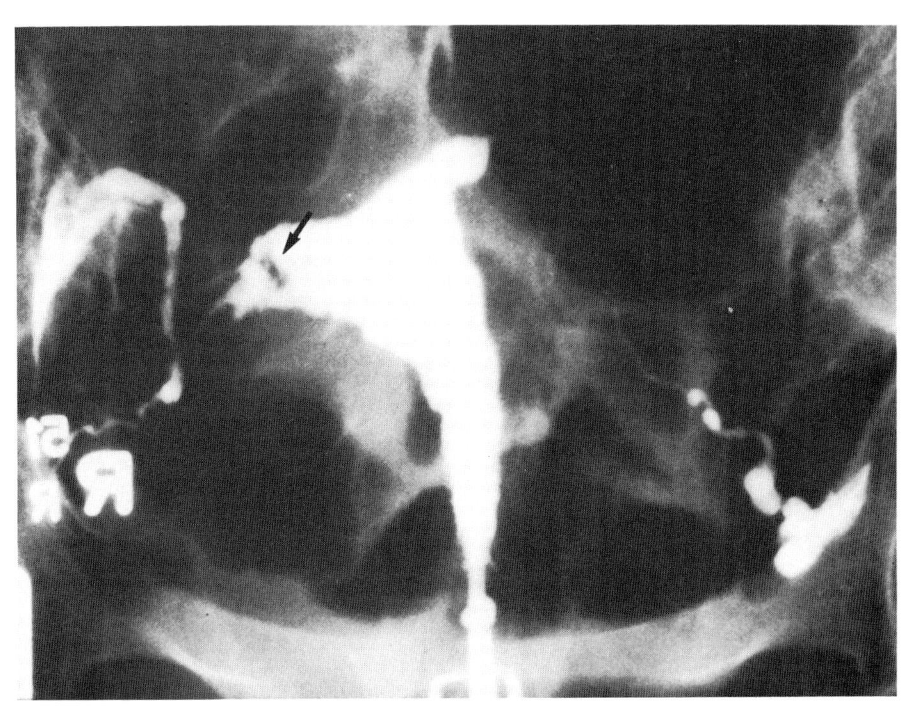

Figure 5.12. Known DES exposure in a multiparous patient. The lumpy and irregular configuration of the endometrial cavity is considered typical. The relatively normal size and overall shape of the cavity may well be a reflection of the previous pregnancies. Note the small synechia in the right cornua (*arrow*).

a lower percentage of desired pregnancies than did controls. Barnes et al. (21), conversely, compared 618 subjects who had prenatal exposure to DES with 618 control subjects, and found that fertility measured in terms of occurrence of pregnancy did not differ between the exposed women and the controls. It is difficult to come to one conclusion concerning the fertility of DES-exposed women because many factors must be considered. Patient populations and the intensity of DES administration vary. Use of control groups and even definitions of pregnancy differ between investigators. There is no one answer, but, clearly, there is concern about the reproductive outcome in these women.

An increased risk of an unfavorable outcome of pregnancy is associated with DES exposure. Among DES-exposed women who become pregnant, 81% had at least one full-term live birth (21), compared to 95% for unexposed women. Spontaneous abortion, ectopic pregnancy, incompetent cervix, and premature labor occur significantly more often in the DES-exposed population than in normal controls, as discussed above.

DES-exposed women have infertility or pregnancy complications due to other causes, and DES-exposure should not be an excuse to postpone or delay the definitive infertility investigation. Stillman and Miller (29–31) reported that 50% of DES-exposed infertile women also had endometriosis, not substantially different from the 39% incidence in non-DES-exposed women, suggesting that endometriosis should be considered in the infertility evaluation and laparoscopy included for definitive diagnosis. Cervical stenosis was identified in 25% of all DES-exposed patients, and could potentially provide an additional diagnosis explaining infertility. Indeed, the incidence of abnormalities unrelated to DES exposure is significant. This would suggest that, in many cases, the presence of these women in infertility groups is due to pathology other than DES exposure.

Conclusion

To conclude, DES exposure is associated with some degree of infertility, fetal wastage, and pregnancy complications. More severe degrees of abnormality appear to be associated with a greater likelihood of reproductive dysfunction. Hysterosalpingography has an important role in elucidating the various degrees of abnormality.

REFERENCES

1. Herbst AL, Ulfelder H, Poskanzer DC: Adenocarcinoma of the vagina: Association of maternal stilbestrol therapy with tumor appearance in young women. *N Engl J Med* 284:878, 1971.
2. Herbst AL, Poskanzer DC, Robboy SJ, et al: Prenatal exposure to stilbestrol: A prospective comparison of exposed female offspring with unexposed controls. *N Engl J Med* 292:334, 1975.
3. Burke L, Antonioli D, Rosen S: Vaginal and cervical squamous cell dysplasia in women exposed to diethylstilbestrol in utero. *Am J Obstet Gynecol* 132:537, 1978.
4. Kaufman RH, Binder GL, Gray PM Jr, Adam E: Upper genital tract changes associated with exposure in utero to diethylstilbestrol. *Am J Obstet Gynecol* 128:51, 1977.
5. Haney AF, Hammond CB, Soules MR, Creasman WT: Diethylstilbestrol-induced upper genital tract abnormalities. *Fertil Steril* 31:142, 1979.
6. Kaufman RH, Adam E, Grey MP, Gerthoffer E: Urinary tract changes associated with exposure in utero to diethylstilbestrol. *Obstet Gynecol* 56:330, 1980.
7. Ben-Baruch G, Menczer J, Mashiach S, Serr DM: Uterine anomalies in diethylstilbestrol-exposed women with fertility disorders. *Acta Obstet Gynecol Scan* 60:395, 1981.
8. DeCherney AH, Cholst I, Naftolin F: Structure and function of the fallopian tubes following exposure to diethylstilbestrol (DES) during gestation. *Fertil Steril* 36:741, 1981.
9. Jefferies JA, Robboy SJ, O'Brien PC, Bergstralh EJ, Labarthe DR, Barnes AB, Noller KL, Hatab PA, Kaufman RH, Townsend DE: Structural anomalies of the cervix and vagina in women enrolled in the Diethylstilbestrol Adenosis (DESAD) Project. *Am J Obstet Gynecol* 148:59, 1976.
10. Prins RP, Morrow CP, Townsend DE, DiSaia PJ: Vaginal embryogenesis, estrogens, and adenosis: *Obstet Gynecol* 48:246, 1976.
11. Ulfelder H, Robboy SJ: The embryologic development of the human vagina. *Am J Obstet Gynecol* 126:769, 1976.
12. Robboy SJ, Taguchi O, Cunha GR: Normal development of the human female reproductive tract and alterations resulting from experimental exposure to diethylstilbestrol. *Hum Path* 13:190, 1982.
13. Robboy SJ: A hypothetical mechanism of diethylstilbestrol (DES)-induced anomalies in exposed progeny. *Hum Path* 14:831, 1983.
14. Kaufman RH, Noller K, Adam E, Irwin J, Gray M, Jefferies JA, Hilton J: Upper genital tract abnormalities and pregnancy outcome in diethylstilbestrol-exposed progeny. *Am J Obstet Gynecol* 148:973, 1984.
15. Antonioli DA, Burke L, Friedman EA: Natural history of diethylstilbestrol-associated genital tract lesions: Cervical ectopy and cervicovaginal hood. *Am J Obstet Gynecol* 137:847, 1980.
16. Siegler AM, Wang CF, Friberg J: Fertility of the diethylstilbestrol-exposed offspring. *Fertil Steril* 31:601, 1979.
17. Berger MJ, Goldstein P: Impaired reproductive performance in DES-exposed women. *Obstet Gynecol* 55:25, 1980.
18. Schmidt G, Fowler WC Jr, Talbert LM, Edelman DA: Reproductive history of women exposed to diethylstilbestrol in utero. *Fertil Steril* 33:21, 1980.
19. Herbst AL, Hubby MM, Blough RR, Azizi F: A comparison of pregnancy experience in DES-exposed and DES-unexposed daughters. *J Reprod Med* 24:62, 1980.
20. Cousins L, Karp W, Lacey C, Lucas WE: Reproductive outcome of women exposed to diethylstilbestrol in utero. *Obstet Gynecol* 56:70, 1980.
21. Barnes AB, Colton T, Gunderson J, Noller KL, Tilley BC, Strama T, Townsend DE, Hatab P, O'Brien PC: Fertility and outcome of pregnancy in women in utero to diethylstilbestrol. *N Engl J Med* 302:609, 1980.
22. Rosenfeld DL, Bronson RA: Reproductive problems in the DES-exposed female. *Obstet Gynecol* 55:453, 1980.
23. Pillsbury SG Jr: Reproductive significance of changes in the endometrial cavity associated with exposure in utero to diethylstilbestrol. *Am J Obstet Gynecol* 137:178, 1980.
24. Veridiana NP, Delke I, Rogers J, Tancer ML: Reproductive performance of DES-exposed female progeny. *Obstet Gynecol* 58:58, 1981.
25. Sandberg EC, Riffle NL, Higdon JV, Getman CE: Pregnancy outcome in women exposed to diethylstilbestrol in utero. *Am J Obstet Gynecol* 140:194, 1981.
26. Herbst AL, Hubby MM, Azizi F, Makii MM: Reproductive and gynecologic surgical experience in diethylstilbestrol-exposed daughters. *Am J Obstet Gynecol* 141:1019, 1981.
27. Mangan CE, Borow L, Burtnett-Rubin MM, Egan V, Guintoli RL, Mikuta JJ: Pregnancy outcome in 98 women exposed to diethylstilbestrol in utero, their mothers, and unexposed siblings. *Obstet Gynecol* 59:315, 1982.
28. Mansi ML, Goldfarb AF: An analysis of pregnancy salvage in a selective DES population. *Infertility* 5:1, 1982.
29. Stillman RJ: In utero exposure to diethylstilbestrol: An adverse effect on the reproductive tract and reproductive performance in male and female offspring. *Am J Obstet Gynecol* 142:905, 1982.
30. Bibbo M, Gill WB, Azizi F et al: Follow-up study of male and female offspring of DES-exposed mothers. *Obstet Gynecol* 49:1, 1977.
31. Stillman RJ, Miller LRC: Diethylstilbestrol exposure in utero and endometriosis in infertile females. *Fertil Steril* 41:369, 1984.

6 The Uterine Cavity

The uterine cavity is a potential space, defined cephalad by the fundus and caudally by the external cervical os. Anatomically, it is readily subdivided by the internal cervical os into the endometrial cavity and the endocervical canal. Considerable normal variation in both dimensions and configuration may be encountered. Varying degrees of traction on the tenaculum during hysterosalpingography will change these measurements. Increasing amounts of contrast media introduced into the cavity may expand it considerably. Furthermore, both uterine size and length of the cavity may vary with hormonal stimulation, correlating in a positive relationship with circulating estradiol levels. Measurements taken from radiographic films to estimate cavity size are subject to both radiologic and geometric distortion and magnification; therefore, measurements from films should not be used to define normal limits. The introduction of up to 2.0 ml of contrast medium will usually fill the uterine cavity, without significant discomfort to the patient. Increasing volume generally results in greater cramping. Most patients begin to appreciate discomfort by the time 1.5 ml has been introduced, but whether this is due to the volume or simply to the irritating stimulus, which induces uterine cramping, is unclear; almost all women will have marked cramping if 3.0 ml of contrast is introduced.

The endometrium, being soft and spongy, may be compressed by pressure from introduced contrast, so hysterosalpingography may not differentiate abnormalities such as hyperplasia or atrophy, although the appearance of the cavity may vary with the phase of the menstrual cycle. On the other hand, if hysterosalpingography is performed late in the luteal phase of the menstrual cycle, the increased thickness of the endometrial lining of the cavity may obstruct the entrance of contrast medium into the fallopian tubes. This is yet another reason why hysterosalpingography should be performed only in the follicular part of the cycle. Endometrial abnormalities will be detected only if they are of sufficient size to distort the uterine cavity, or present as a mass displacing the contrast medium. They are visualized as either filling defects, outpouchings, or irregularities of contour.

Abnormalities of the uterine cavity that are diagnosed by hysterosalpingography include lesions of the endometrium, such as neoplasms, infoldings, and intrauterine synechiae; abnormalities of the myometrium, including submucous and intramural myomata; space-occupying masses and foreign bodies within the cavity; and congenital anomalies of müllerian development as discussed in Chapter 4.

Filling Defects

Spurious Artifacts

Intrauterine abnormalities often present as filling defects. The most common are spurious artifacts such as air bubbles or displaced cervical mucus (Fig. 6.1). *Air* is recognized as solitary or multiple round mobile defects and may be avoided, in large measure, by carefully filling the introducing instrument with contrast prior to initiating the procedure. Fluoroscopic observation of movement of the rounded defect(s) establishes the diagnosis.

Cervical mucus pushed retrograde through the uterine cavity may appear as an unusual filling defect. Its appearance is that of an amorphous mass, often linear in shape

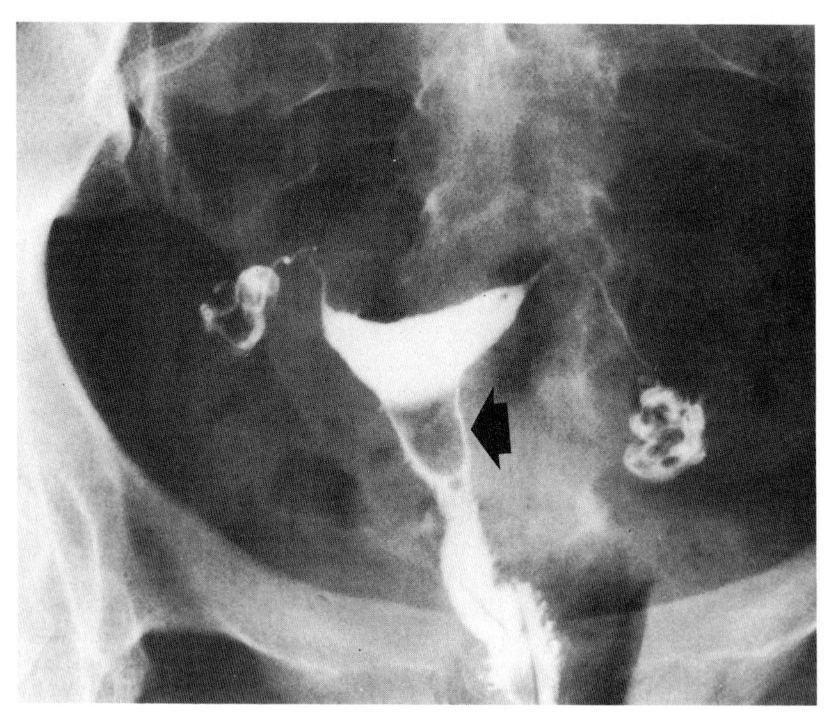

Figure 6.1. Large radiolucent filling defect (*arrow*) in the uterine cavity due to an air bubble. Subsequent films showed disappearance of this mobile lucency.

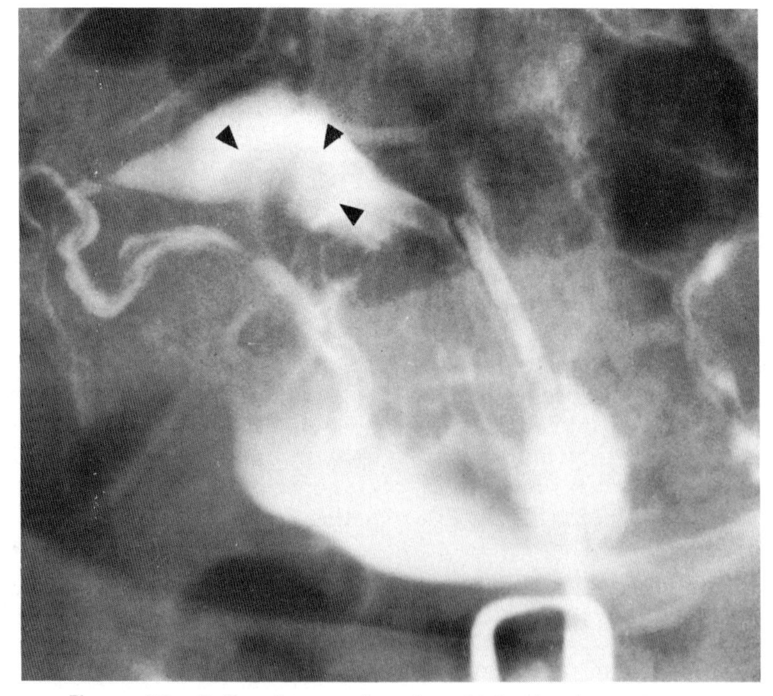

Figure 6.2. Solitary large cell endometrial polyp (*arrowheads*).

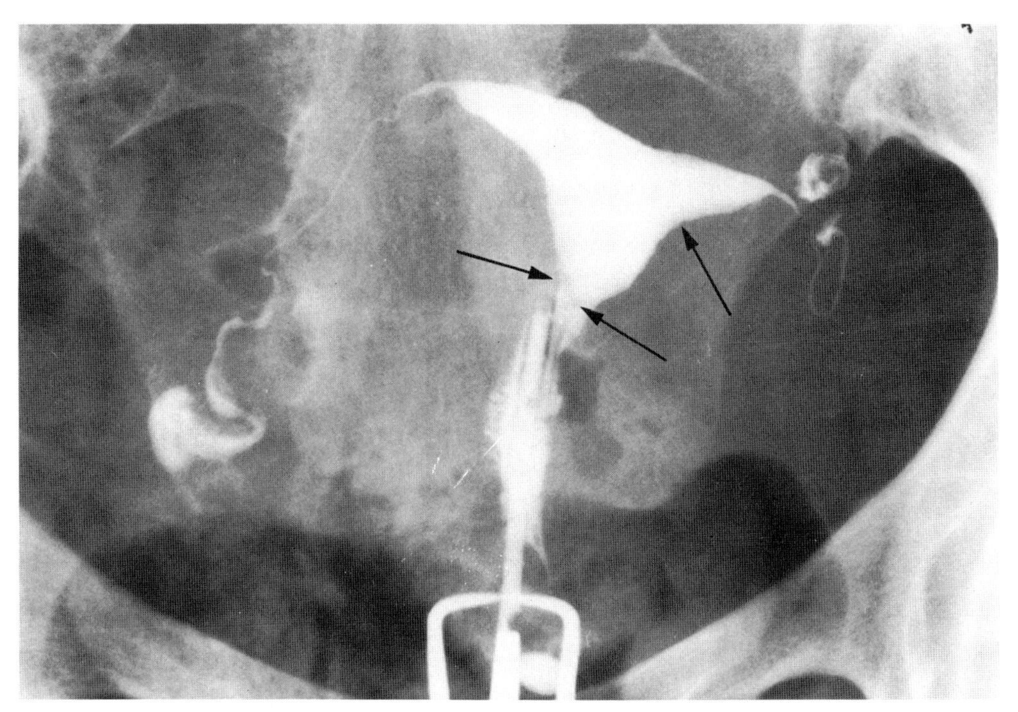

Figure 6.3. Small fixed filling defect (*arrows*) almost completely obscured by the contrast material.

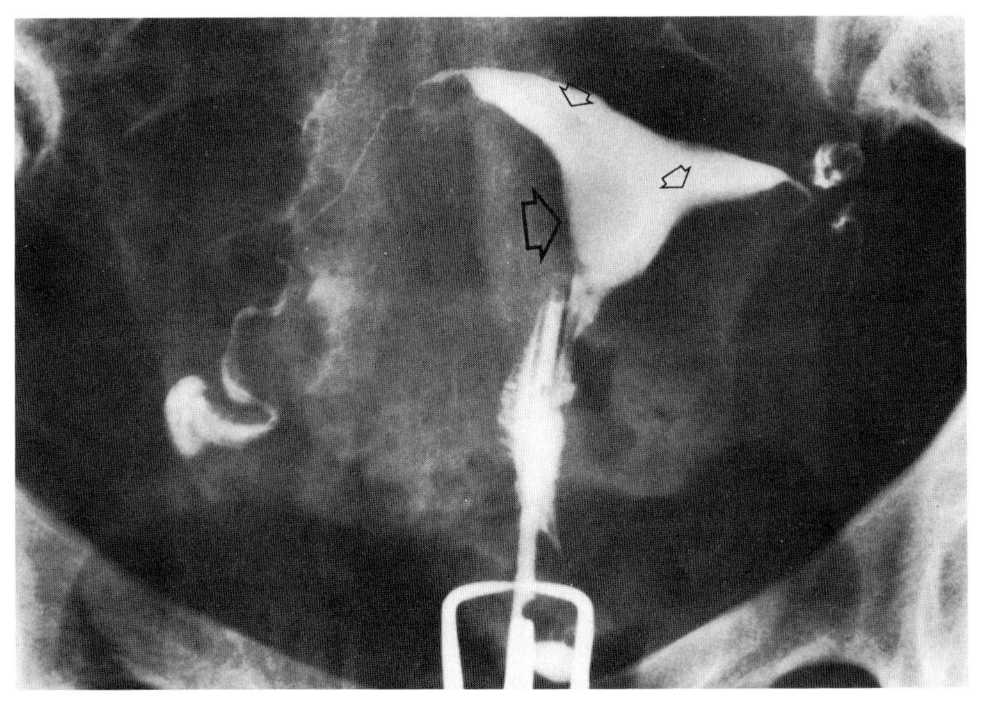

Figure 6.4. Same patient as Fig. 6.3. Reduction in the quantity of contrast material allows demonstration of multiple endometrial polyps of varying size and configuration (*arrows*).

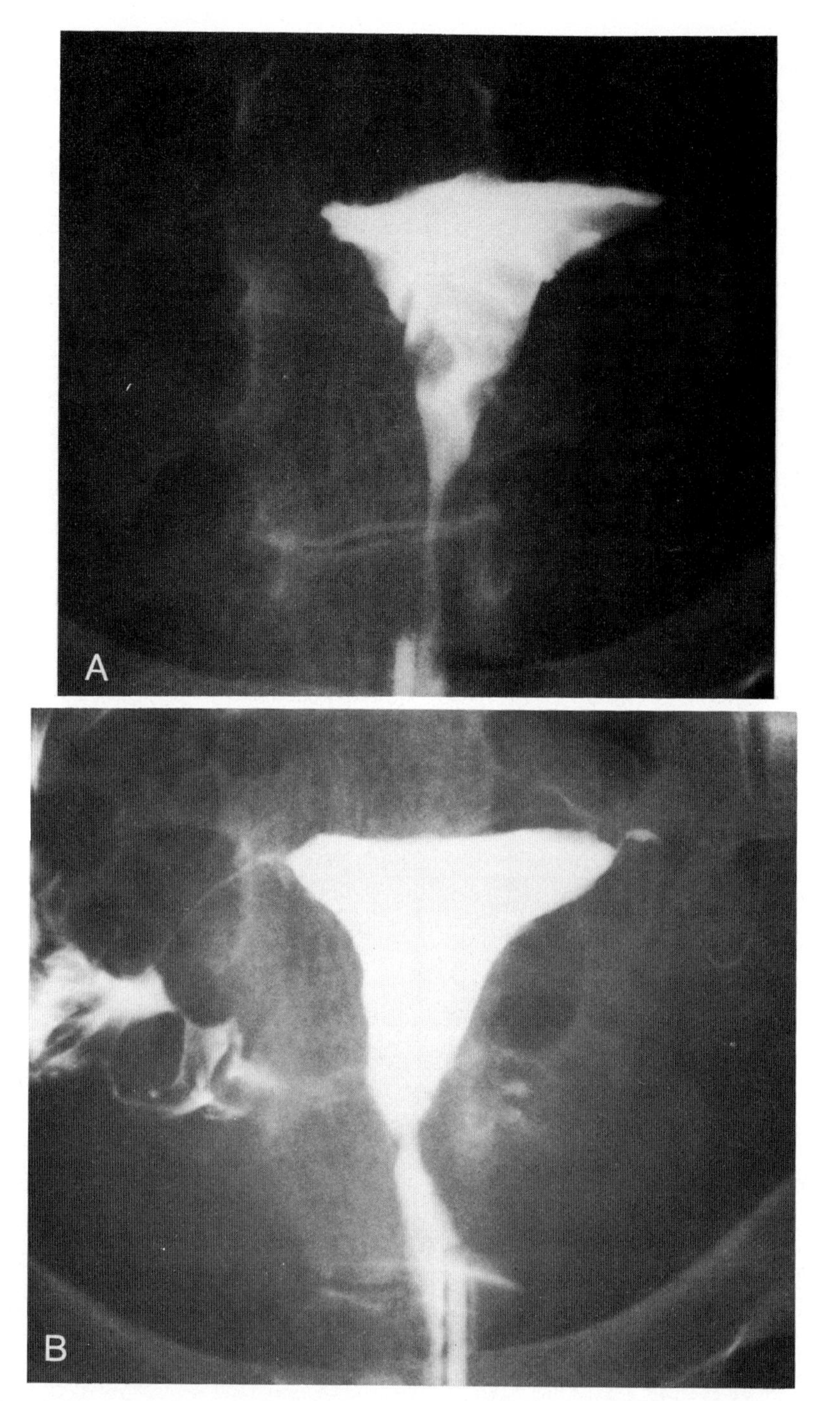

Figure 6.5. *A*: Multiple endometrial polyps of varying size and configuration. *B*: Same patient after hysteroscopy with resection and curettage.

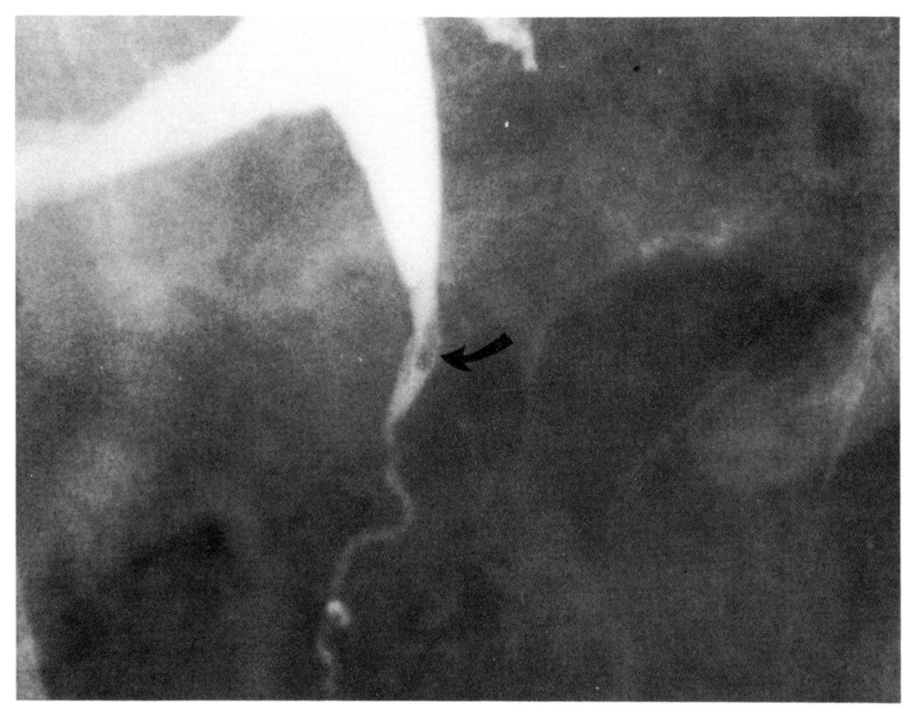

Figure 6.6. Small polypoid filling defects are occasionally seen in the proximal portion of the fallopian tubes (*arrow*).

without clear margins or rounded contour. Blood clots, due either to pre-existing bleeding or the trauma of the instrumentation, are another unlikely but possible cause of such mobile filling artifacts.

Polyps

Persistent filling defects can be caused by polyps, which give a rounded or sessile pattern (Fig. 6.2). Small polyps may be completely obscured by contrast medium, so care must be taken to inspect the configuration of the cavity as contrast is introduced (Fig. 6.3). Polyps may be single or multiple, pedunculated or sessile, and are distinguished from synechiae by their more rounded and regular appearance (Fig. 6.4). Although of varying size, the overwhelming majority of these lesions are smaller than 1 cm in diameter. A recent case at hysterosalpingography had the appearance of multiple grape-like filling defects; at hysteroscopy, the operator likened the appearance to moguls, or humps, on a ski slope (Fig. 6.5). Occasionally, such filling defects, presumed to be polyps, may be seen in the fallopian tubes (Fig. 6.6).

Neoplasms

Malignant neoplasms of the endometrium will rarely be observed at hysterosalpingography during investigation for infertility, but several studies have delineated the characteristics of endometrial carcinoma before and after radiation treatment (1–3). Hysterography was used in the 1950s as a simple, reliable diagnostic technique for menopausal women with abnormal bleeding in whom the diagnosis of cervical cancer had been eliminated (4, 5). A normal uterine cavity was thought to rule out endometrial carcinoma, whereas characteristic defects allowed a presumptive diagnosis permitting hysterectomy without intervening curettage. Endometrial cancer seen on hysterography in this setting was divided by these authors into several types: (a) a single growth in the fundus; (b) multiple circumscribed growths; (c) surface, diffuse, or spreading growths (Fig. 6.7); (d) total

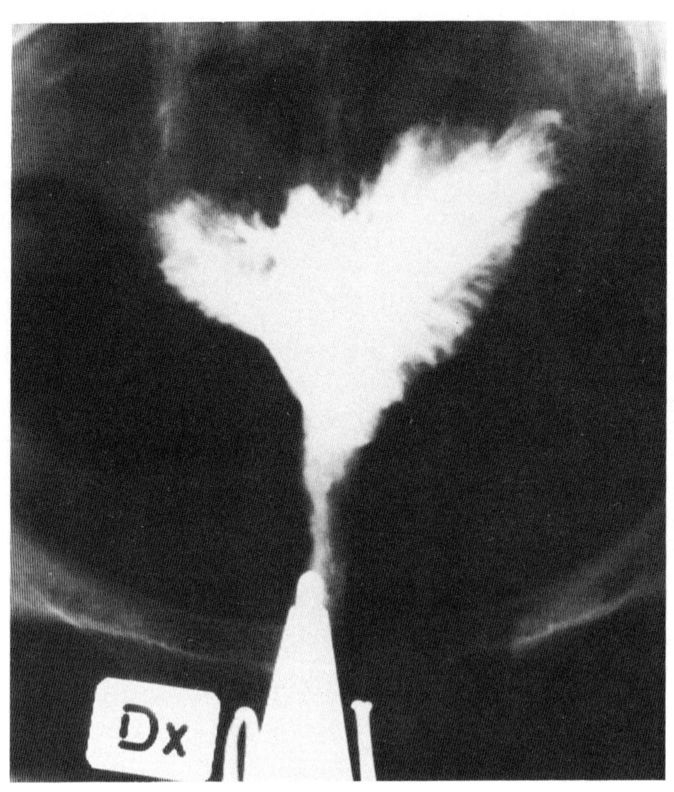

Figure 6.7. Endometrial carcinoma extensively involving the entire endometrial cavity with diffuse infiltration and irregular filling defects. (Courtesy of Leif Ekelund, Lund, Sweden.)

involvement of the uterine cavity (Fig. 6.8); and (e) endometrial cancer coexisting with myomata. Hysterography has largely been replaced in the evaluation of uterine neoplasms, but serendipitous discovery of a neoplasm may occasionally occur.

Filling defects due to *myomata uteri* are usually distinctive. As myomata enlarge, they can stretch and deform the cavity into bizarre configurations, often eliminating usual landmarks (Figs. 6.9 and 6.10). Such distortion may induce crescentric or semilunar contours as contrast outlines the misshapen cavity (Fig. 6.11). A rather unique capability of myomata is to enlarge the uterine cavity markedly, probably relating to the intramural position of the tumor or tumors. On occasion, quantities of contrast exceeding 30 ml may be introduced into the cavity without complete filling (Fig. 6.12).

A myoma located in the cornual area may cause a mechanical block and make it impossible to introduce contrast media into the fallopian tubes. Fibromyomata may demonstrate a relatively typical pattern of coarse calcifications (Fig. 6.13). A fundal location of a myoma may at times create an unusual pattern suggestive of congenital septum formation (Fig. 6.14). Small fibroids may have the appearance of endometrial polyps, particularly if the neoplasm is pedunculated. Under these circumstances, hysteroscopy is necessary for definitive diagnosis.

Pregnancy

Although hysterosalpingography is usually done only in the follicular part of the cycle, some patients may not be taking basal temperature charts, and may mistake spotting in early pregnancy for a menstrual period. For this reason, inadvertent hysterosalpingography in the presence of pregnancy occasionally occurs. A pregnancy may have the appearance of a sessile

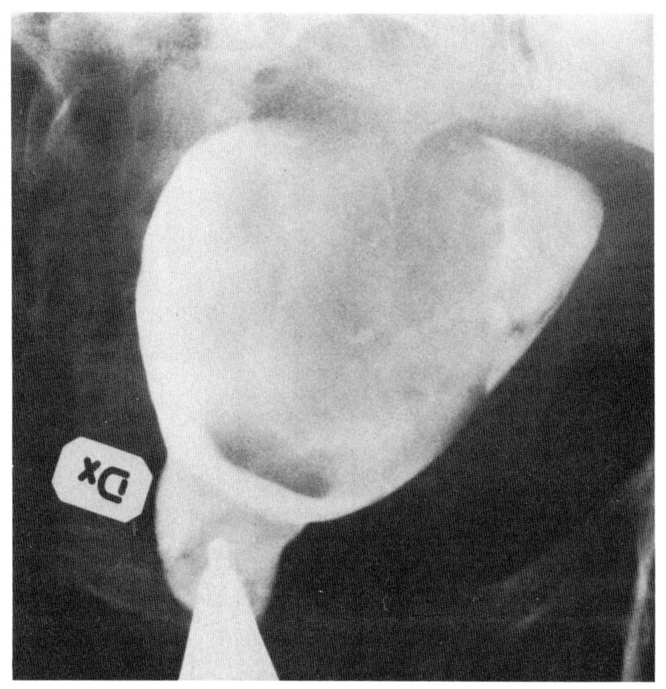

Figure 6.8. Endometrial carcinoma presenting as a large irregular filling defect filling virtually the entire endometrial cavity. (Courtesy Leif Ekelund, Lund, Sweden.)

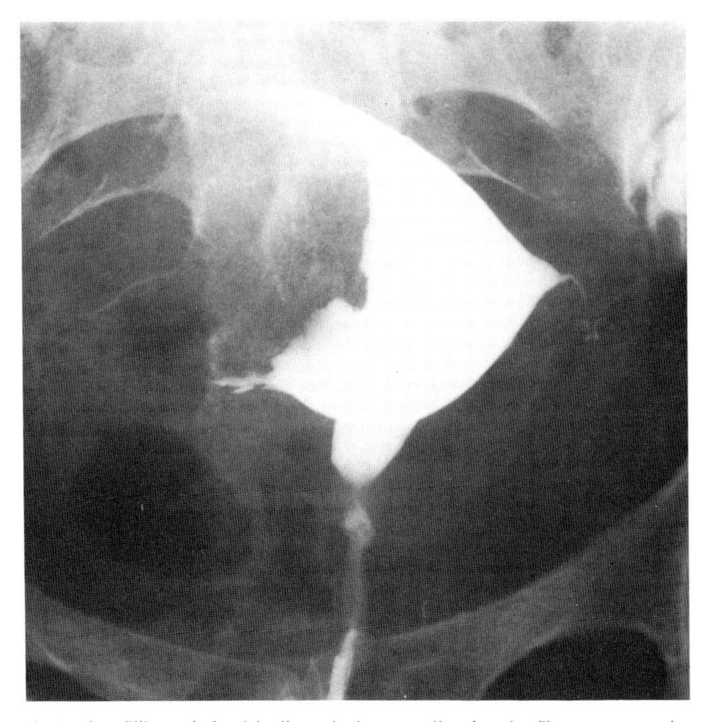

Figure 6.9. Large irregular filling defect in the uterine cavity due to fibromyoma. In addition to deformation of the uterine cavity there is cavity enlargement and failure to fill the right fallopian tube.

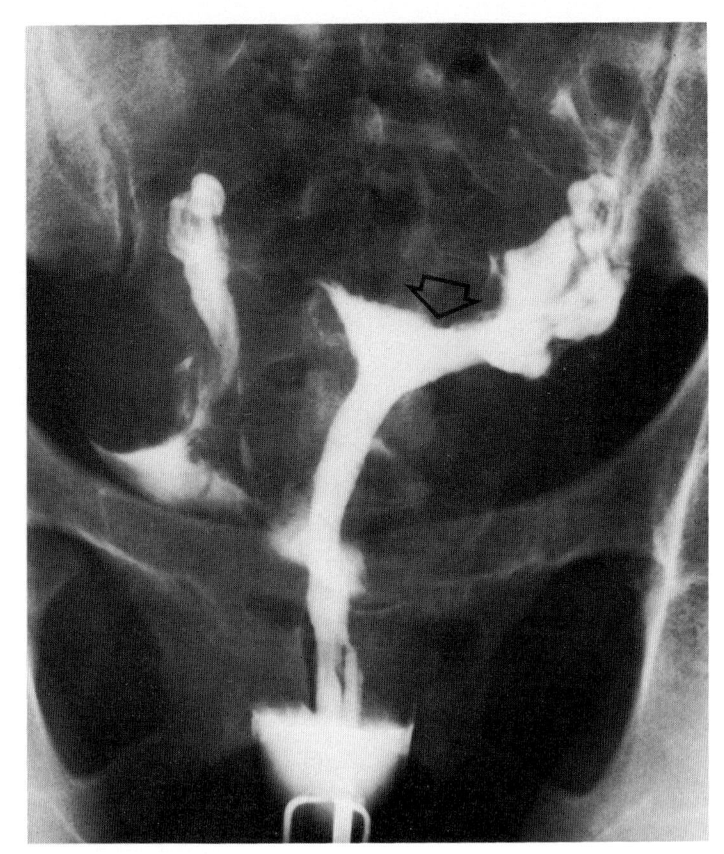

Figure 6.10. Multiple fibromyomata resulting in stretching and elongation of the cervical canal and lower uterine segment. Minimal irregularity along the fundus (*arrow*) secondary to a myoma in this location.

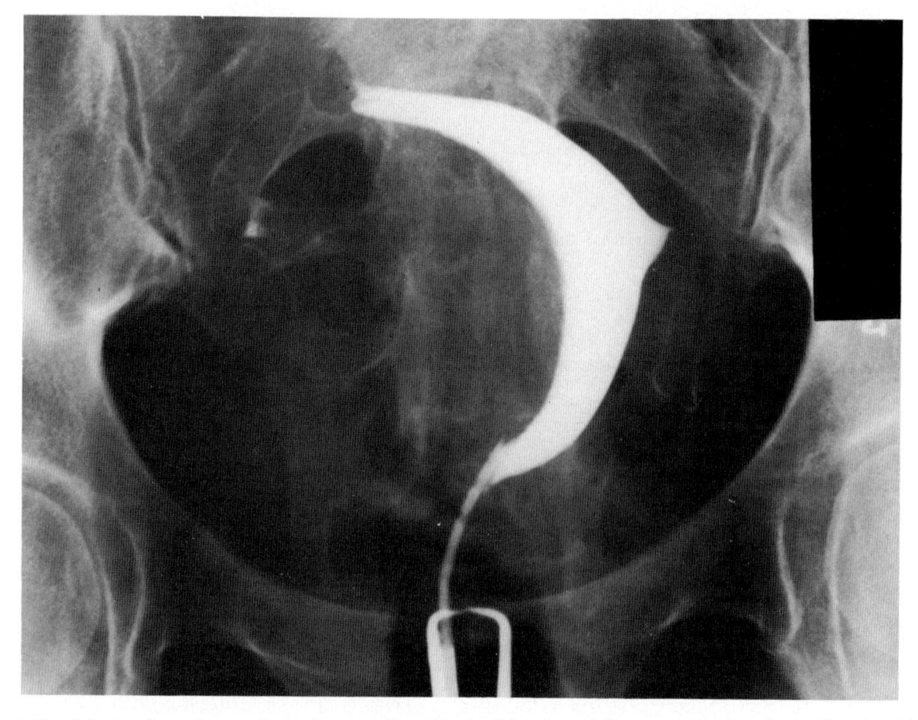

Figure 6.11. Markedly enlarged uterine cavity distorted by large fibromyomata along the right lateral margin of the endometrial cavity. The resultant cavity is crescentic in contour.

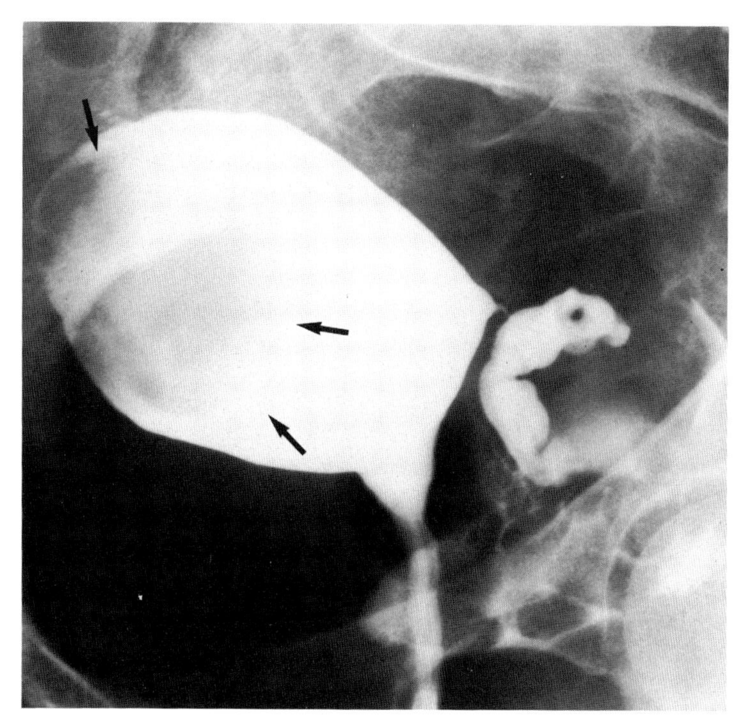

Figure 6.12. Marked stretching of the endometrial cavity by fibromyomata. More than 30 ml of contrast was introduced into the cavity. The large filling defect (*arrows*) is the result of indentation of the lumen by the large myoma.

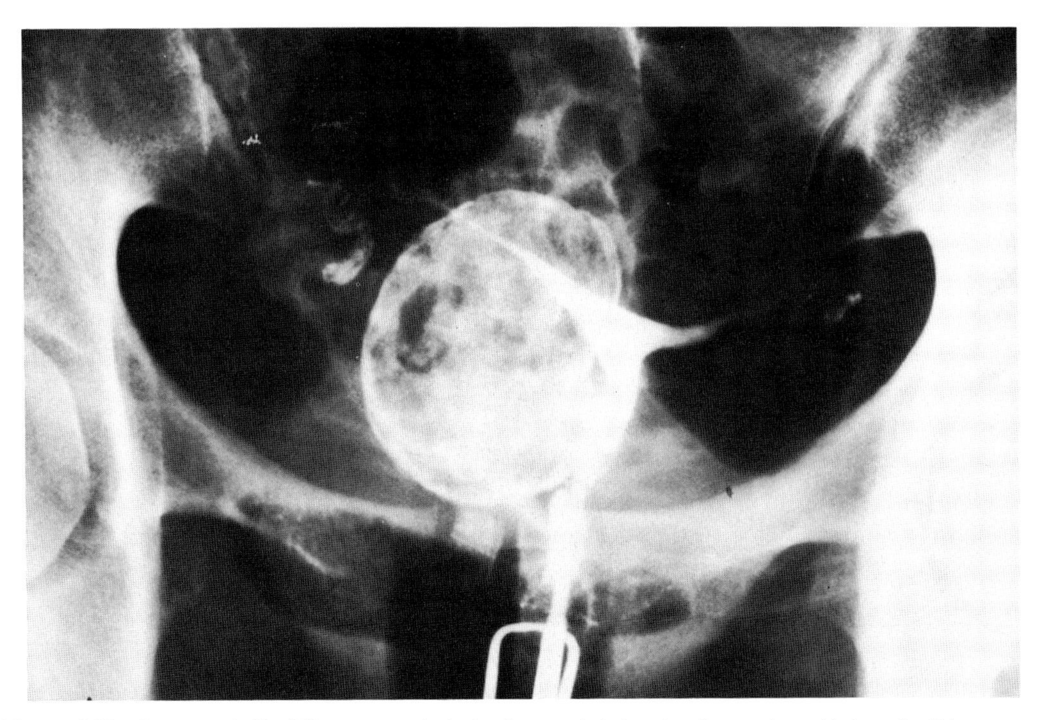

Figure 6.13. Large calcified fibromyomata indenting and deforming the endometrial cavity. This coarse, calcified pattern is relatively typical for fibromyomata.

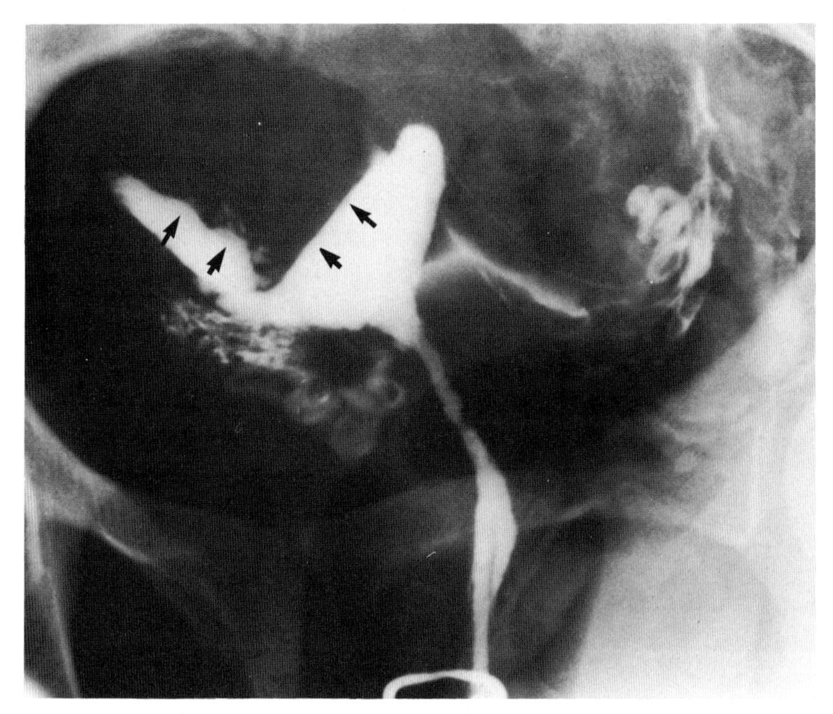

Figure 6.14. Wedge-like deformity (*arrows*) in the uterine fundus is the result of a large solitary myoma. The configuration is somewhat suggestive of congenital septum formation.

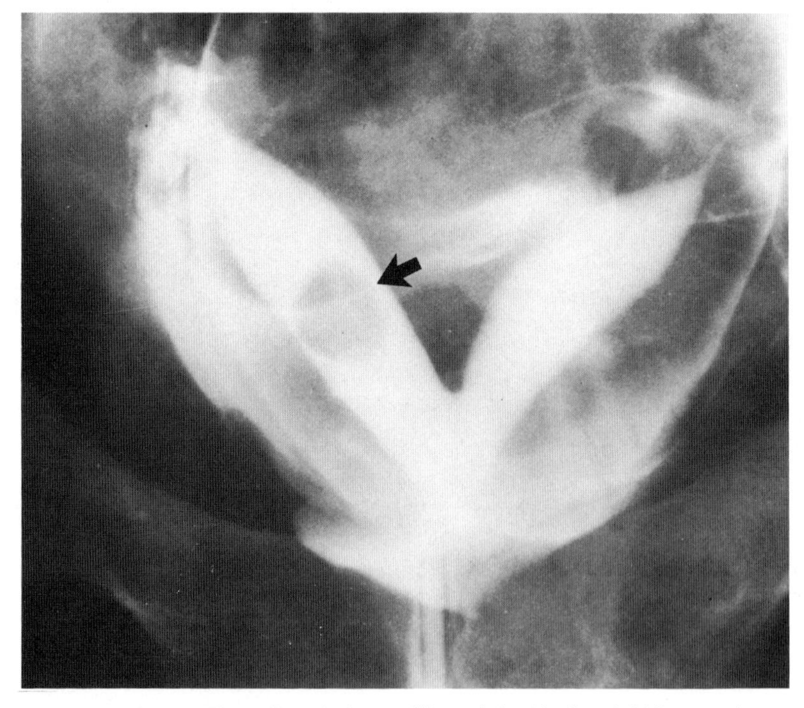

Figure 6.15. Bicornuate uterus with a discrete large filling defect in the right horn, subsequently demonstrated to be an intrauterine pregnancy.

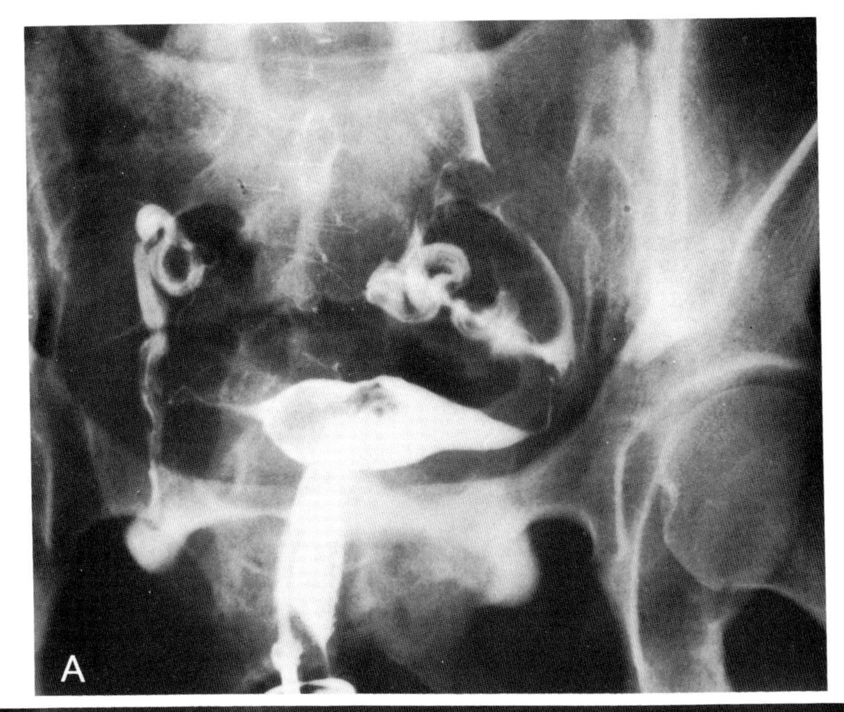

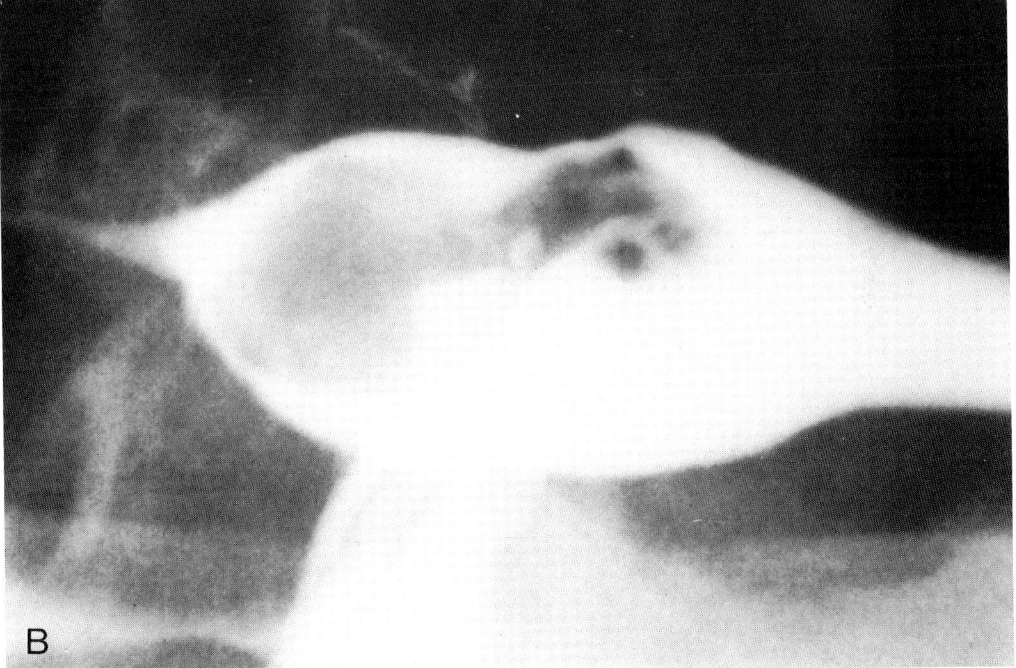

Figure 6.16. *A*: Large polypoid-like filling defect near the fundus of the uterus. Hysteroscopy and resection demonstrated retained product of conception. *B*: Magnification view.

polyp or, at later stages, appear as a large filling defect (Fig. 6.15). Isaacs (6) has reported a characteristic but most unusual pattern, a double outline sign, caused by infiltration of contrast media into the decidua.

The question of both traumatic and radiation injury to the pregnancy then arises. Miscarriage due to the instrumentation may result, with the usual bleeding and cramping. However, if the pregnancy continues, one must determine the appropriate course of action. Is termination of pregnancy warranted because of the radiation incurred during the hysterosalpingogram? Some have suggested that radiation exposure of more than 10 rads to the fetus in the first 6 weeks of pregnancy warrants therapeutic abortion (7). Current opinion, on the other hand, estimates that risk of congenital anomalies increases only from 1% to 3% following radiation exposure of up to 15 rads in the first trimester (8). As mentioned in Chapter 2, estimates of gonadal radiation dosage vary but certainly seem less than 1 rad. The survey published by the American College of Radiology suggests average ovarian radiation exposure to be 590 mrads (9). It seems reasonable to review this slight risk with the couple prior to any course of action. Such risk factors may well be acceptable to many.

Retained Products of Conception

Retained products of conception following fetal loss are a relatively common cause of intrauterine filling defects. These may be of any size and shape but are frequently irregular in contour and of significant size (Fig. 6.16). Their appearance may suggest malignant neoplasm (Fig. 6.17).

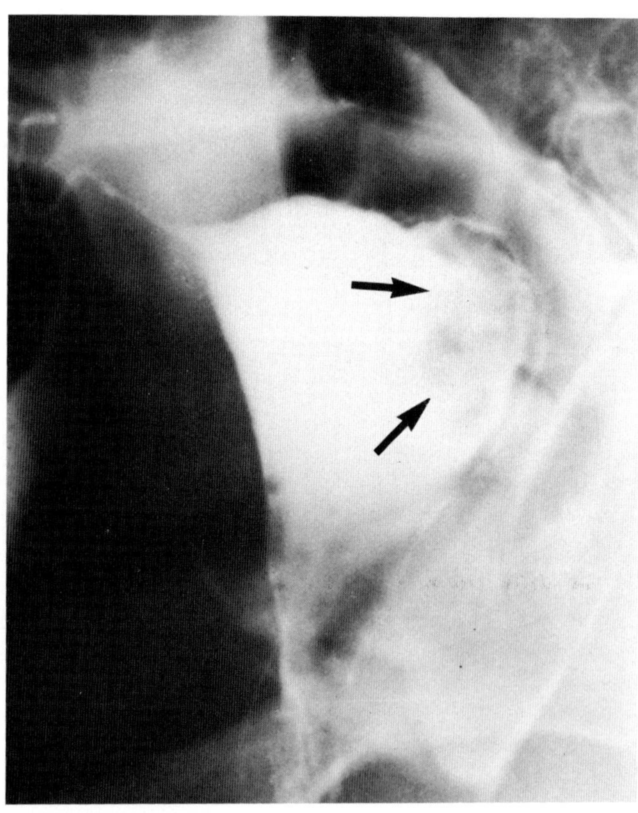

Figure 6.17. Large regular filling defect in the left uterine cornua (*arrows*) representing retained products of conception.

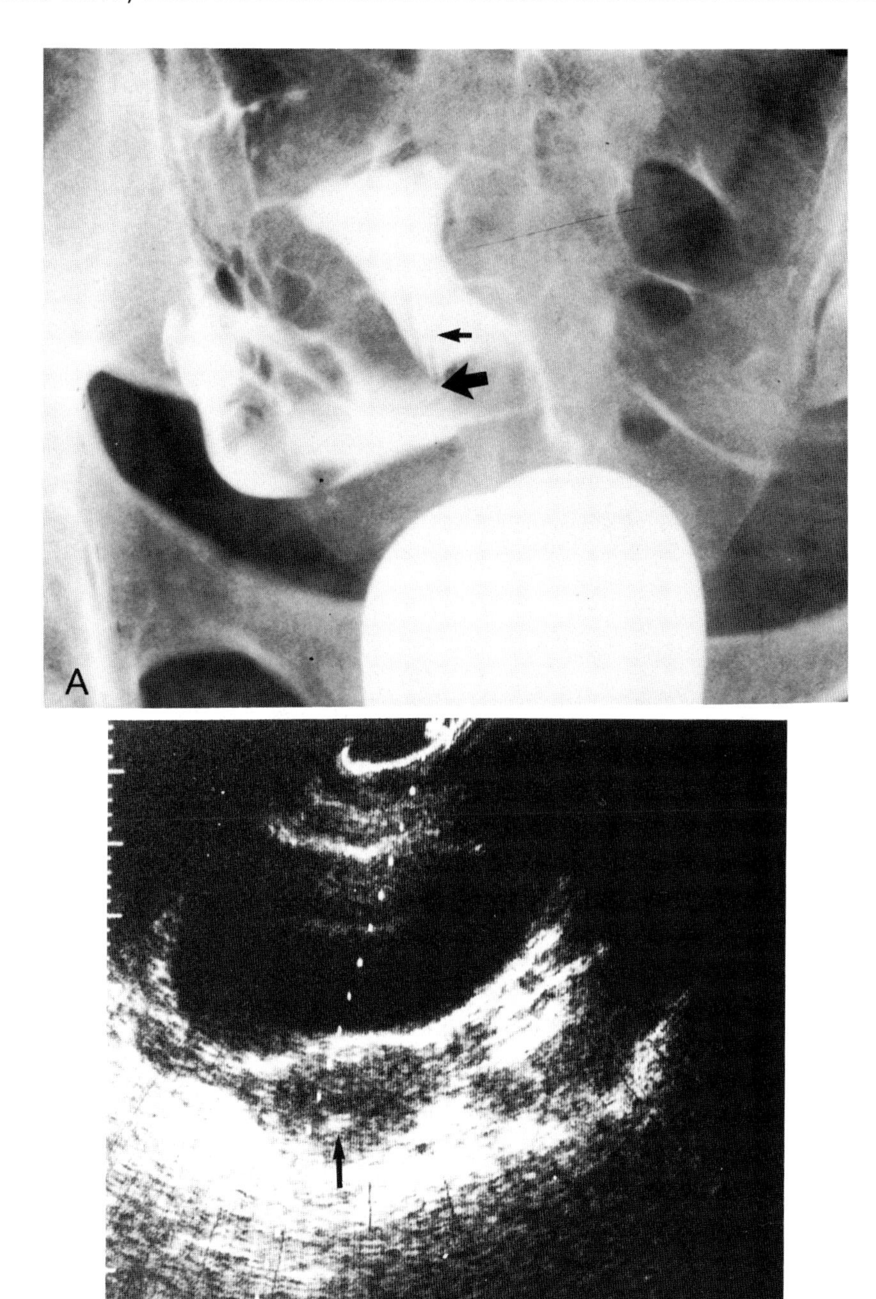

Figure 6.18. *A*: Partial uterine perforation by a Cu-7 IUD. Much of the IUD remains within the uterine cavity (*small arrow*), although the inferior portion of the device extends beyond the lumen and into the myometrium near the lower uterine segment (*large arrow*). *B*: Ultrasound reveals a low position of the IUD but cannot establish the diagnosis of partial perforation. (Courtesy of W. Stern, Bronx, New York.)

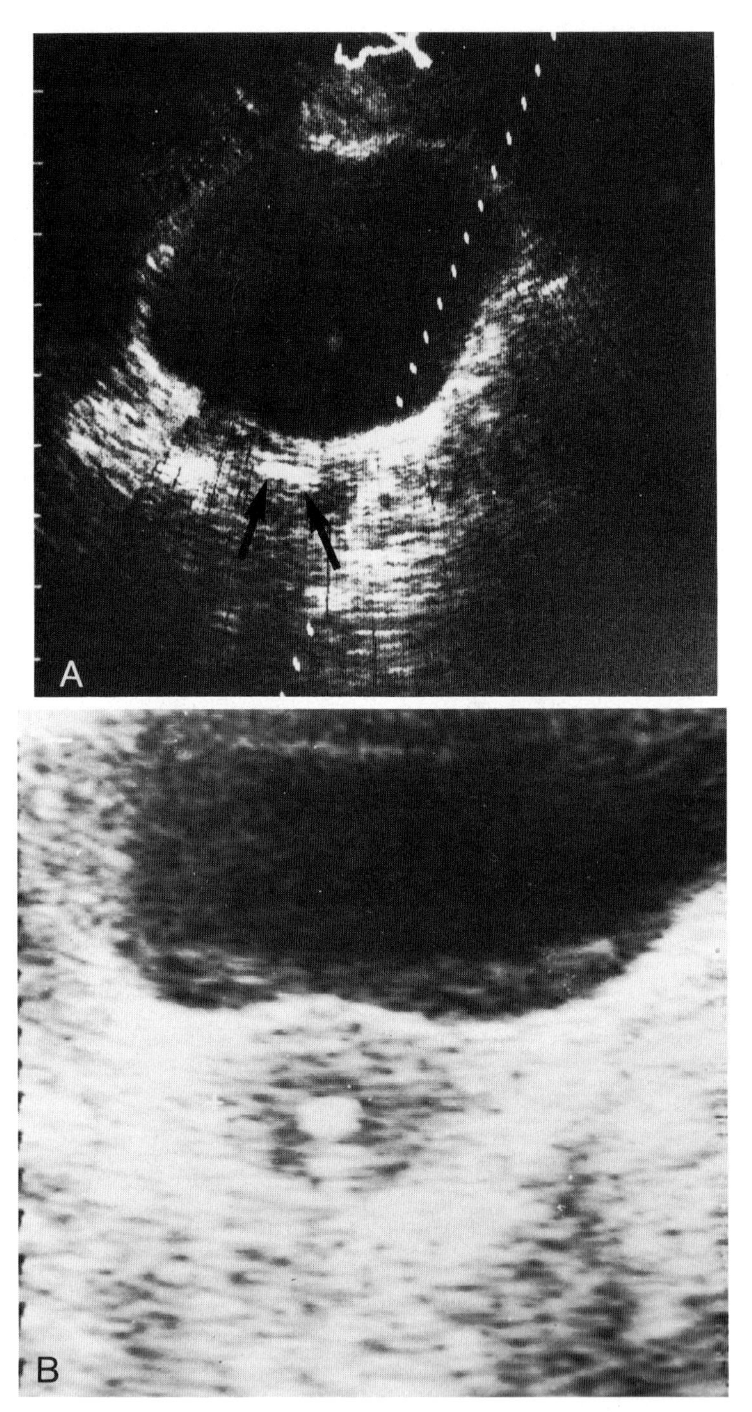

Figure 6.19. *A*: Lippes Loop (IUD) (*arrows*) identified remote from uterine cavity in this demonstration of complete perforation. The IUD lies in the cul-de-sac. *B*: Ultrasound fails to detect IUD within uterine cavity. *C*: IUD (*arrow*) identified free in the cul-de-sac, separated from the uterus. (Reprinted by permission from Rosenblatt R, Zakin D, Stern W, Kutcher R: Uterine perforation and embedding by intrauterine device: evaluation by US and hysterography. *Radiology* 157:765, 1985.)

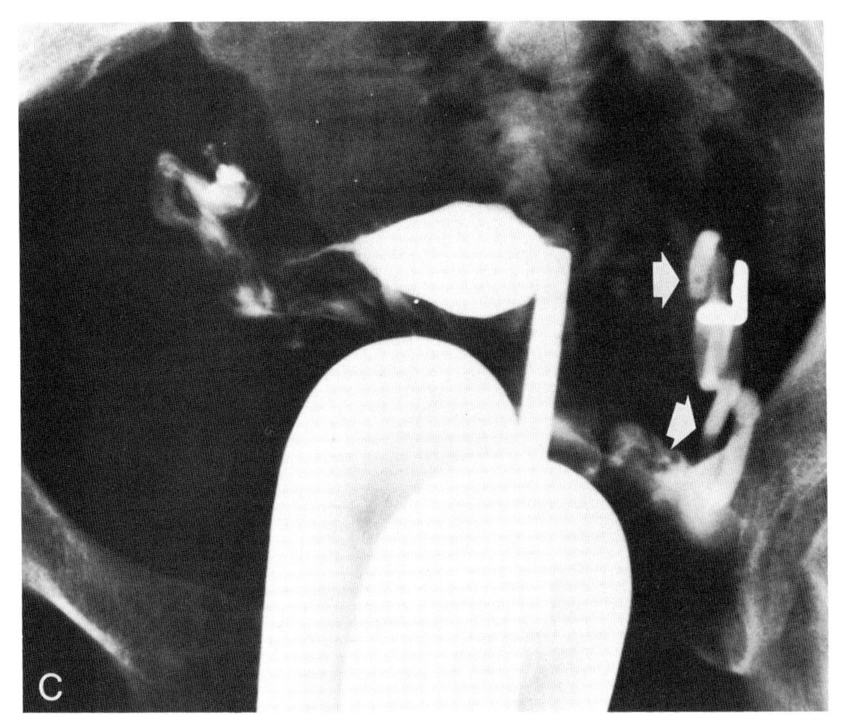

Figure 6.19C

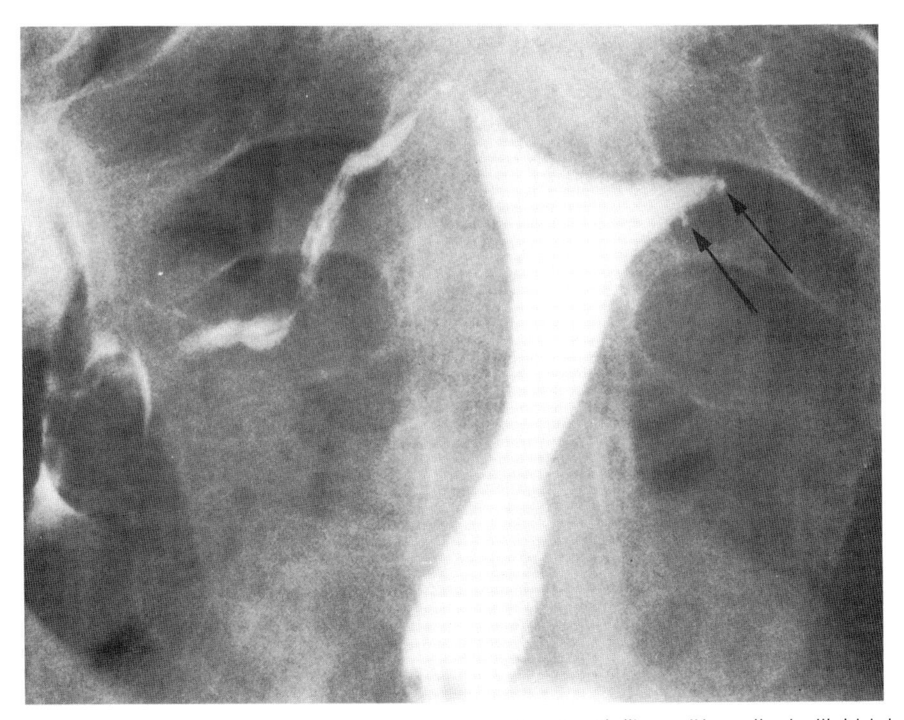

Figure 6.20. Small marginal irregularities (*arrows*) suggesting tiny "lollipops" in patient with histologically proven adenomyosis.

Foreign Bodies

Defects resulting from introduced foreign bodies are occasionally encountered. Unusual elongated or irregular filling defects can tax the diagnostic ingenuity of the observer. History of previous placement of an intrauterine contraceptive device (IUD) is important. Inability to visualize the "tail" or string of the device, resulting in a "lost" IUD, occasionally requires hysterosalpingography, although careful ultrasound examination usually suffices. Elongation of the uterine cavity during pregnancy may result in such displacement of the string of the IUD into the uterine cavity (10).

The IUD, or a fragment of it, may become embedded in the wall of the myometrium, resulting in partial perforation of the uterus, or, progressively, may extrude completely from the uterine cavity and reside in the peritoneal cavity (11). Ultrasonography and hysterography are complementary tools in these situations and are generally diagnostic (Figs. 6.18 and 6.19).

Postabortion Fetal Parts

Finally, prolonged intrauterine retention of fetal parts after spontaneous or induced abortion may cause infertility, leading to the performance of a hysterosalpingogram. Dawood and Jarrett (12) were unable to distinguish the hysterosalpingographic appearance of fetal bones from synechiae, but calcification may occur, making a more obvious and easily diagnosed finding. Fetal bones tend to have a linear appearance. Intrauterine synechiae, or scars, are one of the more common of the filling defects. The usually distinctive angular, irregular appearance is characteristic, but occasionally these are sources of diagnostic confusion. The frequency and importance of these synechiae have warranted their separate discussion (Chapter 7).

Marginal Irregularities

Adenomyosis

Adenomyosis is a process in which endometrial glands are found within the myometrium, and may occur in patients with external endometriosis. Adenomyosis is a diagnosis made only by a pathologic examination of the uterus, and cannot definitely be made by hysterosalpingography, dilatation and curettage, diagnostic laparoscopy, or any other surgical procedure. A clinical hint suggesting adenomyosis is an enlarged uterus and cavity, presumptively due to the thickened and boggy myometrium, and a history of irregular bleeding and spotting. At hysterosalpingography, the presence of adenomyosis may be suggested by the appearance of "lollipops," diverticula-like projections from the uterine cavity into the walls of the uterus (Fig. 6.20). The spicular projections end in small sacs that are thought to be endometrial glands extending into the myometrium and distended with contrast medium.

Postsurgical Abnormalities

Abnormalities of the uterine cavity secondary to previous surgery are discussed more thoroughly in Chapter 9. Contour irregularity, saccular dilatations, and defects simulating neoplasms may present. The most common of these alterations are secondary to previous cesarean section, with an abnormal uterine cavity observed in 36% of patients undergoing a low segment cesarean section (13). Most such deformities are clinically insignificant, but larger defects may dictate elective cesarean section in the event of pregnancy.

REFERENCES

1. Norman O: Hysterosalpingography in cancer of the corpus of the uterus. *Acta Radiol* 34(suppl 79):1, 1950.
2. Norman O: Hysterographically visualized radionecrosis following intrauterine radiation of cancer of the corpus of the uterus. *Acta Radiol* 37:96, 1952.
3. Schwartz PE, Kohron EL, Knowlton AH, Morris JMcL: Routine use of hysterography in endometrial carcinoma and post menopausal carcinoma. *Obstet Gynecol* 45:378, 1975.
4. Beclére C: L'hystérigraphie dans le diagnostic des lésions intrautérines et des métrirragies fonctionelles. *Bull Soc Obstet Gynecol* 22:815, 1933.
5. Beclére C, Fayolle G: *L'Hystéro-salpingographie.* Paris, Masson et Cie, 1966.
6. Isaacs I: Hysterographic double-outlined uterine

cavity: A sign of unsuspected pregnancy. *Am J Roentgenol* 131:305, 1978.

7. Hammer-Jacobsen E: Therapeutic abortion on account of x-ray examination during pregnancy. *Dan Med Bull* 6:112, 1959.
8. Swartz HM, Reichling BA: Hazards of radiation exposure for pregnant women. *JAMA* 239:1907, 1978.
9. *Medical Radiation: A Guide to Good Practice.* American College of Radiology, Washington DC, 1985.
10. Guha-Ray DK: Translocation of the intrauterine contraceptive device: Study of thirty-one cases. *Fertil Steril* 28:9, 1977.
11. Rosenblatt R, Zakin D, Stern W, Kutcher R: Uterine perforation and embedding by intrauterine device: Evaluation by US and hysterography. *Radiology* 157:765, 1985.
12. Dawood MY, Jarrett JC II: Prolonged intrauterine retention of fetal bones after abortion causing infertility. *Am J Obstet Gynecol* 143:715, 1982.
13. Durkan JP: Hysterography after caesarean section. *Obstet Gynecol* 24:836, 1964.

7 Intrauterine Synechiae

Intrauterine scarring first appeared in the medical literature in 1894 when Fritsch reported a 25-year-old who developed amenorrhea following a curettage performed 3 weeks postpartum (1). By 1946, when Asherman described the condition that now bears his name, the literature already contained 61 cases. Asherman described a variable syndrome of intrauterine adhesion formation with scarring and obliteration of the potential space of the uterine cavity resulting clinically in hypo- or amenorrhea (2, 3). At the time, the most common etiologic factor was thought to be genital tuberculosis; it has since been shown that other causative factors are much more prevalent.

The incidence of intrauterine synechiae is unknown and there is no general agreement as to its prevalence or its impact upon fertility. This uncertainty is multifactoral and reflects differences in the experience of investigators, the use of induced abortion throughout the world, the incidence of genital tuberculosis in varied localities, the techniques used to evaluate the infertile couple, and the criteria used to diagnose synechiae.

However, there is a general consensus as to the major etiologic factors responsible for the development of these synechiae. Asherman's syndrome, or endometrial sclerosis, occurs when intrauterine adhesions form and obliterate, either partially or completely, the uterine cavity, cervical canal, internal cervical os, or one or both tubal ostia. The major predisposing factors are pregnancy, infection, and trauma, and synechiae are almost always iatrogenic in origin. From a practical standpoint, adhesion formation should be suspected in any patient who has undergone (a) curettage following pregnancy, (b) induced abortion (Fig. 7.1), or (c) any uterine surgery (including cesarean section) in the presence of pregnancy. However, pregnancy is not essential for the development of synechiae. Trauma without pregnancy, i.e., dilatation and curettage (D&C), can predispose to the condition (4), and, furthermore, the presence of infection seems to increase the likelihood of synechiae formation. Rabau and David (5) maintained that infection, frequently low grade or subclinical, is almost always associated with scarring. However, other investigators report no evidence of infection (6–8) and still others have been unable to document evidence of inflammatory cells or endometritis in the surgical specimens obtained in patients with synechiae (9, 10).

Asherman's syndrome offers a continuum of symptomatology and a variable effect on fertility. Patients may present with secondary amenorrhea, hypomenorrhea, and/or severe dysmenorrhea; less commonly they present with infertility but a normal menstrual history. The diagnosis is usually made by hysterosalpingography. This inconstant clinical presentation is the result of the marked variability of both the extent and the location of the synechiae. The lesions of Asherman's syndrome may be found in any portion of the uterus and may involve the entire uterine cavity or only a small area. Synechiae located in the cornua may result in tubal occlusion. Stenosis or atresia of the internal cervical os may cause partial or complete obstruction to menstrual flow and significant dysmenorrhea. Since endometriosis is more likely in these patients, fertility may be severely hampered. If implantation occurs, continued

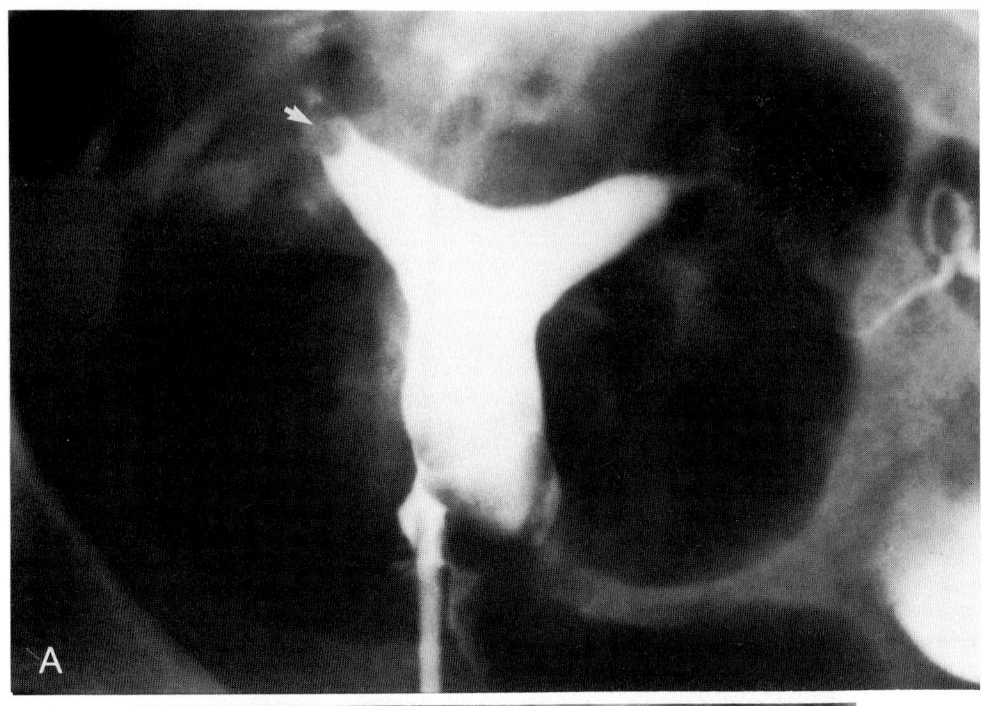

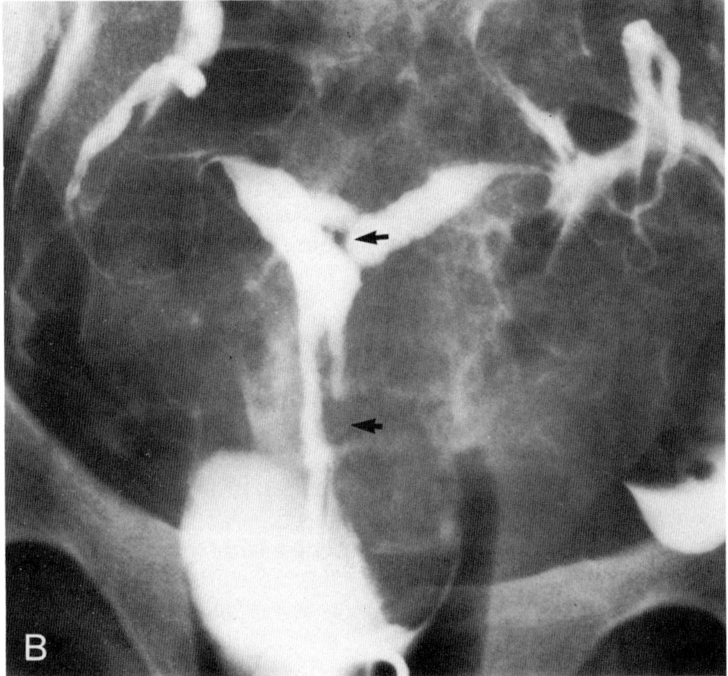

Figure 7.1. *A*: Baseline hysterosalpingogram during infertility workup. Cornual occlusion on the right is noted (*arrow*). *B*: Subsequent hysterosalpingogram on the same patient. In the interim between examinations, a pregnancy terminated in a first trimester abortion, which was followed by dilatation and curettage. Note the synechiae (*arrows*) in the midzone of the uterine body and in the lower uterine segment.

development of the placenta may be impaired and result in first or second trimester abortion. Other obstetric complications or mishaps may occur, including premature delivery, malpresentation, premature separation of the placenta, and placenta accreta, a condition in which the placenta becomes markedly adherent to the myometrium, resulting in incomplete postpartum separation (11–13).

Classification of Intrauterine Adhesions

A grading system to describe the extent of intrauterine synechiae is useful as an objective means to evaluate improvement after a therapeutic attempt as well as to provide prognostic criteria for the physician and the patient. The grade of severity has a positive correlation with menstrual dysfunction, infertility, and complications of pregnancy. However, minor adhesion formation that partially occludes the internal cervical os may induce severe dysmenorrhea. Toaff (14, 15) has proposed a useful classification based on the extent of the cavity obliteration, that categorizes the condition into four grades (15).

Grade 1. A single, small filling defect, frequently well inside the uterine cavity, occupying up to about one-tenth of the uterine area (Fig. 7.2).

Grade 2. A single, medium-size filling defect occupying up to one-fifth of the uterine area, or several smaller defects adding up to the same degree of involvement, located inside the uterine cavity, whose outline may show minor indentations but no gross deformation (Fig. 7.3).

Grade 3. A single large, or several smaller, filling defects involving up to about one-third of the uterine cavity, which is deformed or asymmetrical because of marginal adhesions (Fig. 7.4).

Grade 4. Large-size filling defects occupying most of a severely deformed uterine cavity (Figs. 7.5 and 7.6).

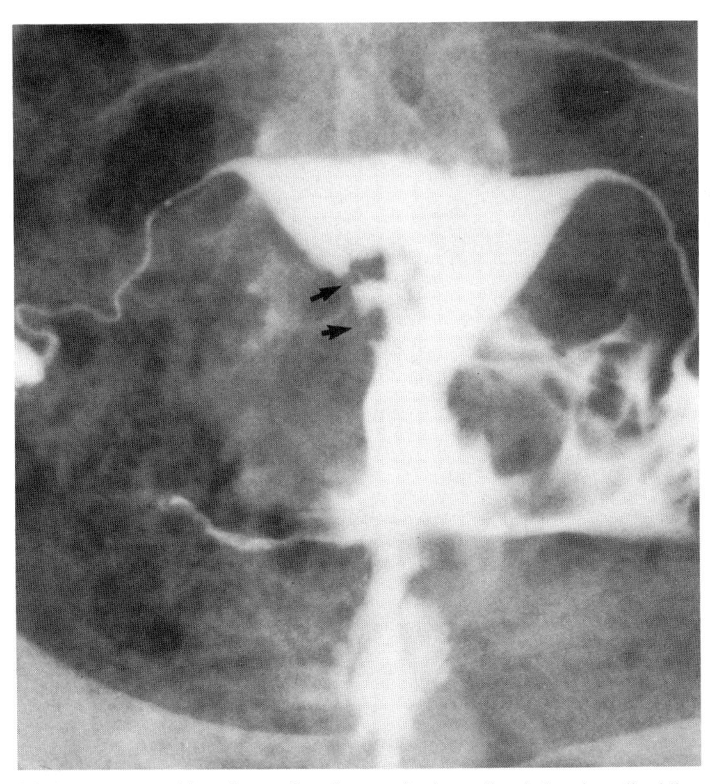

Figure 7.2. Minimum synechiae formation (*arrows*) along the lateral wall of the uterine body.

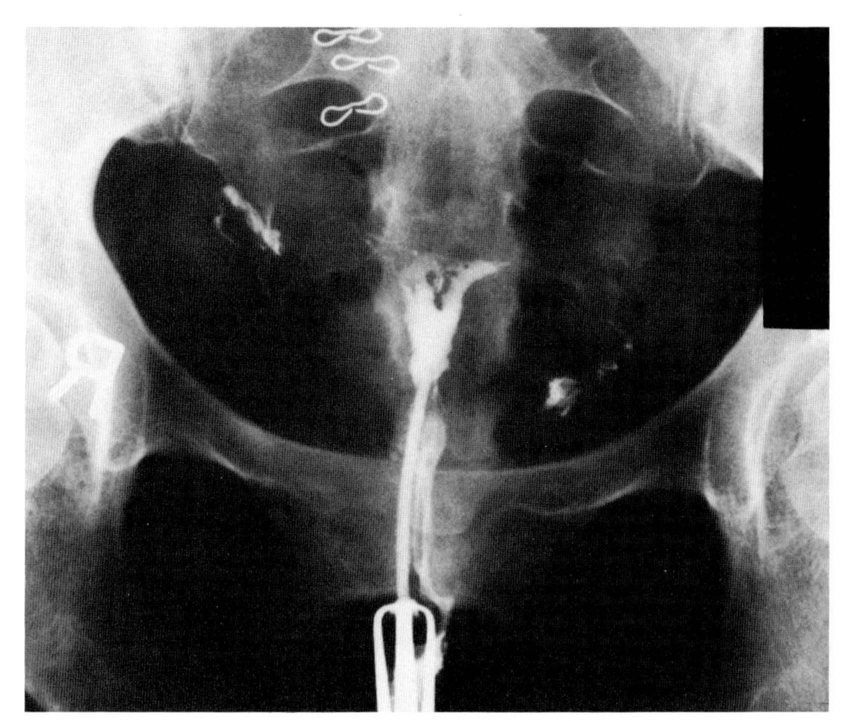

Figure 7.3. Several irregular synechiae localized to the fundus and upper portion of the uterine body in this patient with a history of abortion and dilatation and curettage.

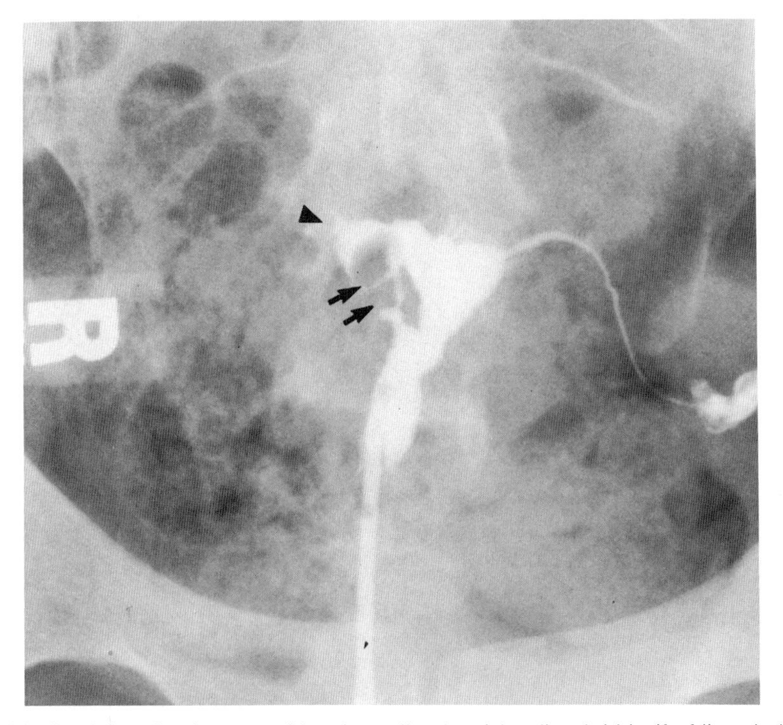

Figure 7.4. Moderately extensive synechiae formation involving the right half of the uterine body (*arrows*). In addition, adhesions have obliterated the ostium of the right fallopian tube (*arrowhead*).

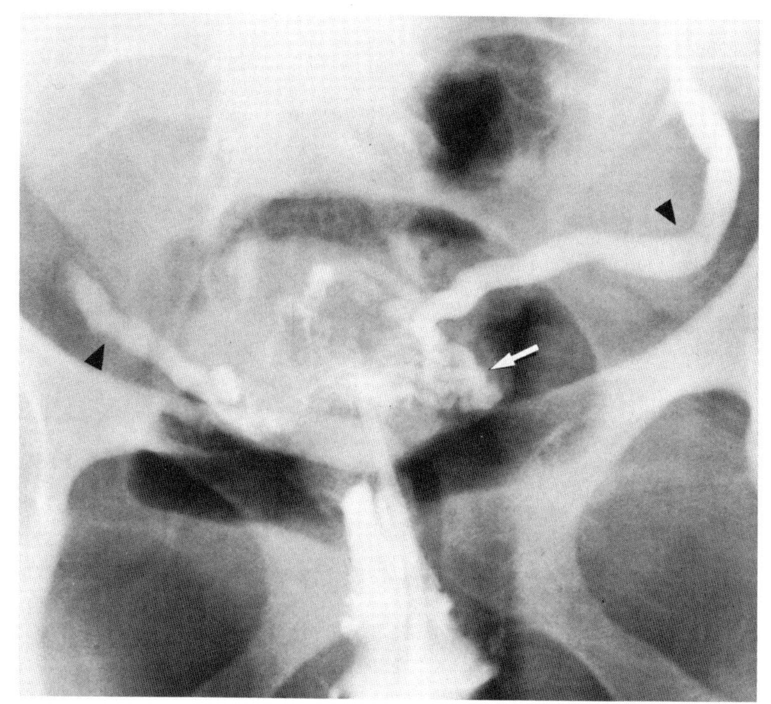

Figure 7.5. Extensive synechiae formation obliterating virtually the entire cavity of the uterine body. Myometrial intravasation (*arrow*) is observed, and there is obvious filling of major uterine and the ovarian veins (*arrowheads*).

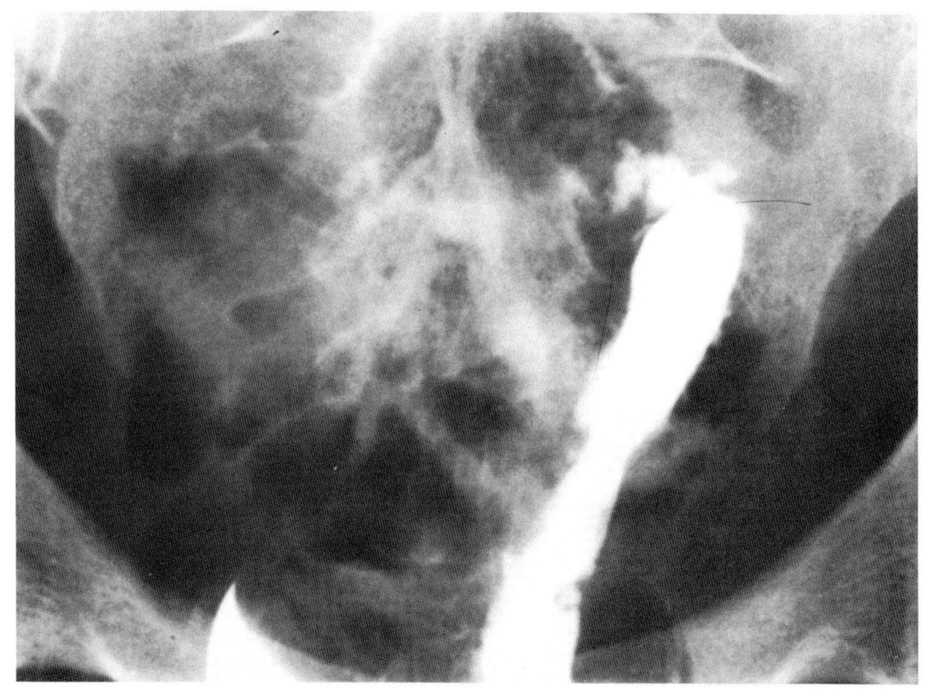

Figure 7.6. Extensive deformity of the uterine cavity secondary to extensive synechiae formation as demonstrated at hysteroscopy. Adhesive changes have occluded both fallopian tubes at their ostia. The appearance simulates a unicornuate uterus.

Diagnosis of Intrauterine Adhesions

Synechiae are scars that have formed as part of the healing process in the potential space of the uterine cavity when the traumatized uterine walls are held in apposition. The history may suggest the diagnosis, particularly since the otherwise unusual condition hypomenorrhea is common. There is frequently difficulty in sounding the uterine cavity or obtaining tissue for an endometrial biopsy. A "gritty sensation" may be detected when the biopsy is attempted.

More often, however, the diagnosis is unsuspected and is made at hysterosalpingography. Synechiae are frequently listed along with polyps, myomata, and uterine septum as "filling defects" within the uterine cavity as seen at hysterosalpingography. Generally speaking, however, the appearance of the scars is characteristic (Fig. 7.7). Neoplastic filling defects of the uterine cavity and retained products of conception

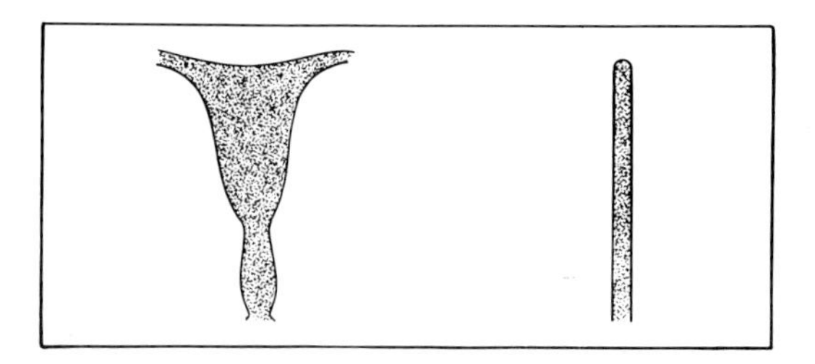

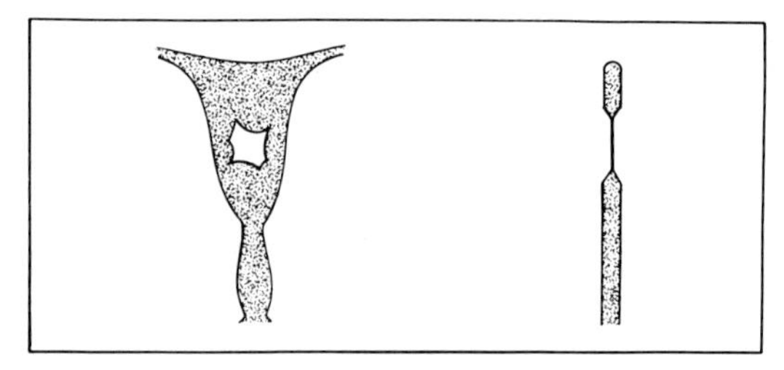

Figure 7.8. Artistic conception of synechia. *Upper panel* demonstrates normal uterine cavity (distended with contrast) in anteroposterior and lateral projection. *Lower panel* reveals that the irregular filling defect seen in anteroposterior projection is really an area of apposition of the anterior and posterior endometrial surfaces rather than a mass lesion.

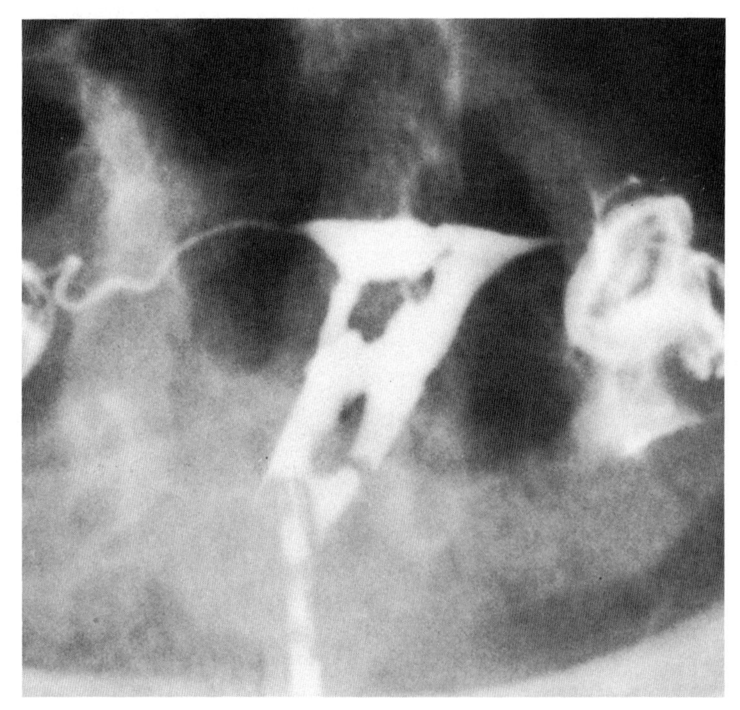

Figure 7.7. Classic appearance of synechiae presenting as irregular, angular, well-defined filling defects.

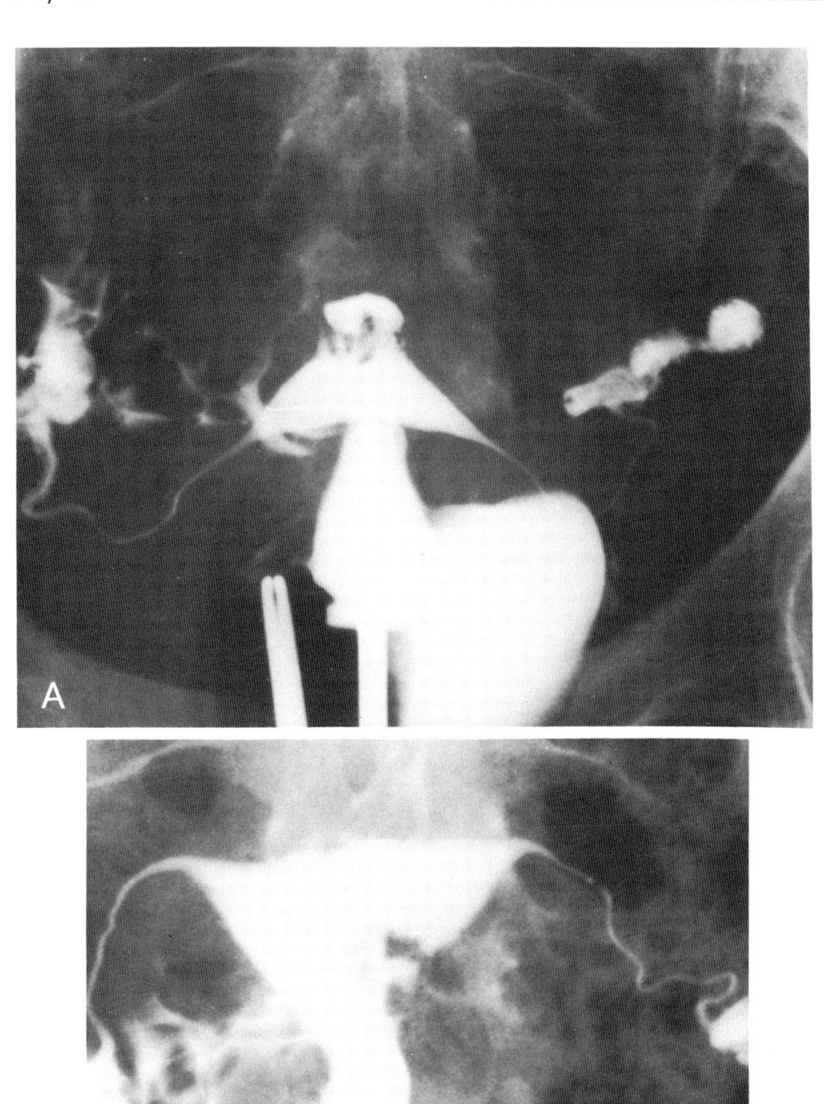

Figure 7.9. *A*: Hysterosalpingogram of patient with multiple small synechiae. One cannot appreciate the presence of the filling defect because of the marked anteflexion of the uterus. *B*: Adequate traction permits better evaluation of the uterine cavity. Multiple small, angular filling defects are noted.

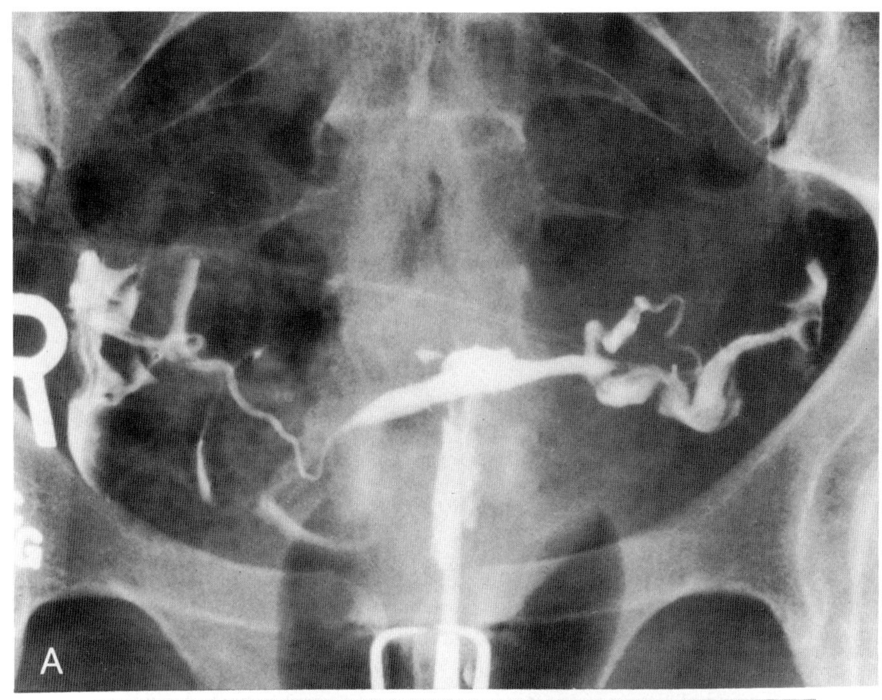

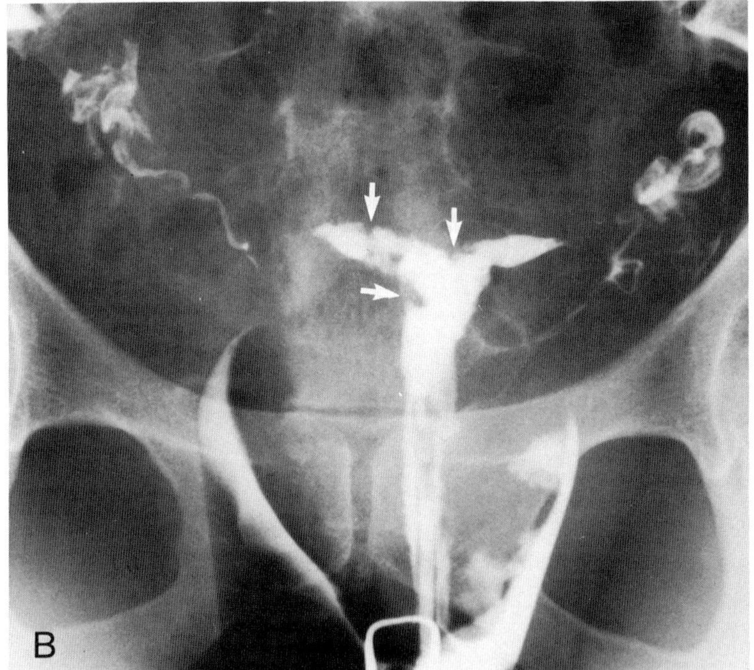

Figure 7.10. *A*: Moderately flexed uterus. No abnormality can be identified. *B*: Traction optimizes visualization of the uterine cavity. Numerous small synechiae are demonstrated (*arrows*).

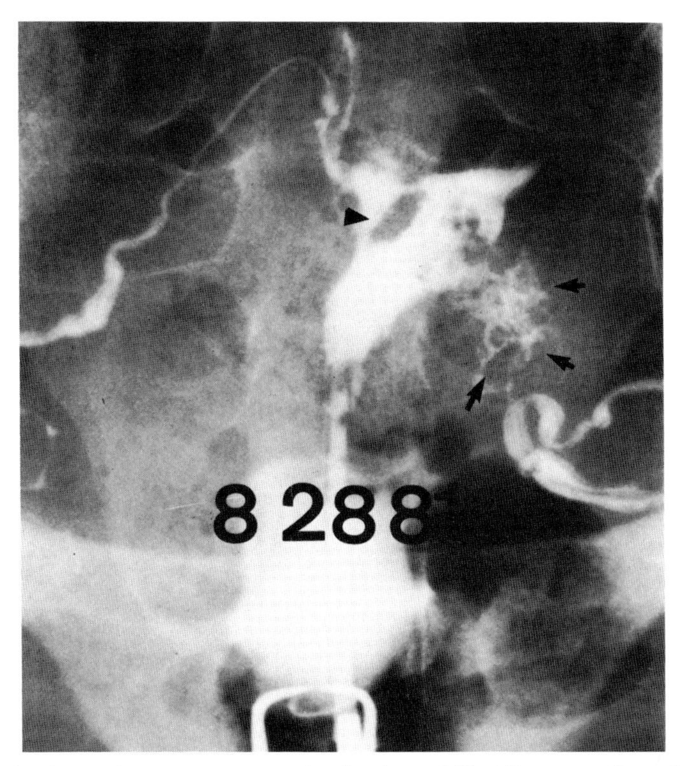

Figure 7.11. Classical angular appearance of adhesions obliterating a portion of the uterine cavity (*arrowhead*). Cavity is also partially obliterated along the right lateral wall with resultant extravasation of contrast into the myometrium (*black arrows*).

are characteristically round with smooth edges and homogeneous in density. Synechiae, on the other hand, are stellate, irregular, and frequently inhomogeneous if contrast enters pockets of no or little adherence. Indeed, the apparent filling defect at hysterosalpingography is not due to a mass lesion within the uterine cavity but rather reflects a constant area of apposition of the anterior and posterior walls of the uterus with failure of distension of the cavity in the area of synechiae formation (Fig. 7.8).

If the diagnosis of intrauterine adhesions is suspected prior to the radiographic examination, a short cannula is wisely employed so as not to disrupt adhesions in the cervical canal or lower uterine segment. It is, of course, important to assume that the uterine cavity is adequately assessed. This requires sufficient traction to counteract any appreciable anteflexion or retroflexion of the uterus and to permit visualization of the uterine cavity in its entirety (Figs. 7.9 and 7.10). Undue pressure should not be used to introduce the contrast agent. In some cases of severe intrauterine adhesion formation, contrast may enter the vascular or lymphatic system (Fig. 7.5). These episodes are usually uncomplicated and without difficulty to the patient. The frequency of intravasation and vascular embolization, however, is such that oil-soluble contrast media utilization may be ill advised (Figs. 7.5 and 7.11).

The intrauterine defects that are characteristic of synechiae are typically irregular, angulated, sharply contoured, and immobile. The position, shape, and number of these adhesions may vary considerably from patient to patient. If localized to the cervical canal, the adhesions may appear quite similar to mesonephric (Gartner) duct remnants (Fig. 7.12). These latter, also described in Chapter 4, present what appear to be linear filling defects in the endocer-

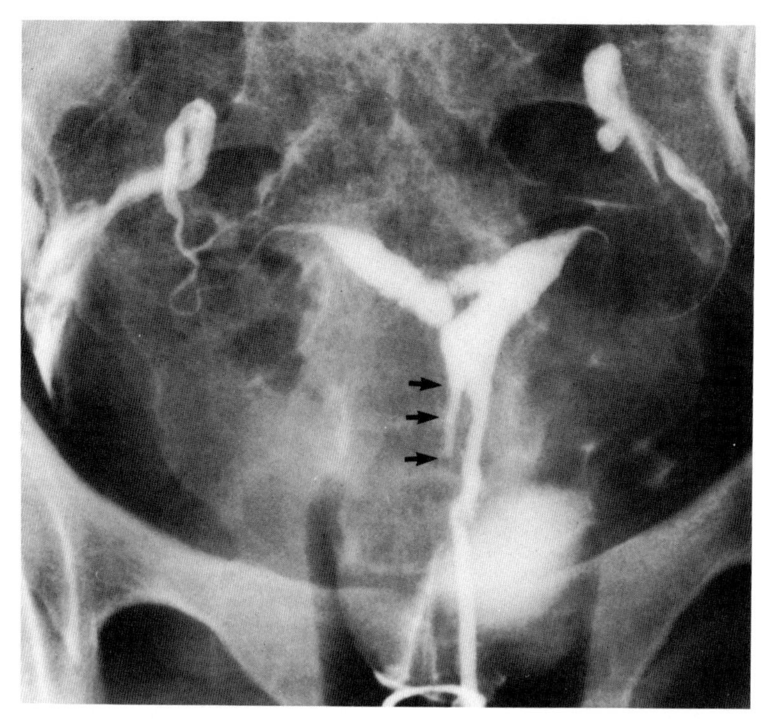

Figure 7.12. Synechiae formation in uterine body and lower uterine segment. The appearance in the lower uterine segment simulates the radiographic appearance of mesonephric duct remnant (*arrows*).

vical canal and may be virtually impossible to distinguish from synechiae; hysteroscopy is needed to make a definitive diagnosis.

Although the diagnosis of intrauterine scarring is usually established at hysterosalpingography, hysteroscopy is necessary for confirmation and further evaluation of the extent of the pathology. The apparent lack of correlation that sometimes occurs when the two procedures are compared is understandable when the differences in the two techniques are considered. The technique of hysteroscopy introduces a viscous medium under pressure, resulting in distension of the uterine cavity and frequently disruption of some of the adhesions. Synechiae that appear large and well defined at hysterosalpingography may present at hysteroscopy only as filmy, band-like adhesions because of the distension produced during the procedure of hysteroscopy. Perfect correlation between the two procedures is therefore not to be expected. Further, synechiae observed at hysterosalpingography may not be seen at hysteroscopy, which is not meant to imply a false-positive radiographic finding.

Treatment and Prognosis

Although there is some controversy surrounding the proper approach to the patient with intrauterine adhesion formation, nevertheless the essential components of therapy are lysis of adhesions at hysteroscopy or at D&C, placement of some form of device to keep the uterine walls apart, and administration of hormones to induce endometrial proliferation and menstruation (16–22). Most of the controversy revolves around the necessity for antibiotic administration, which ordinarily is employed in the therapy, and corticoid administration, to prevent subsequent adhesion formation. The recommended devices for intrauterine placement have included intrauterine contraceptive devices (IUDs) of several types, an inflated Foley catheter balloon, and a form made of distensible material molded to fit the uterine cavity. The Foley catheter

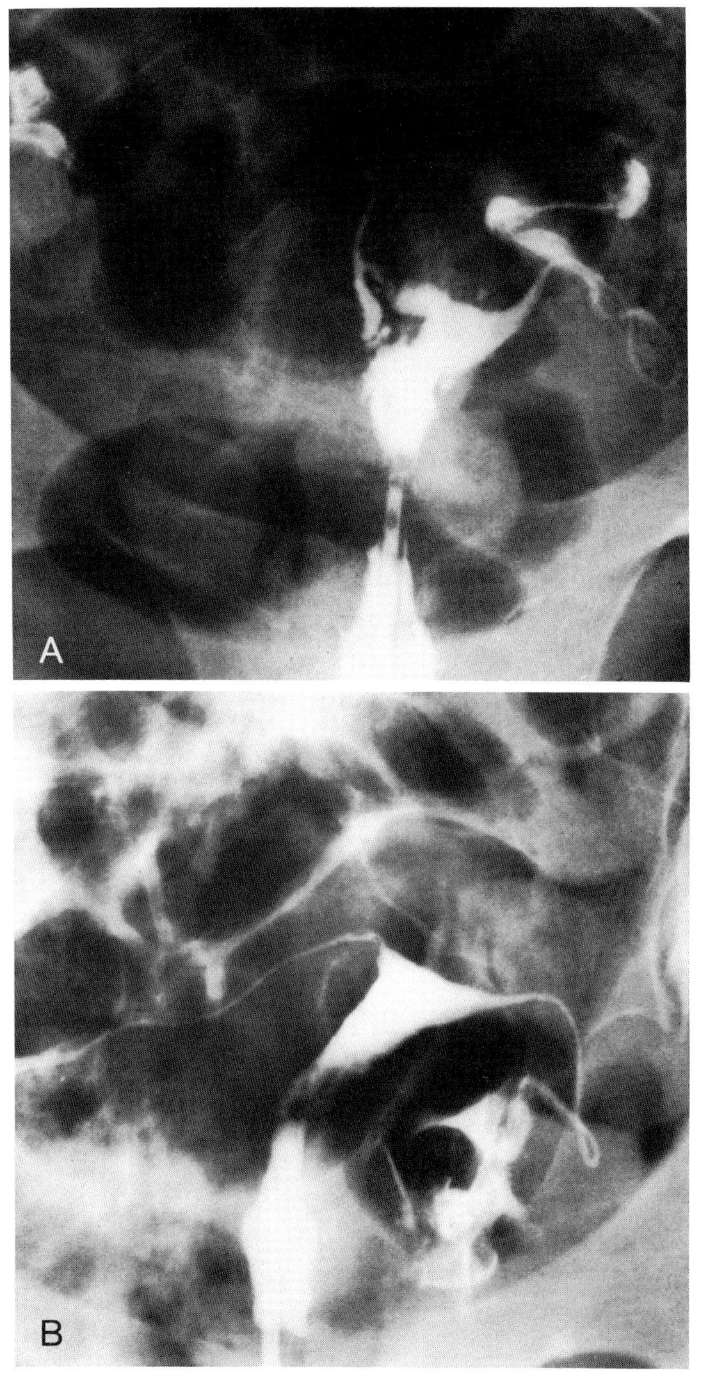

Figure 7.13. *A*: Moderate synechiae formation. *B*: Postoperative hysterosalpingogram demonstrates a normal-appearing uterine cavity.

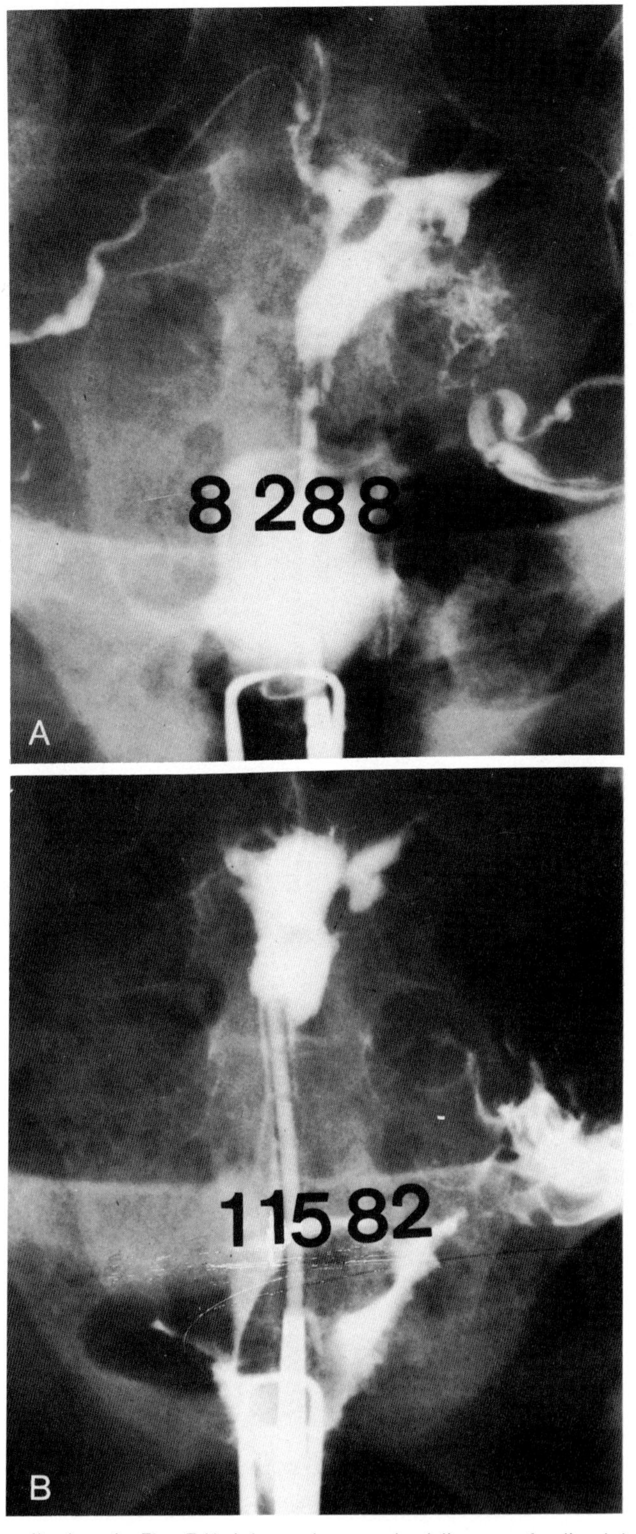

Figure 7.14. Same patient as in Fig. 7.11. Intervening surgical therapy for the intrauterine synechiae formation resulted in partial improvement. Considerable uterine cavity deformity due to residual and/or recurrent synechiae is noted.

can be left in place for up to 2 weeks, but these patients require hospitalization and may have considerable discomfort. Advantages of the use of the IUD include its capacity for retention for a longer time, its acceptability for the patient, and the lysis of adhesions that must occur at its removal; if considerable trauma occurs, however, adhesions may reform.

With complete cervical stenosis or adhesion formation at the level of the internal os, hysterotomy may be the only approach to lysis of intrauterine adhesions (9, 12). Using this approach, the uterus is incised in the fundus, and a route of dissection toward the cervix is employed, using either blunt or sharp dissection. Neither has been proven superior to the other. An intrauterine device is left within the uterine cavity to be removed later through the external cervical os, and the surgical incision in the fundus is closed. The IUD is removed 4–8 weeks later.

Postoperative hysterosalpingography is essential and is usually performed in the next proliferative phase after IUD removal and discontinuation of the hormonal treatment, which results in a "withdrawal" menstrual bleed (Fig. 7.13). It is not unusual to find recurrent adhesion formation, which it is hoped will be less than at the initial procedure (Fig. 7.14).

Results of present-day treatment are excellent in terms of symptomatic relief and correction of menstrual disorders, but remain far from ideal with respect to restoration of fertility (8, 12, 23–25). Most women regain normal menstruation, although about 11% continue to describe hypomenorrhea. In contrast, about half of women desiring fertility conceive, and among these pregnancies, about 25% abort. Placenta accreta occurs in 10% of deliveries, and other obstetric complications are encountered more frequently than normal. Of particular concern is the high incidence of placenta accreta, most probably caused by an abnormal adherence of placental tissue to sites of the severed adhesions adjacent to the myometrium where lack of decidua basalis is most evident. Thus, restoration of menstruation after treatment of intrauterine adhesions does not necessarily imply normal fertility, and the ensuing pregnancy may be subject to a number of complications. Pregnancies in women who have had treatment for intrauterine adhesions are potentially hazardous and should therefore be regarded as high risk for complications.

REFERENCES

1. Fritsch J: Einfall von volligen Schwund der Gebermutterhohle nach Auskratzung. *Zentralbl Gynaekol* 52:337, 1894.
2. Asherman J: Amenorrhea traumatica (atretica). *J Obstet Gynaecol Br Emp* 55:23, 1948.
3. Asherman J: Traumatic intrauterine adhesions. *J Obstet Gynaecol Br Commonw* 57:892, 1950.
4. Taylor PJ, Cumming DC, Hill PJ: Significance of intrauterine adhesions detected hysteroscopically in eumenorrheic infertile women and role of antecedent curettage in their formation. *Am J Obstet Gynecol* 139:239, 1981.
5. Rabau E, David A: Intrauterine adhesions: Etiology, prevention and treatment. *Obstet Gynecol* 22:626, 1963.
6. March CM, Israel R: Intrauterine adhesions secondary to elective abortion. Hysteroscopic diagnosis and management. *Obstet Gynecol* 48:422, 1976.
7. Polishuk WZ, Anteby SO, Weinstein D: Puerperal endometritis and intrauterine adhesions. *Int Surg* 60:418, 1975.
8. Schenker JG, Margalioth EJ: Intrauterine adhesions: An updated appraisal. *Fertil Steril* 37:593, 1982.
9. Jensen PA, Stromme WB: Amenorrhea secondary to puerperal curettage (Asherman's syndrome). *Am J Obstet Gynecol* 113:150, 1972.
10. Schenker JG, Yaffe H: Induction of intrauterine adhesions in experimental animals and in women. *Israel J Med Sci* 14:261, 1978.
11. Georgkopoulos P: Placenta accreta following lysis of uterine synechiae (Asherman's syndrome). *J Obstet Gynaecol Br Commonw* 81:730, 1974.
12. Jewelewicz R, Khalaf S, Neuwirth RS, Vande Wiele RL: Obstetric complications after treatment of intrauterine synechiae (Asherman's syndrome). *Obstet Gynecol* 47:701, 1976.
13. Breen JL, Neubecker R, Gregori CA, Franklin JE Jr: Placenta accreta, increta and percreta. A survey of 40 cases. *Obstet Gynecol* 49:343, 1977.
14. Toaff R: Amenorrea e ipomenorrea traumatics (Sindrome di Asherman) *Atti Soc Ital Ginecol* 49:258, 1962.
15. Toaff R, Ballas S: Traumatic hypomenorrhea-amenorrhea (Asherman's syndrome). *Fertil Steril* 30:379, 1978.
16. Danezis J, Souplis A, Papathanassiou Z: Conservative correction of uterine anomalies in cases of congenital and post-traumatic infertility. *Int J Fertil* 23:118, 1978.
17. Sugimoto O: Diagnostic and therapeutic hysteroscopy for traumatic intrauterine adhesions. *Am J Obstet Gynecol* 131:539, 1978.
18. Badawy S, Nusbaum M: Intrauterine synechiae—

etiological factors and effect of treatment on reproductive function. *Infertility* 2:303, 1979.
19. Ikeda T, Morita A, Imamura A, Mori I: The separation procedure for intrauterine adhesion (synechiae uteri) under roentgenographic view. *Fertil Steril* 36:333, 1981.
20. Siegler AM, Kontopoulos VG: Lysis of intrauterine adhesions under hysteroscopic control. A report of 25 operations. *J Reprod Med* 26:372, 1981.
21. Neuwirth RS, Hussein AR, Schiffman BM, Amin HK: Hysteroscopic resection of intrauterine scars using a new technique. *Obstet Gynecol* 60:111, 1982.
22. Hamou J, Salat-Barous J, Siegler AM: Diagnosis and treatment of intrauterine adhesions by microhysteroscopy. *Fertil Steril* 39:321, 1983.
23. Oelsner G, David A, Insler V, Serr DM: Outcome of pregnancy after treatment of intrauterine adhesions. *Obstet Gynecol* 44:341, 1974.
24. Caspi E, Perpinial S: Reproductive performances after treatment of intrauterine adhesions. *Int J Fertil* 20:249, 1975.
25. Bergquist CA, Rock JA, Jones HW Jr: Pregnancy outcome following treatment of intrauterine adhesions. *Int J Fertil* 26:107, 1981.

8 Hysterosalpingography of the Fallopian Tube: Inflammatory and Congenital Conditions

Tubal factors are thought to be responsible for some 30–50% of all infertility problems and require prompt and accurate assessment. Hysterosalpingography is usually the initial diagnostic tool employed in assessing tubal disease because of its ease, accuracy, and cost effectiveness. Diagnostic laparoscopy, however, frequently plays a significant role in tubal assessment. It is practical to view these two diagnostic procedures as complementary rather than competitive tools in the search for fallopian tube abnormalities (1–5).

Hysterosalpingography can be done easily on an outpatient basis without anesthesia or significant premedication. In addition to visualizing tubal patency and configuration, it permits evaluation of the uterine cavity. Disadvantages include occasional inability to differentiate cornual spasm from occlusion, as well as the inability to diagnose peritubal or adnexal disease.

Diagnostic laparoscopy permits accurate assessment of tubal patency by the retrograde injection of a colored dye, such as dilute methylene blue or indigo carmine, which can be seen to emerge from the fimbriated ends of the tube. Cornual occlusion due to spasm is less likely with laparoscopy because of the general anesthesia needed for the procedure. However, both procedures, laparoscopy and hysterosalpingography, may result in faulty assessment of tubal occlusion. If results from the two procedures contradict one another, any data signifying tubal patency should be accepted. The ability of laparoscopy to find evidence of peritubal and adnexal disease is a distinct advantage of this procedure.

Some data suggest that pregnancy rates may be increased for the 3–4 months following hysterosalpingography (6–8). This seems to hold true whether oil- or water-soluble contrast material is used, although advocates of oil-soluble media report that the incidence of pregnancy is doubled, from 13% with aqueous to 29% with oil-soluble agents (7). In either event, the procedure may possibly have therapeutic as well as diagnostic value.

Tubal dysfunction and disease can take several forms: tubal occlusion, which may be proximal, midposition, distal, or combined; hydrosalpinx, which can be patent or obstructed; intratubal lesions, which can include destruction of ciliated and nonciliated cells of the tubal epithelium, intratubal polyps, and salpingitis isthmica nodosa. Large series have been reported tabulating the abnormalities diagnosed (9, 10); these are of limited use because of population and technique differences.

Abnormalities of the Uterotubal Junction

The normal uterotubal junction at hysterosalpingography has a diamond or wedge shape in the intramural segment of the fal-

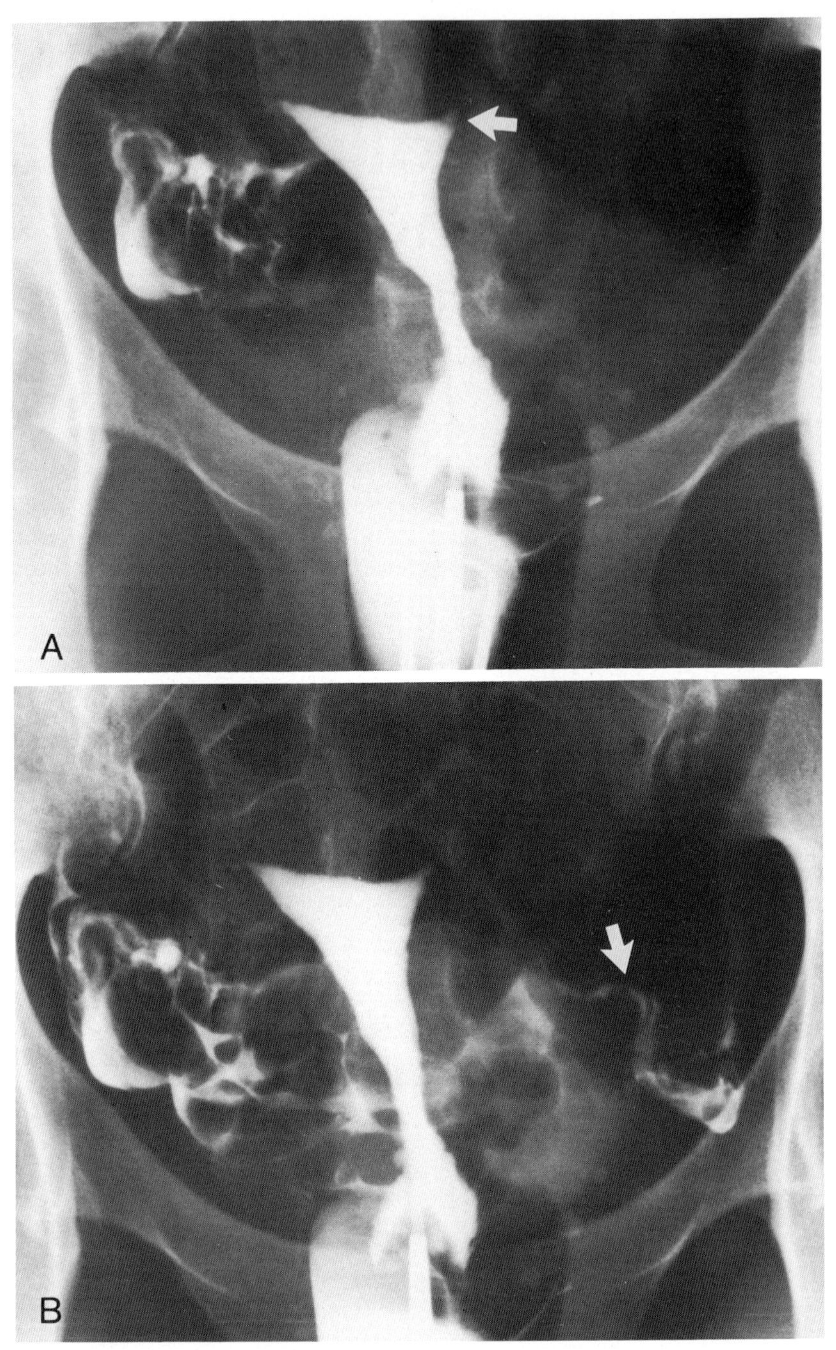

Figure 8.1. *A*: Cornual occlusion on left (*arrow*). *B*: Patency of left tube (*arrow*) established after intravenous administration of 1 ml glucagon.

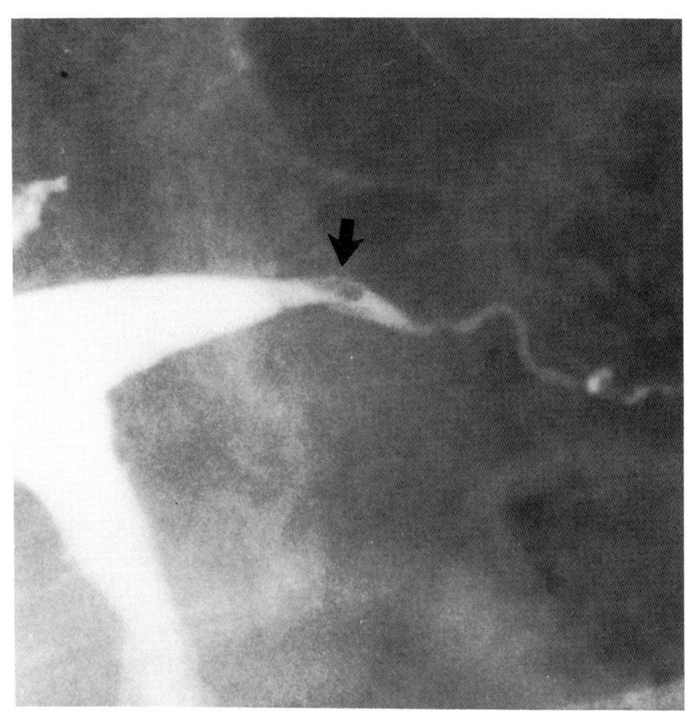

Figure 8.2. Small polyp in the intramural or interstitial segment of the left fallopian tube (*arrow*).

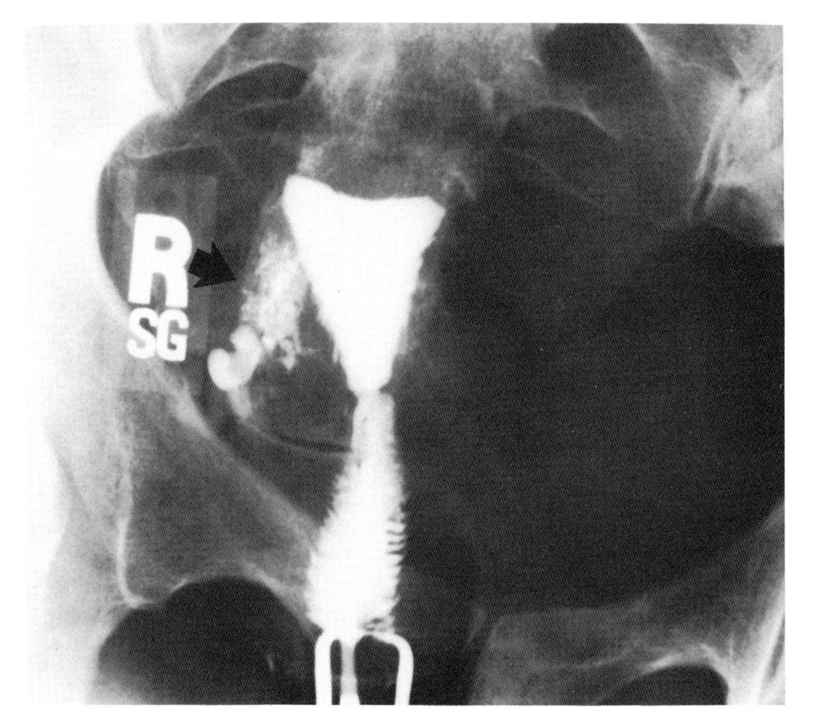

Figure 8.3. Obstructive changes at the uterotubal junction bilaterally secondary to obstructive inflammatory fibrosis. Note extravasated contrast filling abundant venous channels within the myometrium (*arrow*).

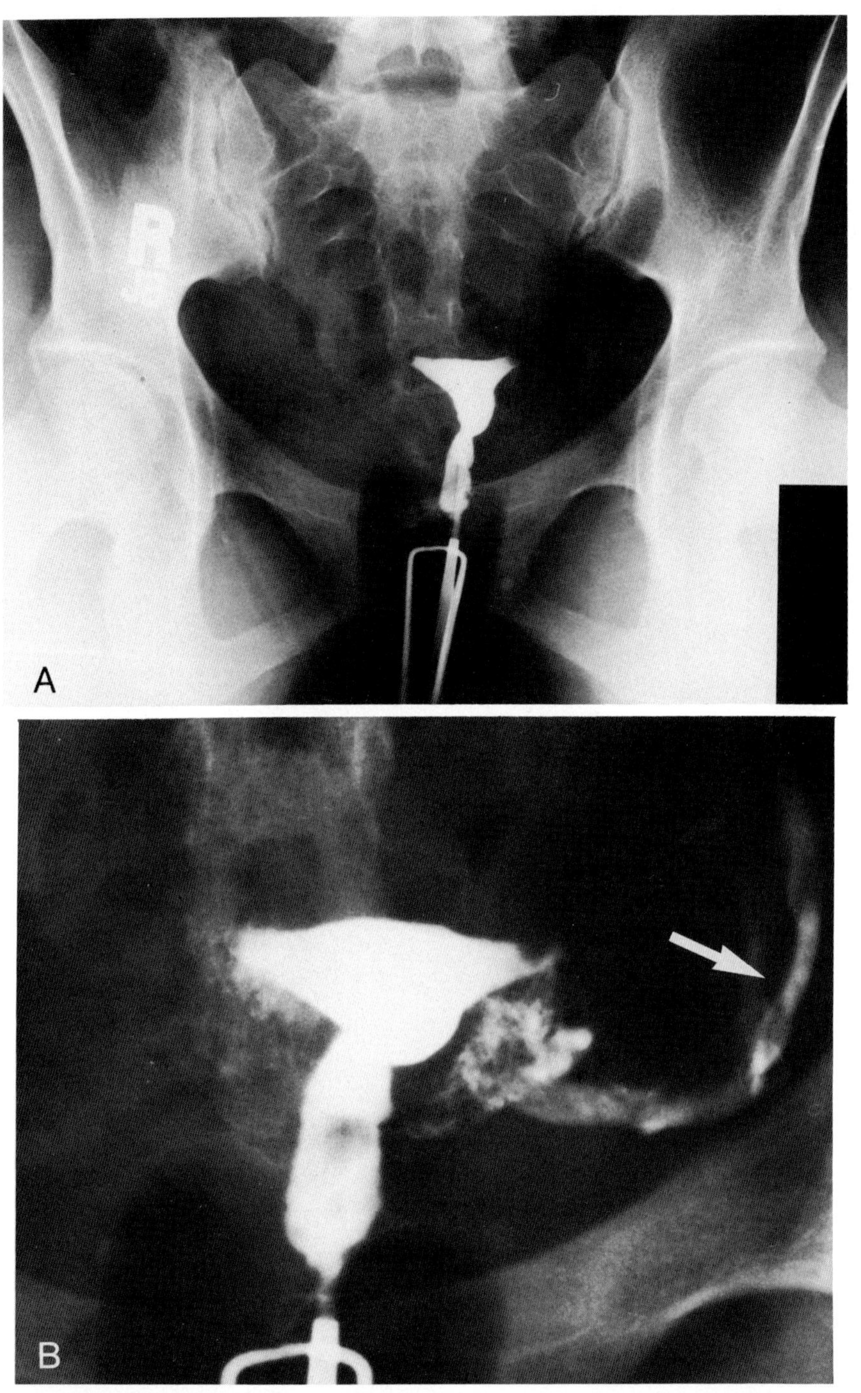

Figure 8.4. *A*: Bilateral obstruction at the cornua due to previous infection. *B*: Continued introduction of contrast results in intravasation of contrast into the myometrium and vascular structures. The ovarian vein (*arrow*) is well filled.

lopian tube (Chapter 3) and has been described as an "endometrial funnel" (11). The length of the intramural tube as measured on extirpated specimens varies from 5 to 14 mm. A true anatomic sphincter is absent; however, the intramural tube has an inner longitudinal muscle layer, which functions with the middle circular layer of muscle to induce a constricting action in this area of the oviduct. Cornual spasm occurring during hysterosalpingography can give the appearance of tubal occlusion; this may or may not be reversed by the intravenous administration of antispasmodics. In our practice, glucagon has been valuable (12) (Fig. 8.1). Cornual occlusion can also occur during the secretory phase of the cycle if protruding secretory endometrium is pushed into the ostium and blocks the opening of the uterotubal junction. True polyps or papillations of the mucosa of the intramural portion of the tube are occasionally observed (Fig. 8.2) and can contribute to the diagnosis of occlusion (13). Similarly, synechia formation at the tubal ostium may cause obstruction. During hysterosalpingography, it is important to move the uterus in relation to the tube, straightening the convoluted intramural portions so that contrast is more likely to pass; pulling down on the tenaculum, then pushing up while injecting contrast will often accomplish this.

Few studies have systematically investigated uterotubal junction pathology. Excised tubal segments from women with uterotubal junction obstruction indicate that the most frequent lesion is obliterative fibrosis (in 38%) (Fig. 8.3), followed by salpingitis isthmica nodosa (24%), intramucosal endometriosis (14%), and chronic tubal inflammation (21%) (14) (Fig. 8.4).

Abnormalities of the Isthmus and Ampullary Portion

Obstruction at the isthmic and ampullary portions of the fallopian tubes can be due to infection, sterilization procedures, and congenital anomalies. Tubal occlusion secondary to infectious processes may occur at any portion of the tube (15). Siegler (16) evaluated 1000 consecutive salpingograms and found unilateral tubal occlusion in 27.7% and bilateral blockage in 11.1%. Cornual occlusion was most common, followed closely by obstruction in the ampullary segment of the tube (Fig. 8.5). Postinflammatory obstruction in the isthmic segment was found in only 5% of his series. Since the myometrial width cannot be distinguished by hysterosalpingography, it is sometimes difficult to differentiate between intramural and isthmic obstruction. Peritubal adhesion formation and tubal tuberculosis involving the isthmic-ampullary segment can cause midportion obstruction. The normal rugal patterns may or may not be seen, and occasionally rugae are observed in markedly dilated diseased tubes. Crohn's disease, or regional enteritis, and ulcerative colitis may induce a significant inflammatory process, and can be associated with tubal occlusion as well as adhesive disease. Chlamydia, which is frequently associated with subclinical infection, is likely to cause fimbrial disease. The use of an intrauterine device (IUD) for contraception has also been associated with adnexal adhesions and fimbrial occlusion. In general, the etiologic agent associated with tuboovarian infection and adhesion formation cannot be distinguished by radiographic or pathologic findings. Pelvic tuberculosis may be an exception.

Poststerilization tubal obstruction is predictable from history, and the purpose of hysterosalpingography is usually to confirm the occlusion or to determine the length of the proximal portion of the tube, if reanastomosis is being considered. Unfortunately, the investigative technique of hysterosalpingography performed in the poststerilization patient to determine successful tubal occlusion may increase the incidence of fistula formation and recanalization, particularly if done early in the postoperative period; this is discussed further in Chapter 9.

Congenital anomalies of the fallopian tube are rare, but may include accessory ostia, multiple lumina, diverticula, and total tubal duplication; complete absence of the fallo-

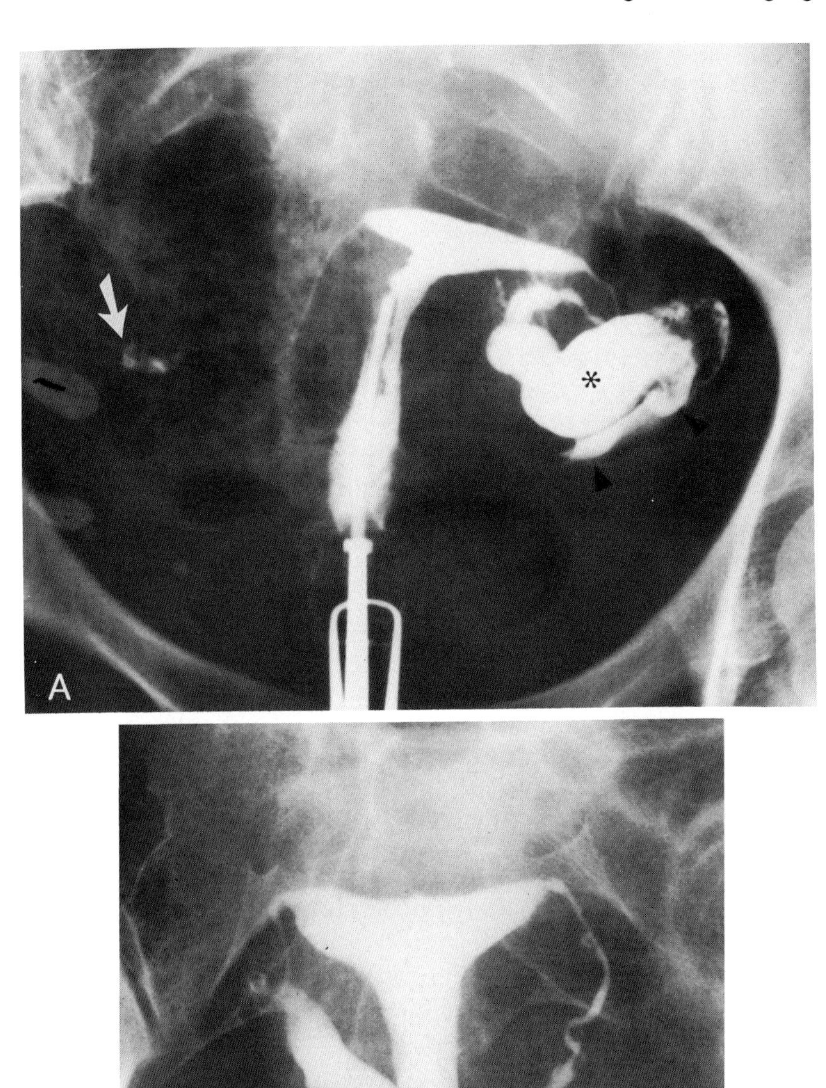

Figure 8.5. *A*: Obstruction in proximal portion of ampullary segment of right fallopian tube (*arrow*). Note peritubal collection of contrast on left (*arrowheads*), reflecting inflammatory changes around the left hydrosalpinx (*asterisk*). *B*: Bilateral tubal occlusions occurring in distal portions of ampullae. Tubes are displaced and fixed deep in the pelvis.

pian tube; and segmental absence of a portion of the fallopian tube (Fig. 8.6). Accessory tubes are thought to contribute to infertility by capturing oocytes that otherwise would have entered the normal tube, and perhaps by increasing the incidence of ectopic pregnancy; they are found in about 0.6% of women (17). Accessory tubes are visualized at hysterosalpingography only if communication with the tubal lumen or fimbrial pick-up of contrast media occurs. Segmental absence may be caused by tubal torsion or twisting with interruption of the blood supply; it supposedly cannot be congenital in origin, so an acquired etiology should be sought (18, 19). Complete torsion of ovary and tube on the pedicle may result in necrosis and sloughing of the tissue; this has been identified later at open surgical procedures as calcified tissue in the cul-de-sac, but the hysterosalpingographic appearance has not been reported.

Other abnormalities of the fallopian tube include polyps in the intramural portion (Fig. 8.2), or less commonly in the proximal isthmic portion. Polyps may be diagnosed in 1.2–2.7% of hysterosalpingograms (20–22), although a tissue diagnosis is usually lacking. In tubes examined for pathology after salpingectomy, 11% reportedly contained polyps (11). Our experience has been minimal because we have seen only one potential case, and this has been without pathology documentation. Infertility has been reported in 27–53% (21) of patients diagnosed at hysterosalpingography to have polyps, but it is difficult to determine that these are not simply incidental findings. Women thought to have tubal polyps probably should be regarded as having relative subfertility and first treated for other causes of infertility. Surgical management has been uniformly poor: removal of the affected segment of tube, with subsequent uterine implantation of the distal portion (20), and microsurgical resection (22) have not led to pregnancy. This suggests that the infertility might better be explained by another etiology, because tubal implantation and reanastomosis procedures ordinarily have a

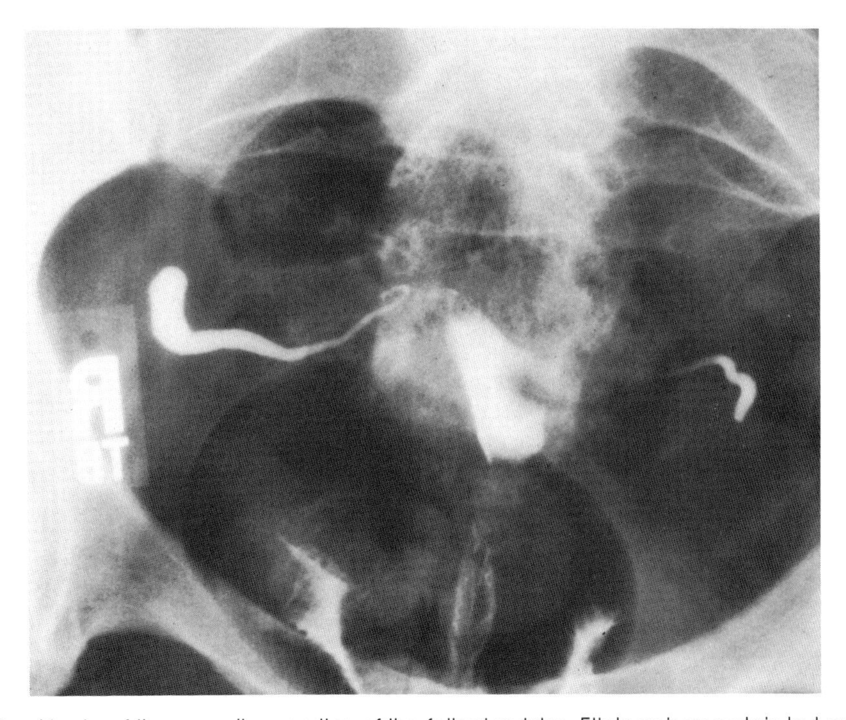

Figure 8.6. Atresia of the ampullary portion of the fallopian tube. Etiology is uncertain but may reflect intrauterine interruption of vascular supply.

higher reported pregnancy rate when performed for frankly obstructing tubal disease.

Distal Fimbrial Obstruction

The finding of a distal hydrosalpinx suggests acute or chronic infectious disease as the etiology of the occlusion, and a history of pelvic inflammatory disease or an IUD for contraception is often obtained. On the other hand, pelvic inflammatory disease may be totally "silent" with no recollection of acute infection; most patients found to have hydrosalpinges at laparoscopy will not recall an episode of pelvic infection (23). Chronic obstruction at the fimbriated end, whatever the etiologic process, leads to dilatation, mucosal damage, and destruction of the ciliated cells normally lining the tube (Fig. 8.7), although on occasion a normal rugal pattern may be preserved (Figs. 8.8 and 8.9). The appearance of hysterosalpingography is characteristic if sufficient contrast is introduced to fill the dilated tube. Partial filling may give the appearance of a normal, nondilated tube, so it is very important to be sure that the salpinx is completely filled (Fig. 8.10). Contrast media must be dispersed freely in the peritoneal cavity to demonstrate patency, and a delayed film may be helpful to clarify peritoneal spill (8.11). If uncertainty exists, continued introduction of more contrast or carbon dioxide is warranted (Fig. 8.12). On occasion, this may result in the appearance of contrast in the lymphatics or the venous vasculature (Fig. 8.7).

Occasionally, a collection of contrast medium surrounding the ampullary portion of the tube is seen associated with a patent or nonobstructed hydrosalpinx. This appearance may persist on the delayed film. Such pooling of contrast medium reflects associated peritubal adhesions and is usually the result of the same process responsible for the formation of the hydrosalpinx.

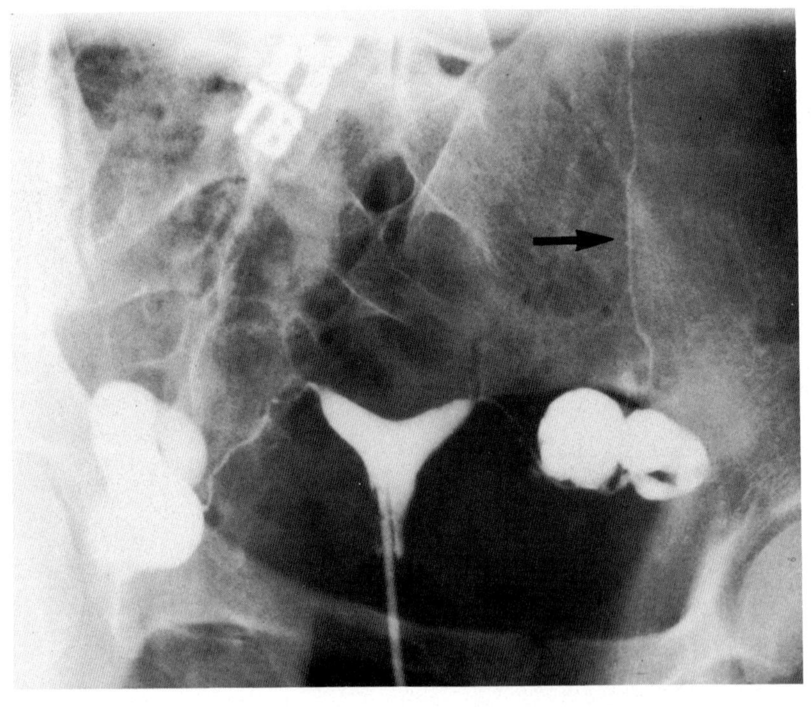

Figure 8.7. Bilateral hydrosalpinges. The tubal dilatation is localized to the ampullary portion of the tubes. Note the filling of a lymphatic channel from the left peritubal region (*arrow*).

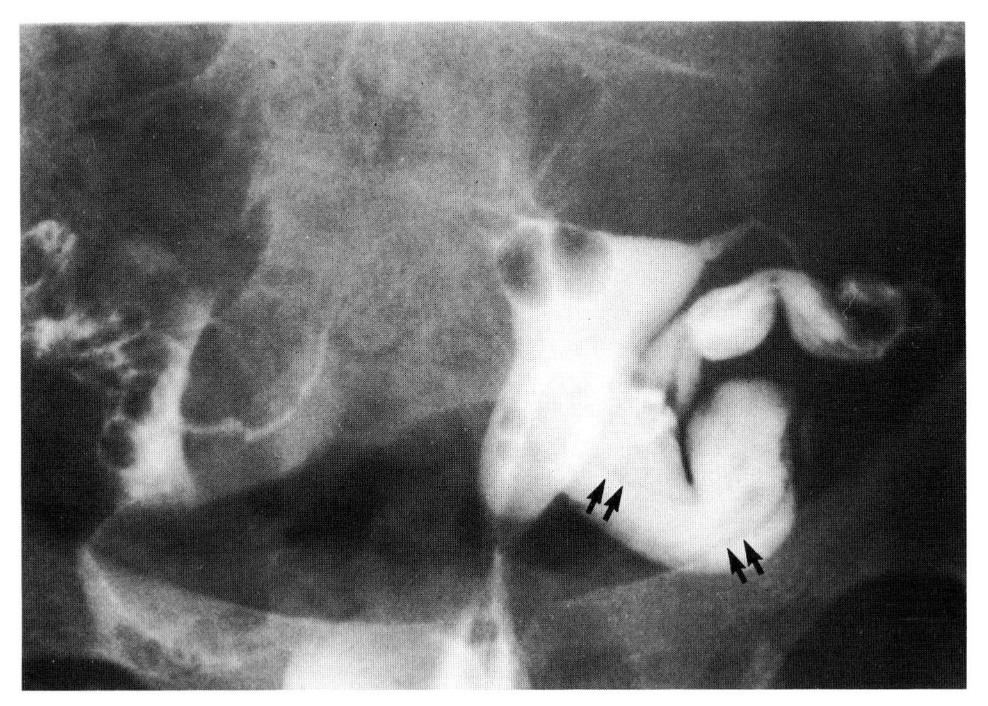

Figure 8.8. Hydrosalpinx on the left. Despite dilatation of the ampullary portion of the tube, rugal folds are intact (*arrows*).

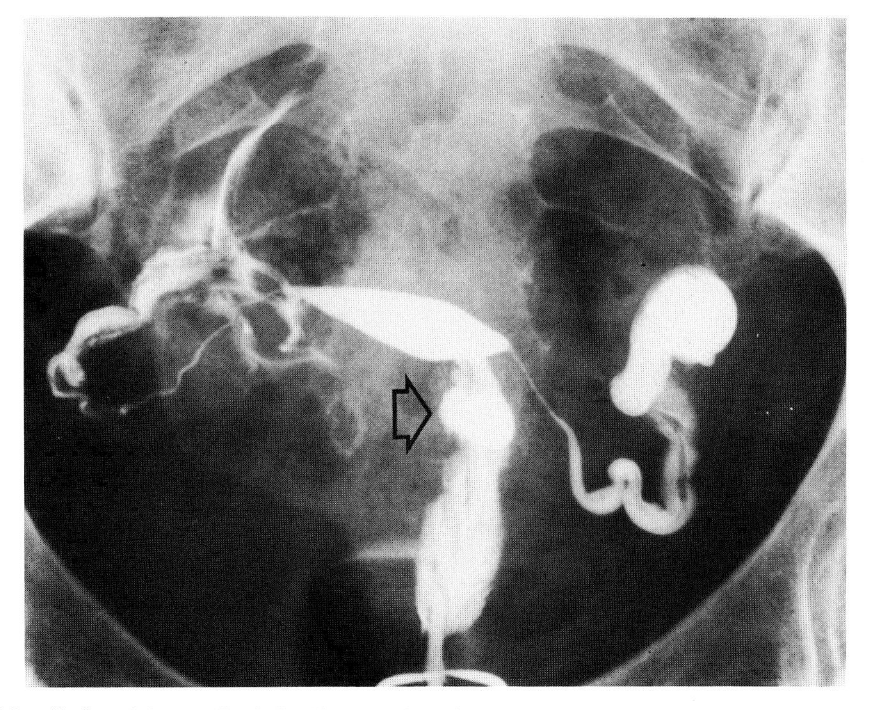

Figure 8.9. Hydrosalpinx on the left with normal and patent tube on right. Both normal and abnormal fallopian tubes demonstrate presence of rugal folds. Incidental note is made of the diverticula-like collection of contrast in the lower uterine segment, the result of a previous cesarean section (*open arrow*).

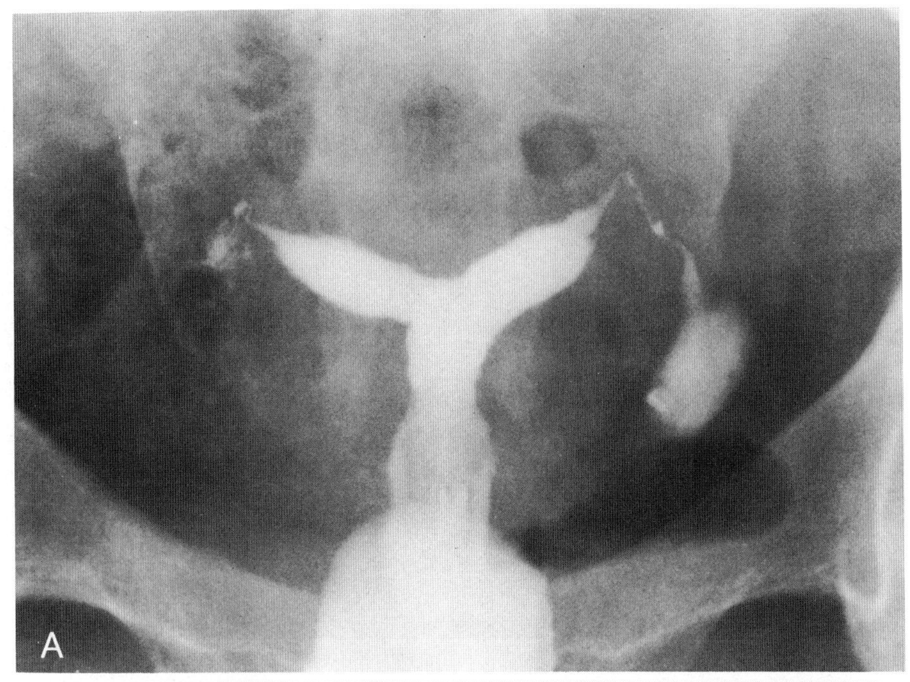

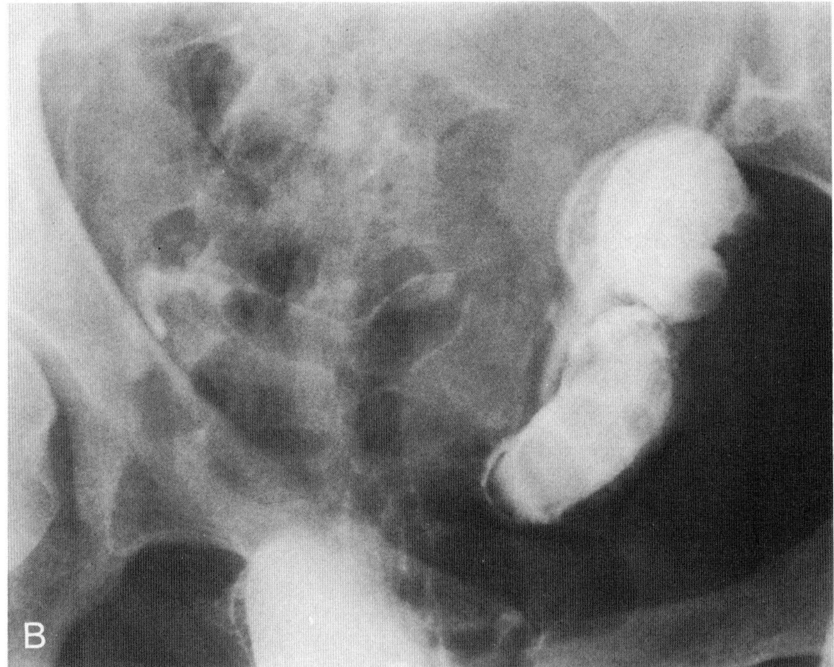

Figure 8.10. *A*: Right tube occluded. Left tube appears occluded and there is a slightly dilated bicornuate uterus. *B*: Further filling demonstrates extent of a very large hydrosalpinx on left.

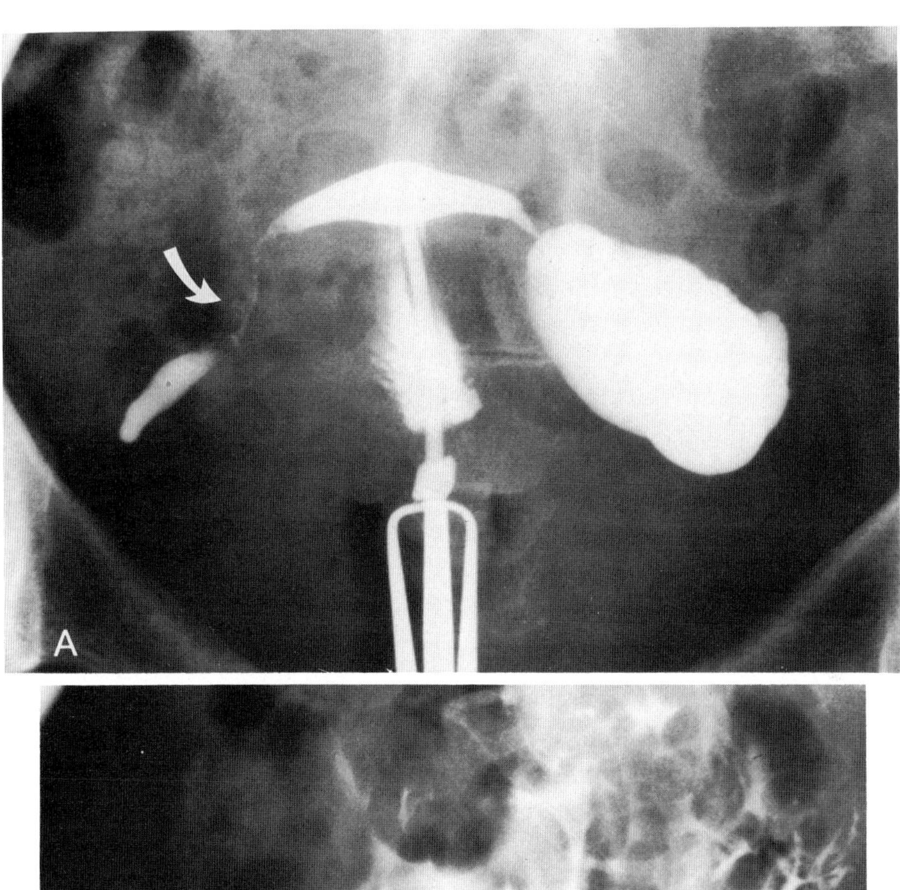

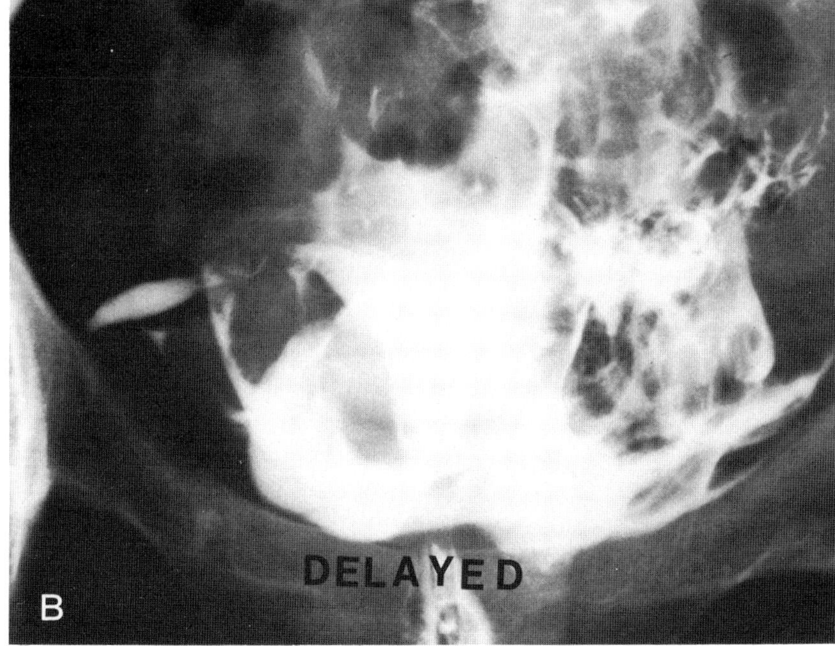

Figure 8.11. *A*: Right tube is occluded in ampulla. Note the salpingitis isthmica nodosa (*arrow*). Left tube is markedly dilated and appears occluded. *B*: Delayed film demonstrates that the left tube is patent as the contrast spills into the peritoneal cavity.

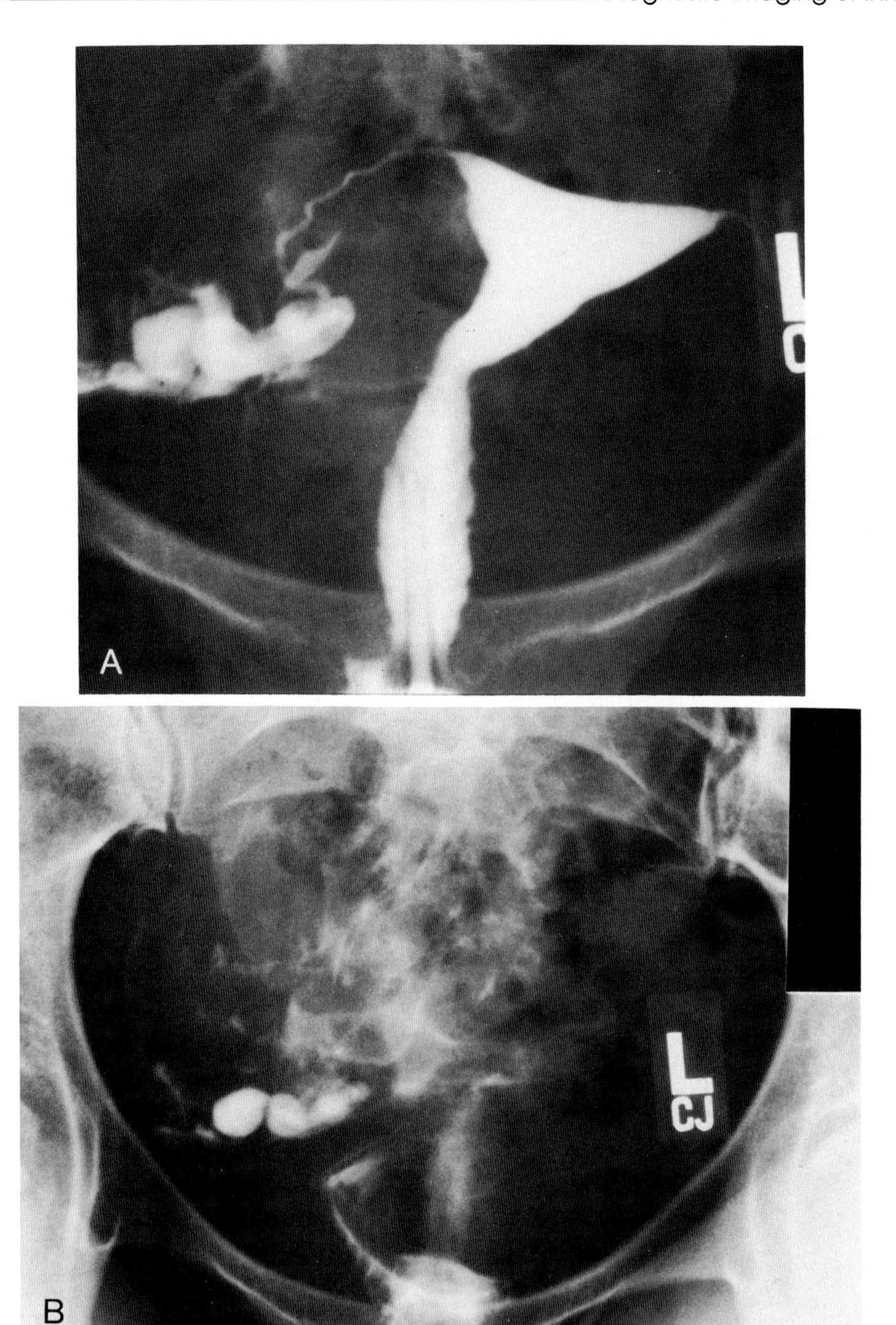

Figure 8.12. *A*: Ampullary portion of the right fallopian tube is minimally dilated and clearly patent. *B*: Delayed film demonstrates retained contrast material within the slightly dilated right fallopian tube. This retention, associated with the dilatation, supports the diagnosis of postinflammatory change.

Salpingitis Isthmica Nodosa

Salpingitis isthmica nodosa (SIN) is the descriptive term for nodular thickenings of the isthmic and occasionally ampullary portion of one or both fallopian tubes (24–26). Radiologically, this condition is characterized by a honeycombed accumulation of radiocontrast material in the wall of the fallopian tube, giving the appearance of multiple small diverticuli (24) (Figs. 8.13, 8.14, and 8.15). These punctate accumulations of contrast medium are distributed in the isthmic, isthmic-cornual, or isthmic-ampullary segments. The area sometimes resembles the branches of a tree or a bush, with a complex maze of opaque channels; more commonly, the image is that of a cluster of diverticula or outpouchings, with persistence of contrast material in these pouches on delayed films.

Pathologically, these nodular areas are characterized by inpouchings and invaginations of normal endosalpinx surrounded by a hypertrophied myosalpinx. There are alveolar-like, slit-like, or cystic irregular spaces in the myosalpinx. These cavities or pockets may be empty or contain a homogeneous coagulum, desquamated epithelial cells, macrophages with or without phagocytized brown pigment, and occasionally a purulent exudate (27). Salpingitis isthmica nodosa appears almost exclusively to involve the isthmic segment of the tube, which has a powerful muscle coat and a very dense adrenergic motor innervation.

The etiology of SIN could be a postinfectious process, a congenital or development defect, or the result of endosalpingeal metaplasia. Salpingitis isthmica nodosa could also be a late morphologic expression of chronic tubal spasm. Some patients with SIN have external endometriosis in other parts of the pelvis, but other studies (25) have excluded endometriosis, suggesting that the two conditions are not identical, but may exist independently. Both acute and chronic infection have been identified in resected nodules, and even the name salpingitis isthmica nodosa, as coined by Chiari (28), denotes the inflammatory origin of the lesion, its predominant location, and its nodular gross appearance. Evidence of previous inflammatory disease, including gonococcal infection, has been reported in the majority of cases diagnosed (26).

Von Recklinghausen (29) in 1896 proposed that the findings of SIN were due to wolffian rests, since embryologically the tubal isthmus is the region where the müllerian and wolffian ducts cross. However, examination of tubal tissue from embryos, term female infants, and older prepubertal girls has failed to demonstrate the lesion (27). Although various etiologies have been recorded for SIN, it seems likely that the process is postinflammatory in origin, with proliferation of normal epithelium in the tubal isthmus, slow penetration into the tubal wall forming a labyrinth of pathways and cysts, and finally secondary muscular hypertrophy.

Salpingitis isthmica nodosa appears to be associated with either subfertility or infertility. Tubal occlusion is not characteristic, and most cases of SIN have tubal patency, with a honeycombed appearance in the isthmic area of the tube, with or without associated hydrosalpinx. As expected, there is an increased incidence of ectopic pregnancy, 75% of which are found in the isthmus or ampulla (26, 30). In many specimens no communication can be demonstrated between the main tubal lumen and the trophoblastic tissue, suggesting implantation in a diverticulum (31). Interestingly, tubal pregnancies are more common in the right than in the left tube; 49% of women with ectopic pregnancies have SIN nodules, with 30% having nodules in the right tube compared to 19% with nodules in the left (30).

Salpingitis isthmica nodosa cannot be detected by pelvic examination but may be observed at diagnostic laparoscopy, and is determined by palpation during open surgical procedures. From a hysterosalpingographic standpoint, the diagnosis of SIN is easily established, is not uncommon, and can be expected in at least 4% of hysterosalpingographies (26). The remainder of the tube may appear normal and patent,

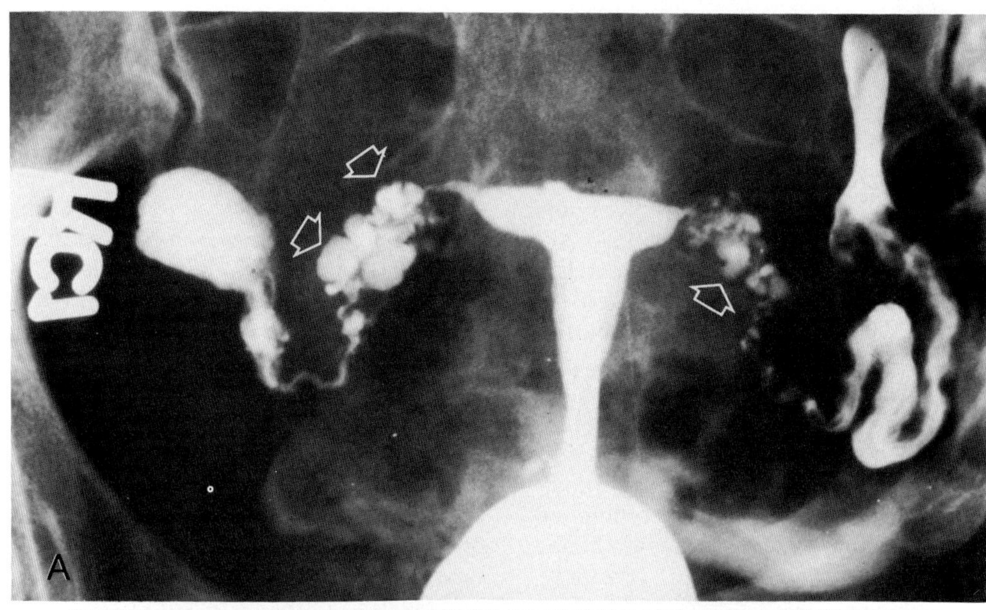

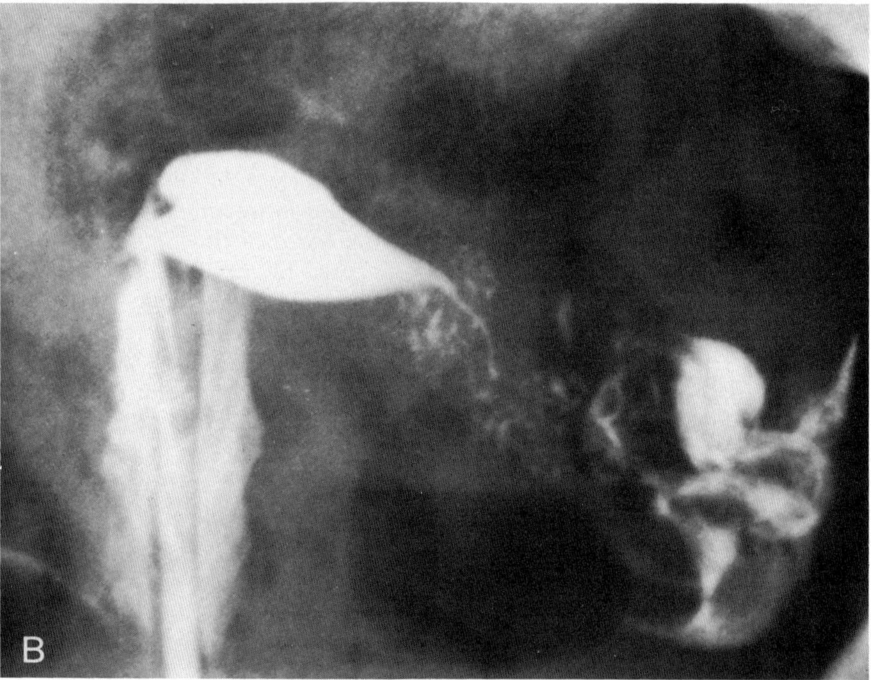

Figure 8.13. *A*: Salpingitis isthmica nodosa. Note the numerous large, diverticula-like collections of contrast in the isthmic segment of the left fallopian tube. (Courtesy of T. A. Baramki, Baltimore, Maryland.) *B*: Another example of salpingitis isthmica nodosa. Collections are smaller but quite numerous and equally characteristic.

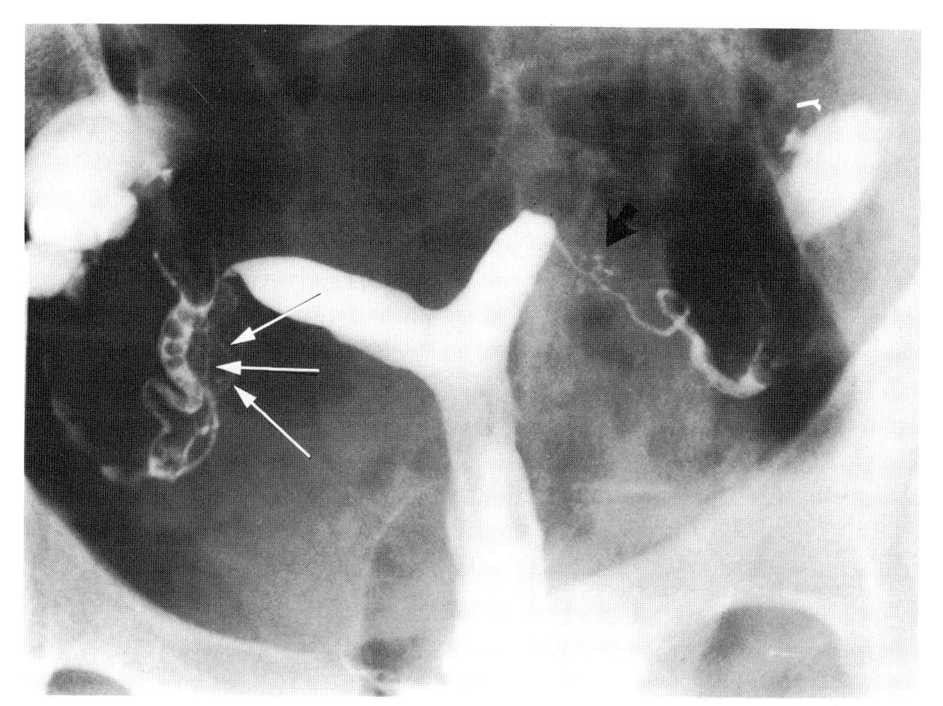

Figure 8.14. Salpingitis isthmica nodosa, bilateral. Both tubes demonstrate ampullary hydrosalpinges. The right fallopian tube shows several diverticula as well as lacy linear tracts of contrast (*arrows*). The left fallopian tube shows the more characteristic appearance of small diverticula-like collections in the isthmic segment of the tube (*black arrow*).

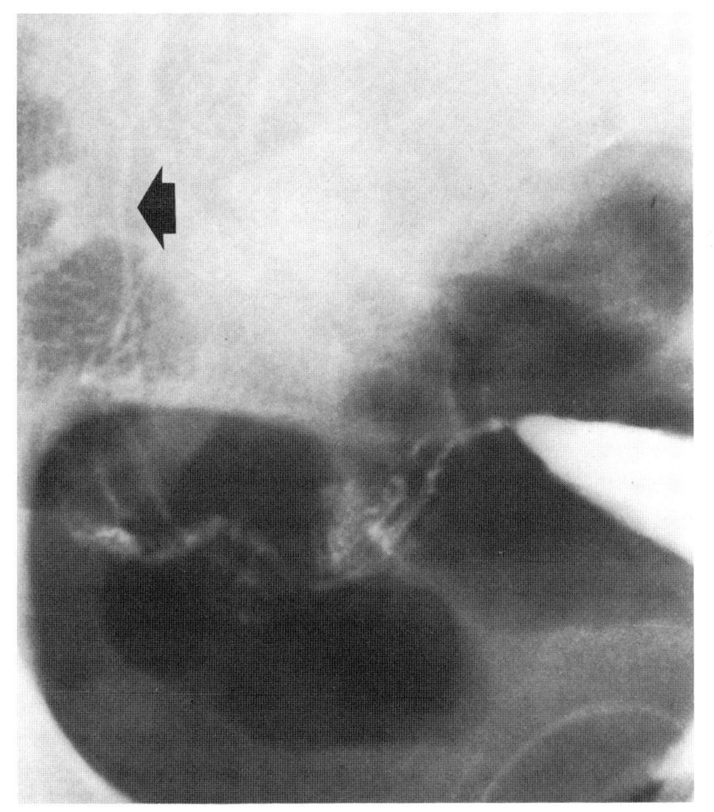

Figure 8.15. Salpingitis isthmica nodosa. Numerous rounded and linear collections of contrast associated with the isthmic segment of the right fallopian tube. In addition to these small diverticula-like collections, some of the linear pattern is believed to reflect lymphatic drainage with a prominent lymphatic channel directed cephalad (*arrow*).

but hydrosalpinx is not uncommon. Approximately 50% of patients with SIN have tubal occlusion distal to the SIN, with frequent evidence of tubal damage on the contralateral side.

Other Findings

The fallopian tube may appear to be coiled on itself, in a corkscrew type of pattern, or have poor mobility, suggestive of adhesive disease. The convoluted tube with a corkscrew appearance on radiographic examination has been described as associated with the finding of endometriosis at laparoscopy (32). Significant peritubal and periovarian adhesions and a laterally shortened and scarred broad ligament are found that contribute to the convoluted pattern (Fig. 8.16). This abnormality may be associated with retrograde menstruation, which in turn could be responsible for the induction of endometriosis and peritubal adhesion formation.

This corkscrew configuration of the tube is not the only way that peritubal disease can present. A beaded appearance suggesting minimal tubal dilatation with a series of short constrictions has similar significance (Fig. 8.17). With peritubal inflammatory disease, the tube may appear crowded and folded on itself. This configuration is persistent even when viewed in varying degrees of obliquity. Generally, peritubal adhesive disease is not associated with tubal occlusion, although exceptions occur.

The classic etiologic factors for peritubal adhesive disease are endometriosis and gonococcal pelvic inflammatory disease (PID). Other processes include nongonococcal PID, IUD with associated subclinical pelvic infection, chlamydia, tuberculosis, and Crohn's disease (Figs. 8.18, 8.19, and 8.20).

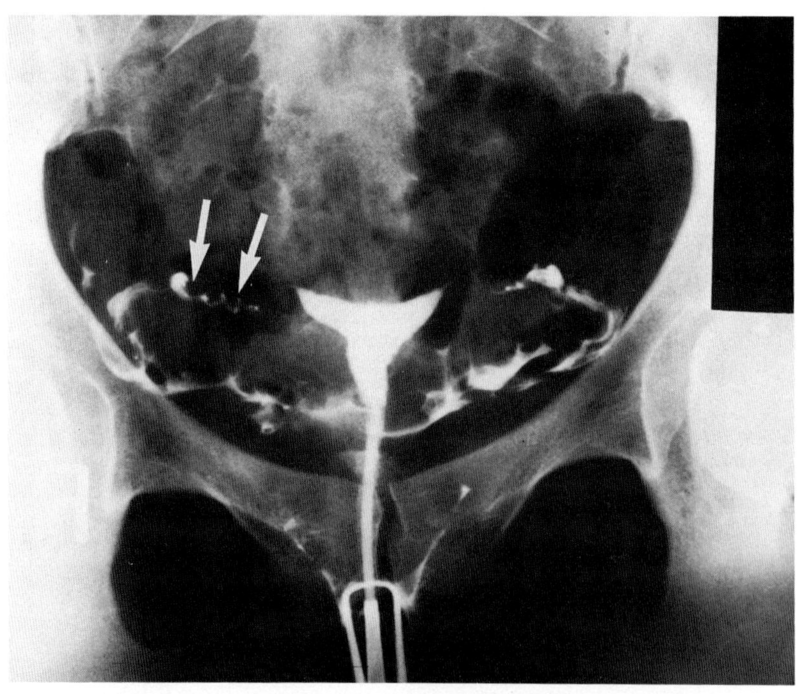

Figure 8.16. Endometriosis proven at laparoscopy. Note the corkscrew-like configuration of the isthmic and proximal ampullary portions of the fallopian tubes (*arrows*), an observation often seen with peritubal inflammatory diseases.

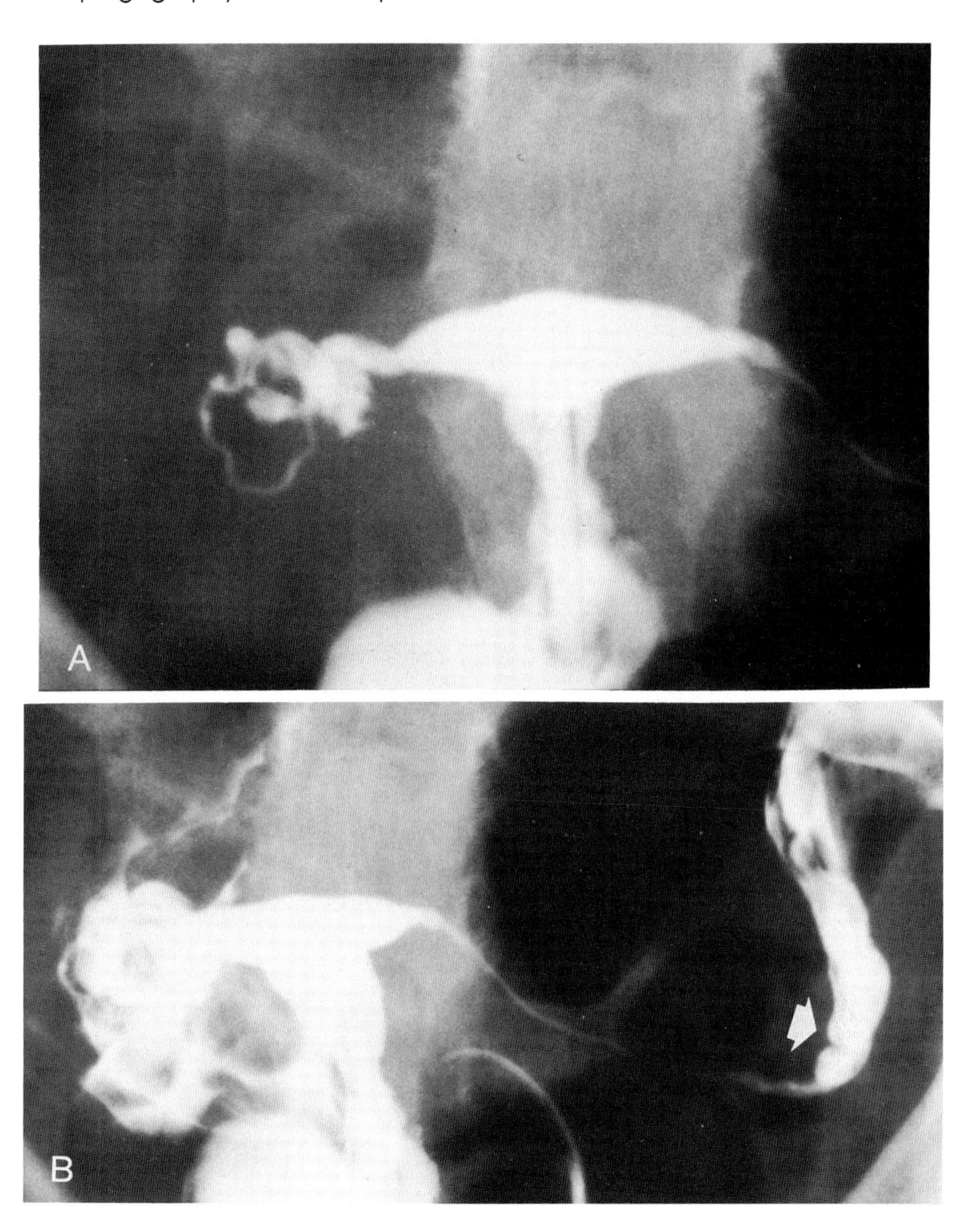

Figure 8.17. *A*: Endometriosis. Fallopian tubes are patent but are distorted and crowded together. *B*: Endometriosis. Ampullary segment of left tube characterized by segmental dilatations not unlike links of sausage (*arrow*).

Hysterographic Diagnosis of Tuberculosis

Genital tuberculosis is now rare, but characteristic abnormalities found at hysterosalpingography suggest that further studies should be done in some women to identify the disease. Genital tuberculosis affects the tube more than the uterus, resulting in destruction of normal tissue, occlusion of the tubes, and partial or complete sclerosis of the uterine cavity, or an ulcerated appearance (33). The tubal ampulla and

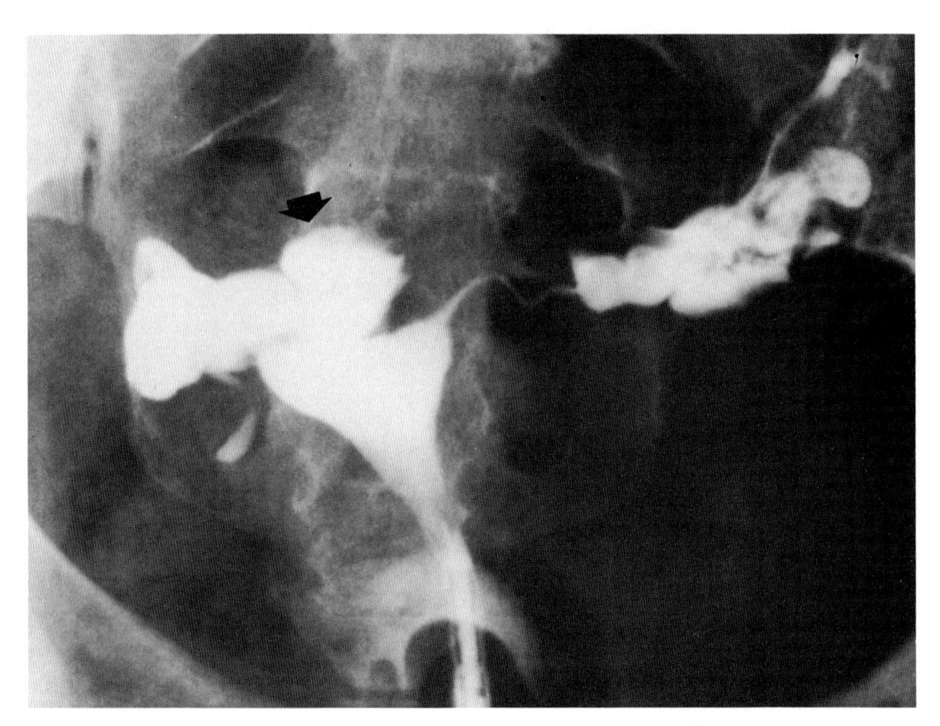

Figure 8.18. Crohn's disease. Essentially normal left adnexa. The right fallopian tube is distorted with a rather crowded, clustered appearance and some dilatation of the distal end of the tube. Peritubal collection of contrast (*arrow*) further reflects the inflammatory process surrounding the adnexae.

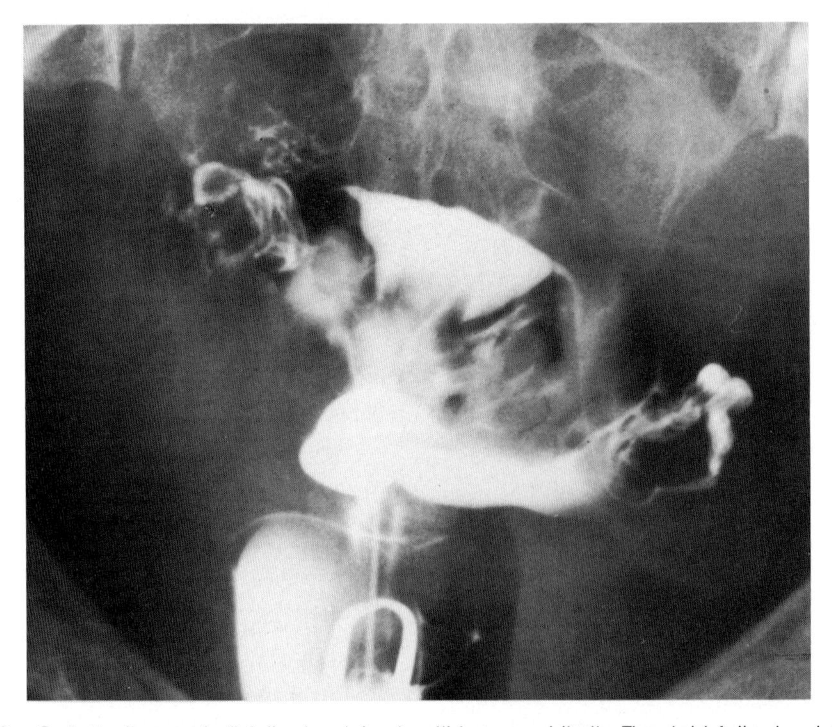

Figure 8.19. Crohn's disease. Left fallopian tube is within normal limits. The right fallopian tube again shows crowded, distorted configuration. The tube is patent and no significant dilatation is evident. Exploration confirmed extensive peritubal inflammatory change with no obstruction.

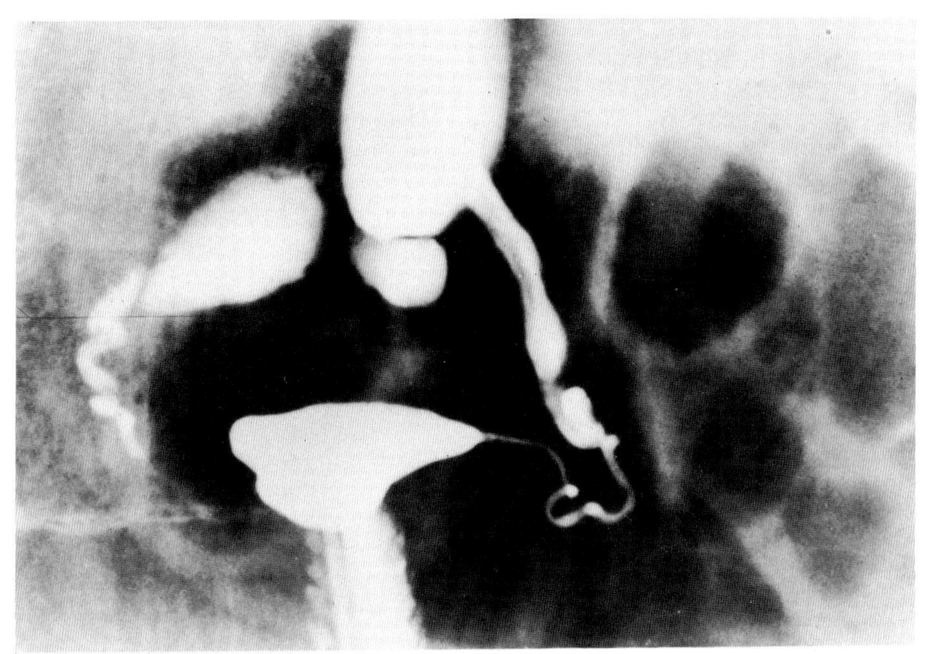

Figure 8.20. Crohn's disease. This pattern reflects bilateral hydrosalpinges and is not typical for peritubal inflammatory change. One must speculate whether this patient had both Crohn's disease and pelvic inflammatory disease.

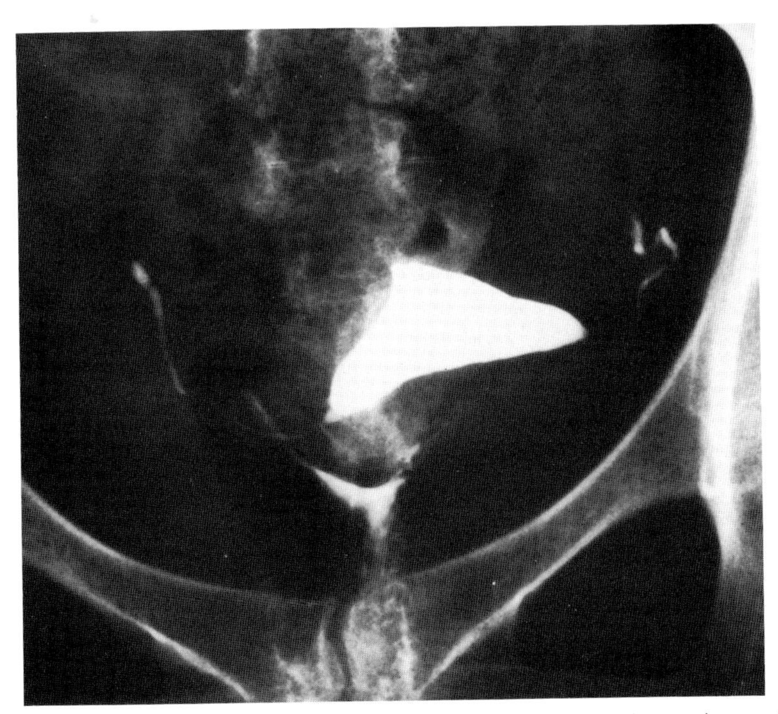

Figure 8.21. Tuberculosis. Essentially normal cavity. Tubes appear narrowed, irregular, and uneven. No significant dilatation despite the bilateral obstruction.

isthmus become constricted and rigid, resulting in a sewer pipe appearance with a rounded blunt end, complete obstruction, and the absence of rugae (Fig. 8.21). Alternating dilatation and constriction result in a "rosary bead" pattern. Calcified pelvic lymph nodes and smaller areas of calcified material in the region of the tubes may also be observed. Fistulous extension from the tubes, appearing like diverticula, and vascular or lymphatic intravasation of contrast are occasionally noted.

Hysterosalpingography performed in the patient with active genital tuberculosis or acute salpingitis of any origin can spread infection to the peritoneal cavity or to extragenital organs. When the condition is suspected at hysterosalpingography, culture of the tubercle bacillus and/or histologic examination of endometrial tissue is mandatory.

Klein and coworkers (34) established criteria for the diagnosis of pelvic tuberculosis:

1. Calcified lymph nodes or smaller, irregular calcifications in the adnexal area.
2. Obstruction of the fallopian tube in the zone of transition between the isthmus and the ampulla.
3. Multiple constrictions along the course of the fallopian tube.
4. Endometrial adhesions and/or deformity or obliteration of the endometrial cavity, in the absence of a history of curettage or abortion.

Although the disease is rare, its radiographic appearance is characteristic, and, if observed, mandates further diagnostic evaluation.

Prevention of Acute Pelvic Inflammatory Disease after Hysterosalpingography

Hysterosalpingography provides useful information about the uterine cavity and fallopian tubes in the evaluation of the infertile patient, and not infrequently may uncover significant pathology, with the potential of activating an infectious process. The examination is therefore not without some risk of complication. The development of acute pelvic inflammatory disease (PID) following hysterosalpingography and the possibility of resultant diminished fertility are of serious concern. The incidence of severe PID after hysterosalpingography has been reported to be as low as 0.3% and as high as 3.4% (35–39). This obviously depends on many factors, not the least of which is a reflection of the patient population studied. Risk factors for infection after hysterosalpingography have been examined, but a proposed scoring system of these risks successfully predicted only 8 of the 14 identified infections (36). Prophylactic ampicillin and tetracycline were reportedly ineffective in preventing serious infections in women having hysterosalpingography. On the other hand, Pittaway and coworkers (35) studied 278 women undergoing hysterosalpingography for infertility, and identified four (1.4%) who developed acute pelvic inflammatory disease. All cases of PID following hysterosalpingography occurred in women with dilated tubes; among those women with dilated tubes, 11.4% developed PID. Patent hydrosalpinx signified greater risk than tubal obstruction, because three of the four patients afflicted had dilated but nonobstructed tubes despite the fact that 75% of all hydrosalpinges in this series were occluded.

Antibiotics may be indicated before performance of hysterosalpingography if the possibility of activating infection is suspected. If such an abnormality is identified at hysterosalpingography, antibiotics should be instituted at the time of the procedure. All patients with tubal dilatation should therefore receive antibiotics. We have advocated the use of doxycycline 100 mg, two tablets on diagnosis, and one twice daily for a total of 5 days. To our knowledge, this regimen has successfully prevented the development of such an acute inflammatory process.

REFERENCES

1. Hutchins CJ: Laparoscopy and hysterosalpingography in the assessment of tubal patency. *Obstet Gynecol* 49:325, 1977.

2. Taylor PJ: Correlation in infertility: Symptomatology, hysterosalpingography, laparoscopy and hysteroscopy. *J Reprod Med* 18:339, 1977.
3. Philipsen T, Hansen BB: Comparative study of hysterosalpingography and laparoscopy in infertile patients. *Acta Obstet Gynecol Scand* 60:149, 1981.
4. Duff DE, Fried AM, Wilson EA, Haack DG: Hysterosalpingography and laparoscopy: A comparative study. *Am J Roentgenol* 141:761, 1983.
5. Snowden EU, Jarrett JC II, Dawood MY: Comparison of diagnostic accuracy of laparoscopy, hysteroscopy, and hysterosalpingography in evaluation of female infertility. *Fertil Steril* 41:709, 1983.
6. Mackey RA, Glass RH, Olson LE, Vaidya R: Pregnancy following hysterosalpingography with oil and water soluble dye. *Fertil Steril* 22:504, 1971.
7. DeCherney AH, Kort H, Barney JB, DeVore GR: Increased pregnancy rate with oil-soluble hysterosalpingography dye. *Fertil Steril* 33:407, 1980.
8. Soules MR, Spandoni LR: Oil versus aqueous media for hysterosalpingography: A continuing debate based on many opinions and few facts. *Fertil Steril* 38:1, 1982.
9. Pontifex G, Trichopoulos D, Karpathios S: Hysterosalpingography in the diagnosis of infertility (statistical analysis of 3437 cases). *Fertil Steril* 23:1972.
10. Lapido OA: An evaluation of 576 hysterosalpingograms on infertile women. *Infertility* 2:63, 1979.
11. Lisa JR, Gioia JD, Rubin IC: Observation on the interstitial portion of the fallopian tube. *Surg Gynecol Obstet* 99:159, 1954.
12. Winfield AC, Pittaway DE, Maxson W, Daniell J, Wentz AC: Apparent cornual occlusion in hysterosalpingography: Reversal by glucagon. *Am J Roentgenol* 139:525, 1982.
13. Merchant RN, Prabhu SR, Chougale A: Uterotubal junction—morphology and clinical aspects. *Int J Fertil* 28:199, 1983.
14. Fortier KJ, Haney AF: The pathologic spectrum of uterotubal junction obstruction. *Obstet Gynecol* 65:93, 1985.
15. Nordenskjold F, Ahlgren M: Laparoscopy in female infertility. *Acta Obstet Gynecol Scand* 62:609, 1983.
16. Siegler AM: *Hysterosalpingography*, ed 2. New York, Medcom Press, 1974.
17. Beyth Y, Kopolovic J: Accessory tubes: A possible contributing factor in infertility. *Fertil Steril* 38:382, 1982.
18. Richardson DA, Evans MI, Talerman A, Maroulis GB: Segmental absence of the mid-portion of the fallopian tube. *Fertil Steril* 37:577, 1982.
19. Silverman AY, Greenberg EI: Absence of a segment of the proximal portion of a fallopian tube. *Obstet Gynecol* 62:90S, 1983.
20. Gordts S, Boechx W, Vasquez G, Brosens I: Microsurgical resection of intramural tubal polyps. *Fertil Steril* 40:258, 1983.
21. David MP, Ben-Zwi D, Langer L: Tubal intramural polyps and their relationship to infertility. *Fertil Steril* 35:526, 1981.
22. Stangel JJ, Chervenak FA, Mouradian-Davidian M: Microsurgical resection of bilateral fallopian tube polyps. *Fertil Steril* 35:580, 1981.
23. Rosenfeld DL, Seidman SM, Bronson RA, Scholl GM: Unsuspected chronic pelvic inflammatory disease in the infertile female. *Fertil Steril* 39:44, 1983.
24. Tulandi T, Wilson RE, Arroent GH, McInnes RA: Fertility aspect of women with tubal diverticulosis: A 5-year follow-up. *Fertil Steril* 40:260, 1983.
25. Honore LH: Salpingitis isthmica nodosa in female infertility and ectopic tubal pregnancy. *Fertil Steril* 29:164, 1978.
26. Creasy JL, Clark RL, Cuttino JT, Groff TR: Salpingitis isthmica nodosa: Radiologic and clinical correlates. *Radiology* 154:597, 1985.
27. Kontopoulos VG, Wang CF, Siegler AM: The impact of salpingitis isthmica nodosa on infertility. *Infertility* 1:137, 1978.
28. Chiari H: Zur pathologischen anatomie des eileitercatarrhas. *Zeittschr Heilkl* 8:457, 1887.
29. Von Recklinghausen F: *Die Adenomyome und Cystadenome der Uterus and Tubenwandung*. Berlin, A. Hirschwald, 1896, p 247.
30. Persaud V: Etiology of tubal ectopic pregnancy: Radiologic and pathologic studies. *Obstet Gynecol* 36:257, 1970.
31. Stewart DB, Skinner SM: Tubal ectopic pregnancy. *Manitoba Med Rev* 47:552, 1967.
32. Cohen BM, Katz M: The significance of the convoluted oviduct in the infertile woman. *J Reprod Med* 21:31, 1978.
33. Nogales-Ortiz F, Tarancon I, Nogales FF: The pathology of female genital tuberculosis. *Obstet Gynecol* 53:422, 1979.
34. Klein TA, Richmond JA, Mischell DR: Pelvic tuberculosis. *Obstet Gynecol* 48:99, 1976.
35. Pittaway DE, Winfield AC, Maxson W, Daniell J, Herbert C, Wentz AC: Prevention of acute pelvic inflammatory disease after hysterosalpingography: Efficacy of doxycycline prophylaxis. *Am J Obstet Gynecol* 147:623, 1983.
36. Stumpf PG, March CM: Febrile morbidity following hysterosalpingography: Identification of risk factors and recommendations for prophylaxis. *Fertil Steril* 33:487, 1980.
37. Moller BR, Allen J, Toft B, Hansen KB, Taylor-Robinson D: Pelvic inflammatory disease after hysterosalpingography associated with *Chlamydia trachomatis* and *Mycoplasma hominis*. *Br J Obstet Gynaecol* 91:1181, 1984.
38. Marshak RH, Roole CS, Goldberger MA: Hysterography and hysterosalpingography. *Surg Gynecol Obstet* 91:182, 1950.
39. Measday B: An analysis of the complications of hysterosalpingography. *J Obstet Gynaecol Br Emp* 67:663, 1960.

9 Postoperative Findings at Hysterosalpingography

Operative procedures performed upon the cervix, uterus, or fallopian tubes may leave characteristic findings at hysterosalpingography. However, there are a limited number of operative procedures that are performed on uterus and tubes, and not all of these will be detected by a hysterosalpingogram performed in the immediate or delayed postoperative period.

Surgical Procedures on the Uterus

CONGENITAL ANOMALIES

Only perhaps 25% of women with a "double uterus" have reproductive problems, and present with recurrent pregnancy wastage or premature delivery. Experience suggests that each subsequent pregnancy is likely to be carried for a longer duration until finally a viable pregnancy results. In women who have aborted early, progesterone supplementation may permit the pregnancy to carry. Therefore, an operative procedure to correct either a septate or bicornuate uterus may be unnecessary in most women. However, if accomplished, a postoperative hysterogram is necessary, primarily to diagnose intrauterine adhesion formation.

Unification procedures performed on bicornuate or septate uteri do result in characteristic findings at hysterosalpingography. The Jones procedure (1), used in the septate uterus, is a wedge metroplasty and removes the septum entirely (Fig. 9.1). The postoperative hysterosalpingogram shows a narrow fundus with a small wedge-like septum or arcuate pattern, a smaller uterine cavity, and two small lateral "dog ears" (Figs. 9.2, 9.3, and 9.4). The Tompkins procedure splits the septum by bivalving the uterus in an anteroposterior direction, dissects the septum away from the fundus, and closes the defect, leaving a wider fundus and larger uterine cavity (Figs. 9.5 and 9.6). The Strassman procedure, which splits the septum across the fundus, also results in a wider cavity but with a different line of incision, and is used only with a bicornuate and not with a septate uterus. Full-term delivery rates of about 80% are to be expected after surgical unification, and are essentially the same for all the procedures described. Synechiae may form after any of these surgical procedures (Fig. 9.7).

The uterus didelphys, which is duplicated throughout, does not lend itself to surgical correction, nor is this indicated. However, excision of a rudimentary or perhaps normal-size cervix has occasionally been attempted. Occasionally, in the past, surgical establishment of communication between the two chambers was carried out (Fig. 9.8). A vaginal septum may be removed for patient comfort or obstetric convenience. The reproductive performance of the didelphic uterus is good, and any approach surgically would likely result in mutilation of a functional organ.

TUMOR RESECTION

The leiomyoma is the most common uterine tumor. Resection may involve removing it from a subserosal location, which should cause no characteristic hysterosalpingographic findings, or a more major proce-

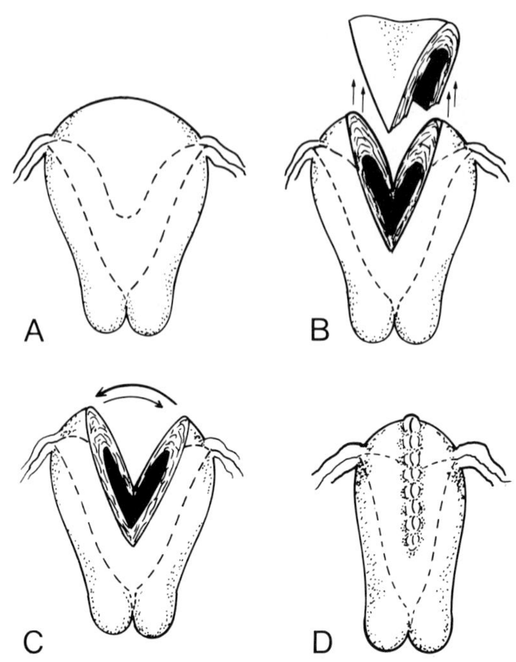

Figure 9.1. Jones procedure. The septum is removed in its entirety (*B*) and the cavity closed (*C*). The cornua are brought closer together and the fundus is narrowed (*D*).

dure might be required. Myomata in intramural or submucous locations may cause considerable distortion of the uterine cavity.

Following myomectomy, with removal of single or multiple fibromas, the uterine cavity may remain as misshapen as it was before the operation was performed (Fig. 9.9). On the other hand, the cavity may revert to a normal appearance (Fig. 9.10) or a new abnormality, intrauterine synechiae, may be diagnosed at follow-up hysterosalpingography. Resection of myomata may also cause tubal obstruction, and their removal may also induce mechanical problems and blockage in the cornual areas (Fig. 9.11).

After resection of endometrial polyps, deformity is rare. The cavity usually is normal in appearance (Fig. 9.12). Similarly, resection of synechiae may restore a normal appearance to the cavity (Fig. 9.13), or result in reformation of the scarring.

The most common of all postoperative findings is the scar left by a lower segment cesarean section. This is ordinarily a jagged, wedge-like, sharp-contoured defect in the area of the lower segment (Fig. 9.14), and is characteristic of a prior operative procedure during pregnancy. Some scars are large saccular defects and others have a wide base suggestive of poor healing of the uterine incision (Fig. 9.15). Dog ear-like diverticula may also be observed (Fig. 9.16).

There should be no hysterosalpingography changes induced by dilatation and curettage, although on occasion a widened internal os may be the result of overdistension during dilatation and evacuation. When this procedure is performed, the internal os must be dilated sufficiently to permit removal of a second trimester fetus.

Uterine perforation by a rigid injection cannula or a probe is occasionally diagnosed. More often, the instrument is buried in the myometrial wall (Fig. 9.17). The procedure should be immediately terminated and no contrast injected. When there is a strong suspicion of perforation, the patient should be admitted for observation, with close attention to the development of symptoms of blood loss, fever, and/or lower abdominal pain; antibiotic administration may be required.

Surgical Procedures on the Fallopian Tube

HYSTEROGRAPHIC FOLLOW-UP OF STERILIZATION PROCEDURES

Sterilization procedures and techniques have changed over the years for the purpose of improving success rates (Fig. 9.18). There has been modification and improvement of older methods of tubal occlusion such as ligation and fulguration. A recent development has been the application of clips and bands to the tubes, and the efficacy of various chemicals and plugs introduced into the tubes is under current investigation.

Tubal ligation to prevent passage of sperm and ova is one of the oldest forms of tubal occlusion. Various combinations of simple ligation, crushing, division and burial of the

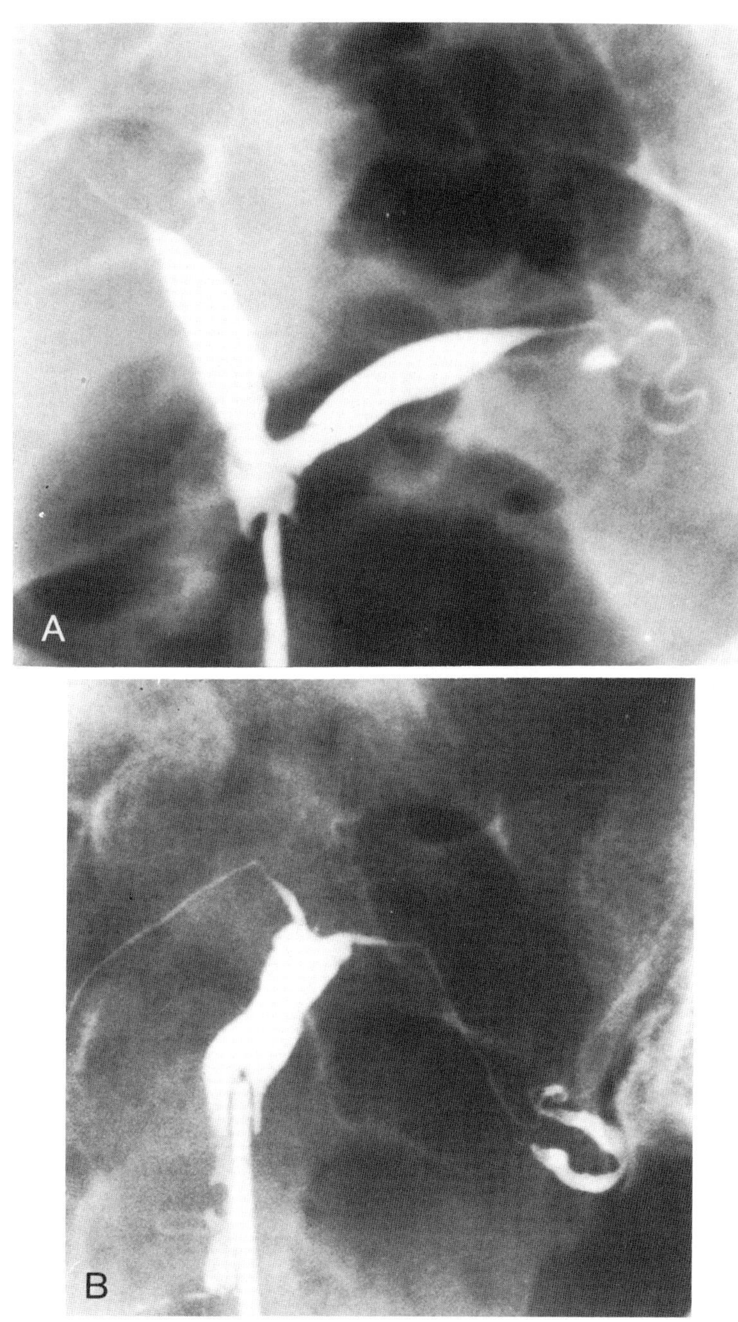

Figure 9.2. Characteristic hysterosalpingographic pattern of a Jones wedge procedure for correction of a septate uterus. Note the narrow fundus and approximated cornua. *A*: Preoperative; *B*: Postoperative. (Courtesy of Howard W. Jones, Jr., Norfolk, Virginia.)

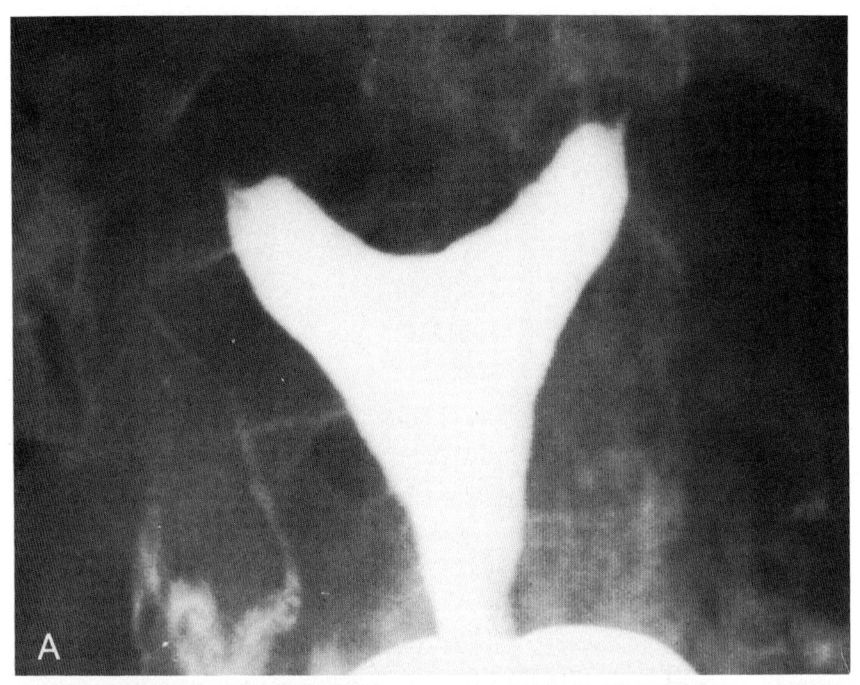

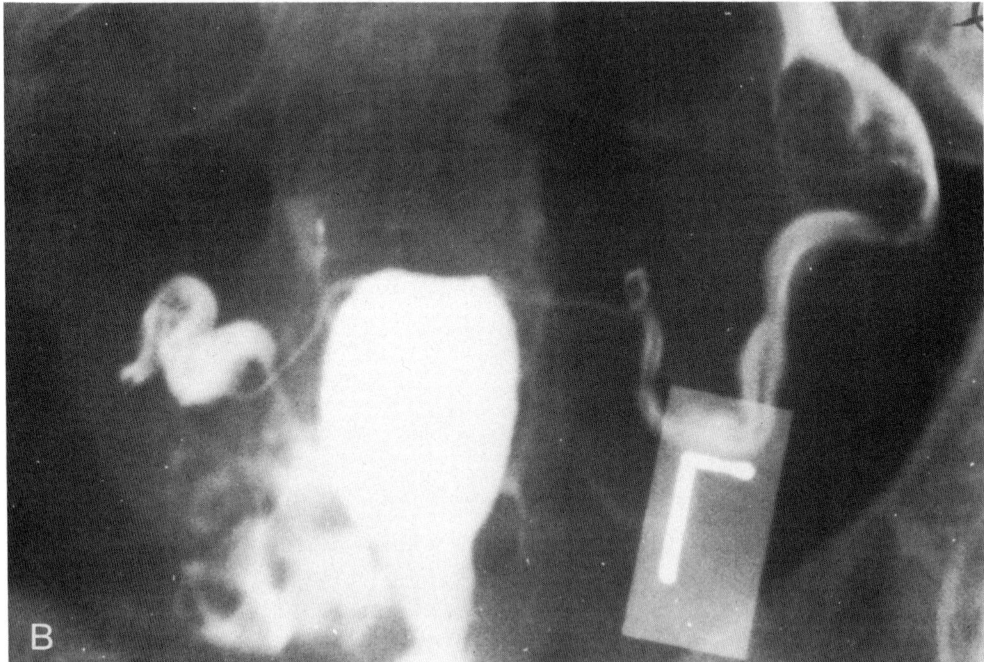

Figure 9.3. Jones procedure for correction of septate uterus. *A*: Preoperative; *B*: Postoperative. (Courtesy of Howard W. Jones, Jr., Norfolk, Virginia.)

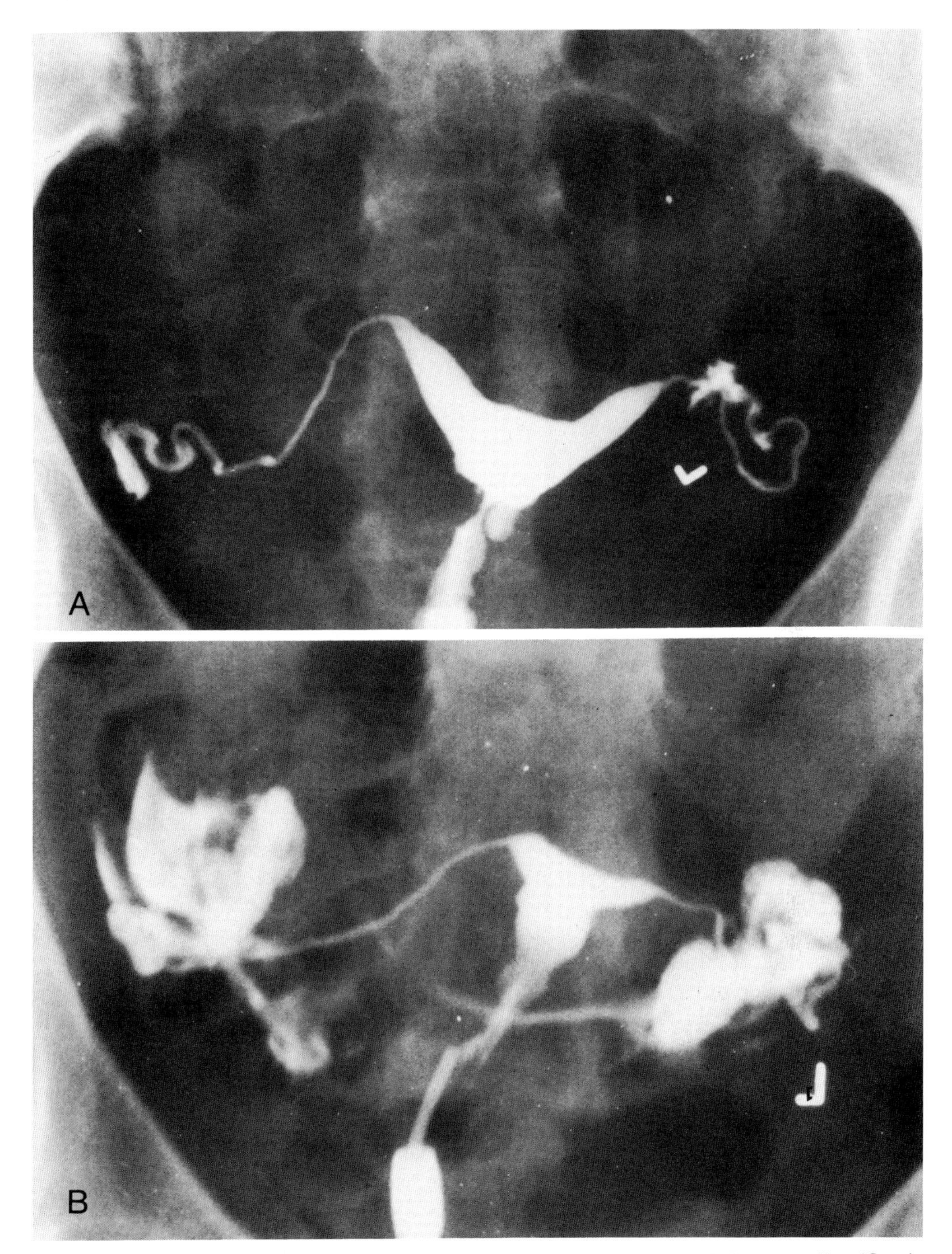

Figure 9.4. Jones procedure for correction of septate uterus. *A*: Preoperative; *B*: Postoperative. (Courtesy of Howard W. Jones, Jr., Norfolk, Virginia.)

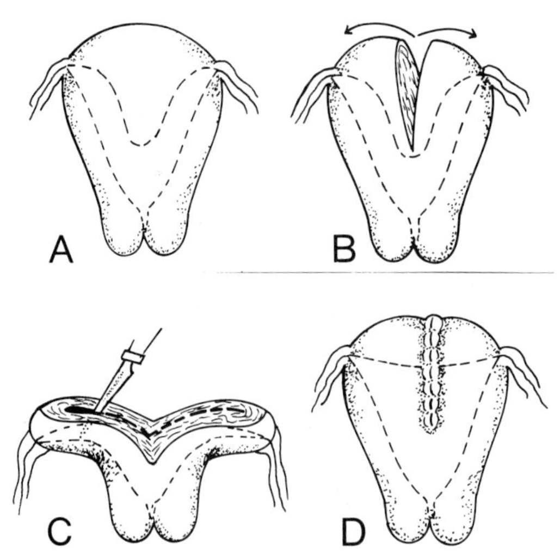

Figure 9.5. Tompkins procedure. The septum is dissected from the fundus (*C*). Closure results in a relatively wide fundal configuration (*D*).

stump, and resection are commonly used (2). The Pomeroy technique is the most frequently performed of all simple ligation techniques, with a low failure rate worldwide. The occluded tubes have a blunt, occasionally dilated ending in the isthmic portion of the tube.

In the Uchida technique, the proximal ligated stump is buried beneath the serosa. With fimbriectomy, the fimbria of each tube is removed. Both the Pomeroy and Uchida techniques result in virtually identical hysterosalpingographic pictures (Figs. 9.19 and 9.20). The film may show a blunted end, fistula formation, or distal patency, and most abnormalities are seen beginning at the midisthmic area (3). With cornual resection, the intramural portion of the tube will not fill with contrast medium.

In the Irving technique, the tubes are divided between two absorbable ligatures, and the proximal stump is buried in the uterine myometrium, usually posteriorly. At hysterosalpingography, the proximal part of the tube will be bent upon itself. The radiographic picture is unique and striking, as contrast is distributed in varying amounts in the uterine myometrium, depending upon the patency of the buried tubal segment (Fig. 9.21). The contrast clears rapidly after the injection.

Most ligation techniques were devised to be done postpartum, with a small incision. In the past 15 years, interval sterilization procedures have been adapted to the laparoscope. Laparoscopic fulguration uses a combination of coagulation and cutting current and attempts to destroy a considerable portion of the tube.

When laparoscopic fulguration techniques were first developed and their efficacy was under investigation, several groups designed studies to test patency using hysterosalpingography in the postoperative period. This was revealed to be a mistake. Jordan et al. (4) initially performed salpingography at 6 weeks, but when 22 of 60 women had spill of contrast, it was deduced that the "depressingly high rate of spill" was because the hysterosalpingogram was being performed before the tube had fully fibrosed, forcing a fistula. When the salpingogram was performed at 12 weeks, only 8 of 383 patients were diagnosed to show spill. Others (5–7) confirmed these findings. Although Sheikh (5) recommended follow-up with hysterosalpingography in teaching institutions and when inexperienced operators were learning the technique of sterilization procedures, routine hysterosalpingography was not, and is not, recommended.

Tantalum clips for sterilization are applied to the tube via specially designed pliers, and failures may result when a clip migrates off a tube, opens slightly renewing tubal patency, cuts through the tube leading to recanalization, or opens subsequent to the pressure produced by a hysterosalpingogram used to test tubal occlusion. Spring-loaded clips have been judged more effective than the Tantalum clip (Fig. 9.22); the hysterosalpingographic pictures may not allow differentiation between the two types of clips.

Bands, including the Falope ring, may be applied by any approach except the transcervical. Electrocautery is not required, and the ring is slipped onto the base of a loop of tube via a special applicator; the blood supply is interrupted and eventually the tube undergoes fibrosis. The characteristic hysterosalpingographic picture is that of tubal

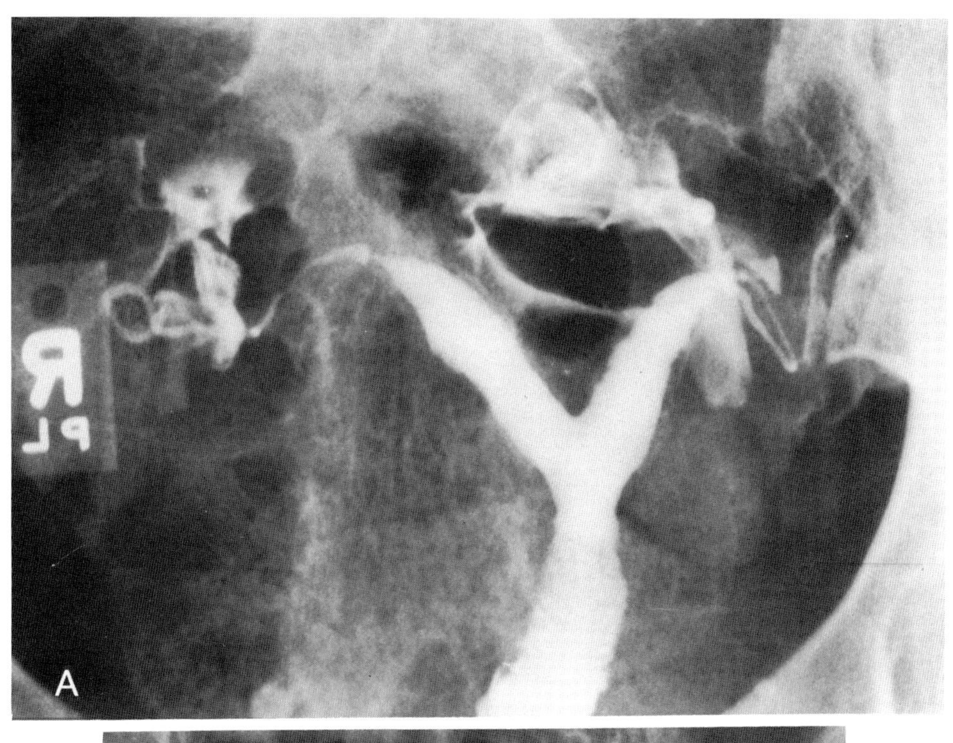

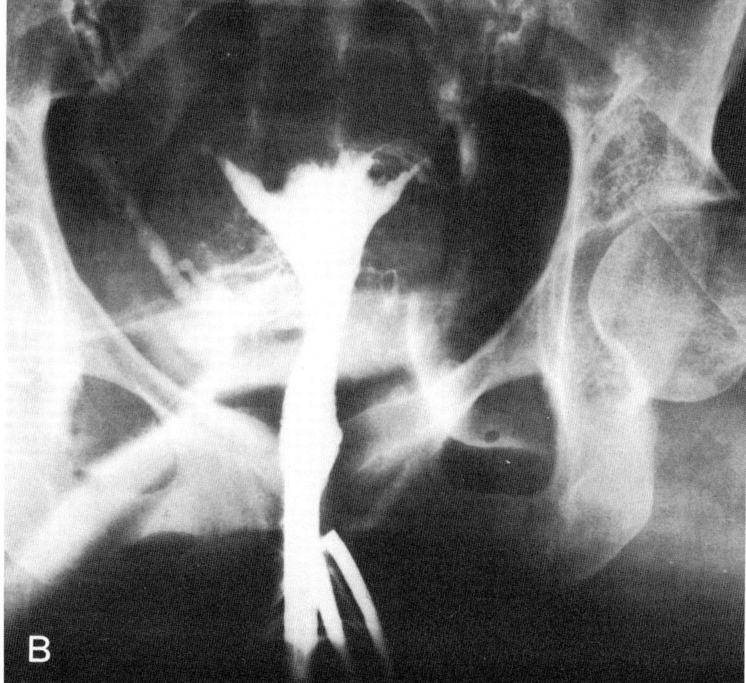

Figure 9.6. *A*: Septate uterus in woman with history of multiple spontaneous abortions. *B*: Postoperative appearance after septum resection via Tompkins procedure. Irregular appearance at fundus is not unusual.

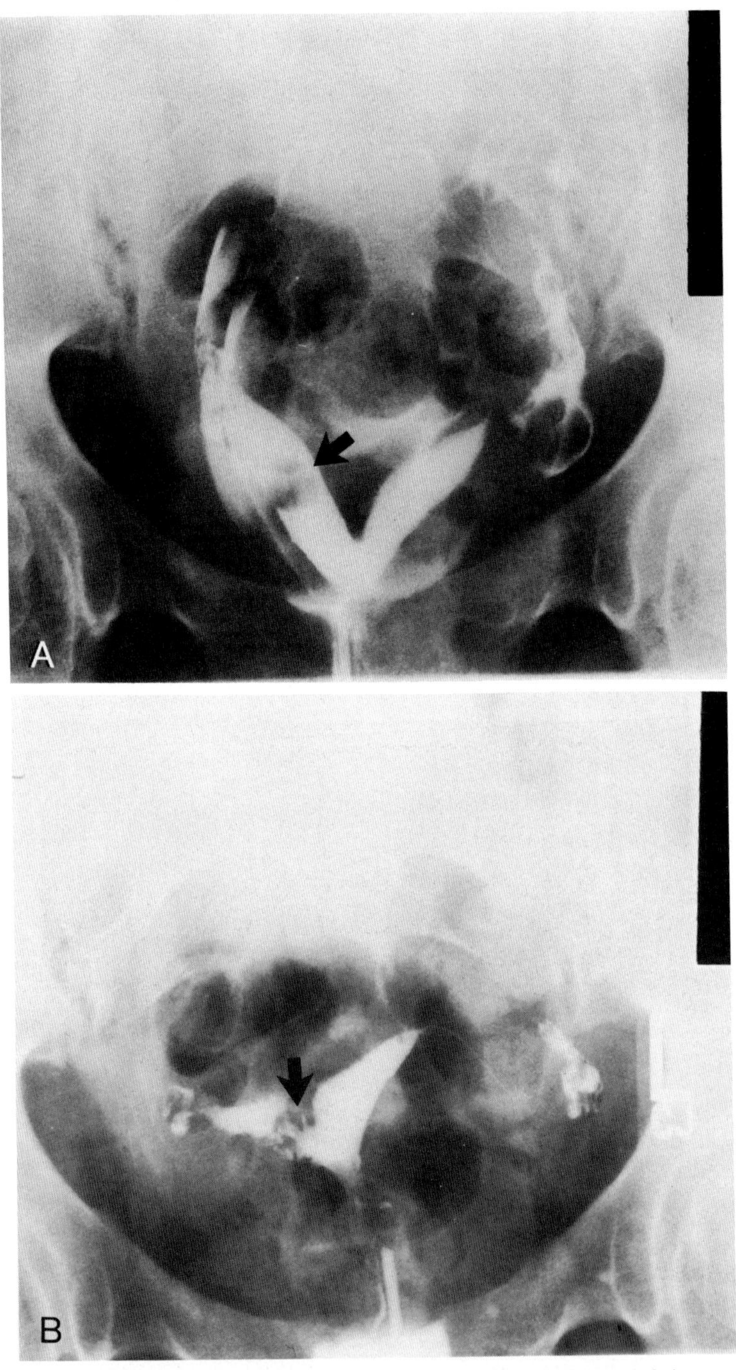

Figure 9.7. *A*: Septate uterus with intrauterine pregnancy in right cornua (*arrow*). Subsequent pregnancy was lost. *B*: Postoperative hysterosalpingogram following reunification procedure. Localized synechia formation in right cornua (*arrow*).

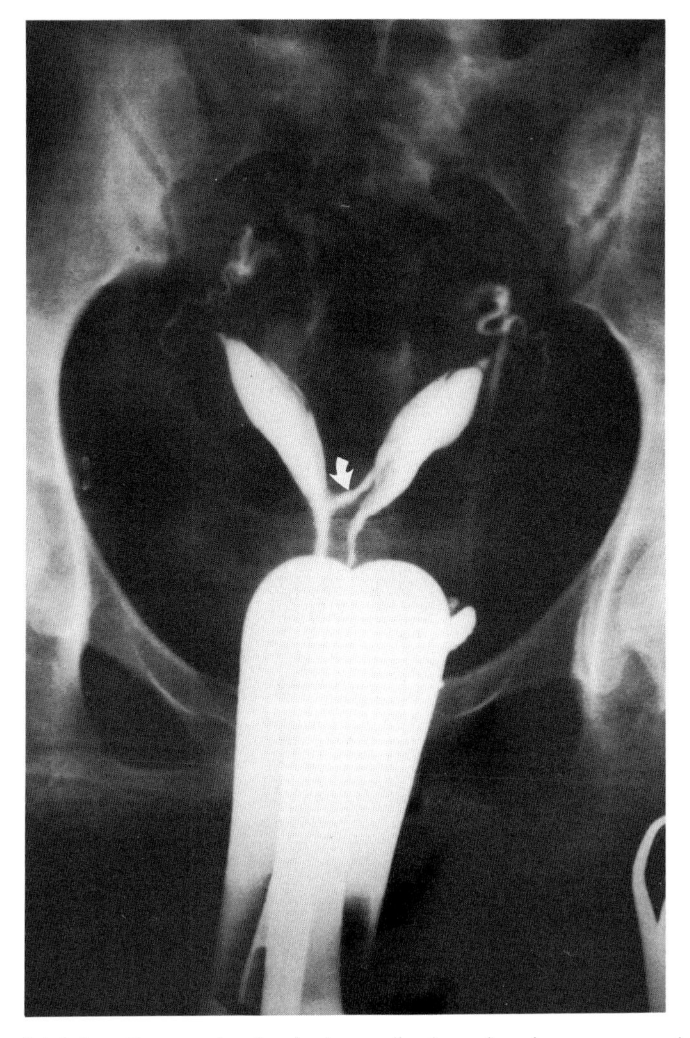

Figure 9.8. Uterus didelphys. Communication between the two chambers was created surgically (*arrow*). This operation is no longer performed.

obstruction adjacent to a radiopaque band, which is the Falope ring (Fig. 9.23).

Tissue adhesives have been used to form a plug, and sclerosing chemicals including quinacrine and silver nitrate have been employed to destroy the inner lining of the tube with subsequent fibrosis. In addition, solid Silastic intratubal devices, polyethylene and Teflon plugs, and a silver-impregnated silicone plug (8) have been instilled to accomplish tubal occlusion. (Fig. 9.24). The major advantage of these techniques is their potential reversibility and the possibility of application by the vaginal or laparoscopic route. The major disadvantages include the specially designed instrumentation and the above-average operator skill required for insertion via laparoscopy. The hysterosalpingographic pictures are characteristic.

TUBAL REANASTOMOSIS AND REIMPLANTATION

A major reason for hysterosalpingography in the sterilized patient is when reanastomosis is being contemplated. Requests for reversal are not uncommon. The tubal

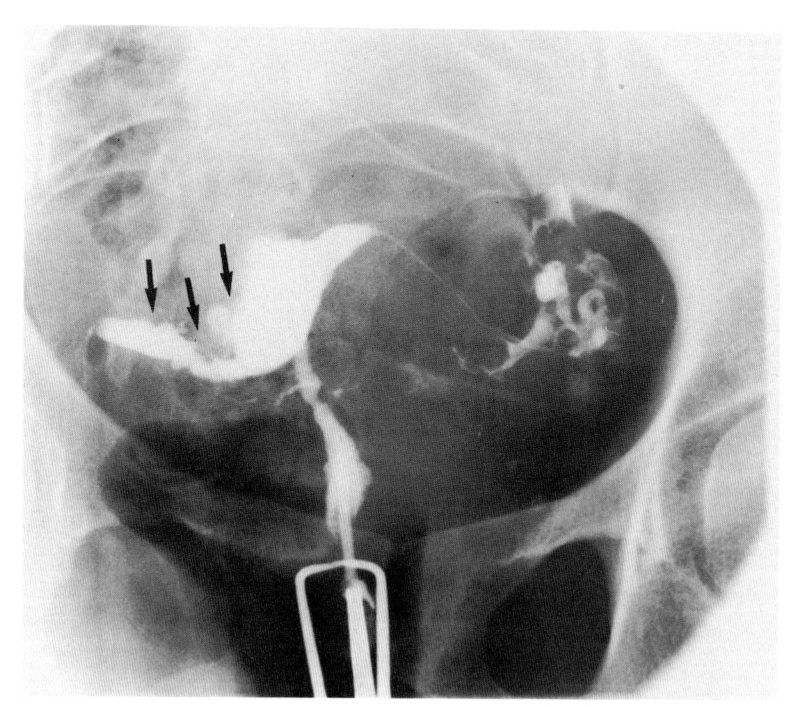

Figure 9.9. Appearance following resection of fundal myoma. Note the marked irregularity of the fundus (*arrows*), the result of cavity distortion and scarring.

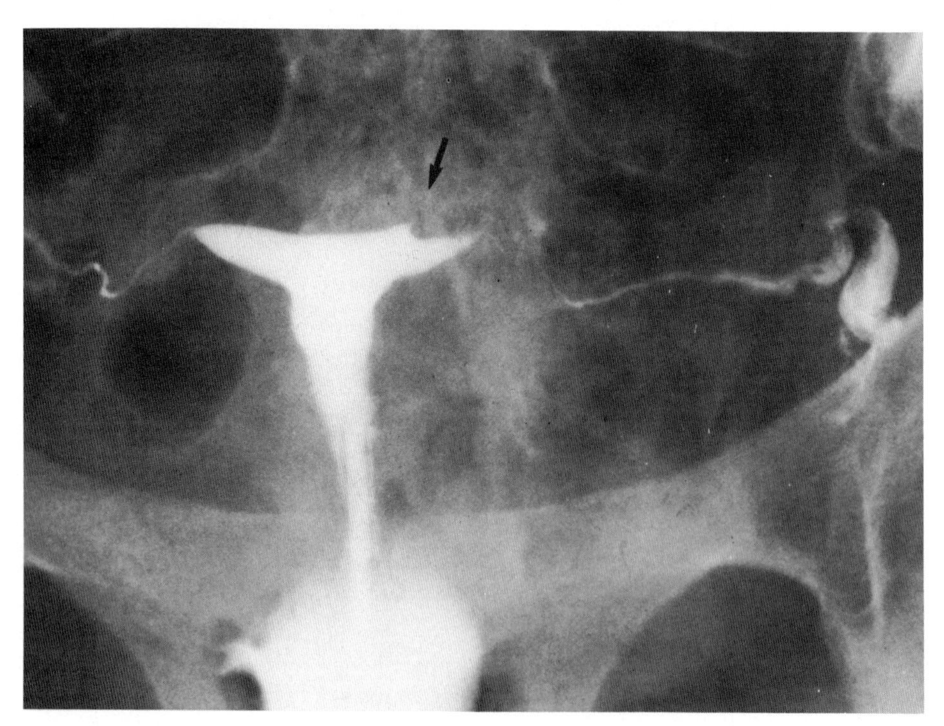

Figure 9.10. Virtually normal appearance of uterine cavity following the resection of large fundal myoma. Note the small deformity near the left cornua (*arrow*).

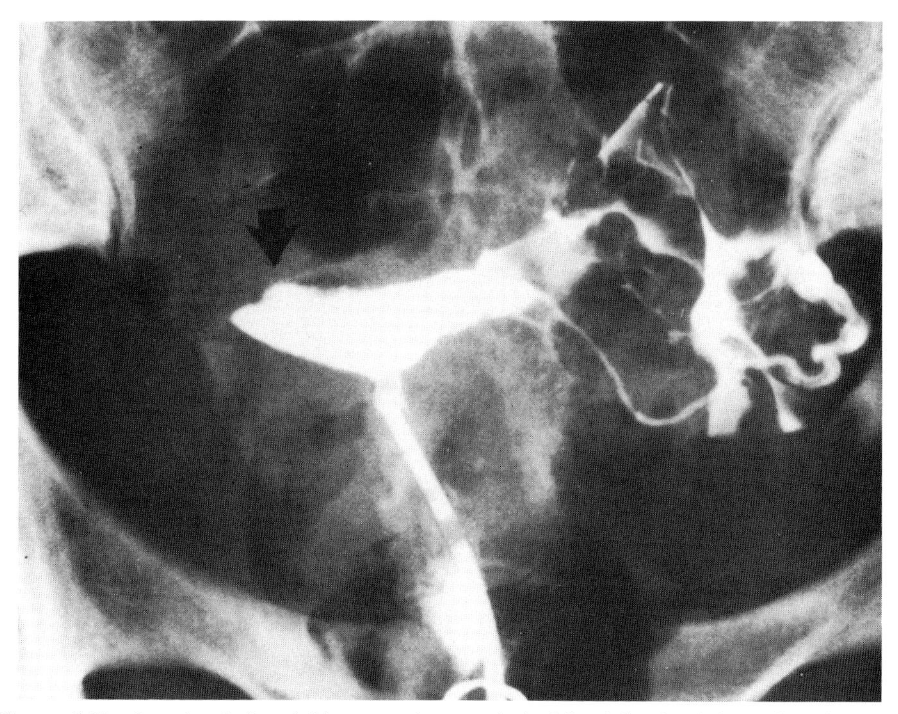

Figure 9.11. Scarring in the right cornua has occluded the fallopian tube ostium (*arrow*).

segments may not remain normal after ligation, so evaluation must be accomplished before anastomosis is undertaken (Figs. 9.20 and 9.25). The length of the proximal segment is important, and the prospect for a normal pregnancy is directly proportional to the length of the remaining tube (9). About 50% of patients with 3–4 cm of tube on the longest side are successful, and few become pregnant after the surgical reanastomosis with less than 3.0 cm of tube. For pregnancy, presence of fimbria and the ampullary segment are both important. At least 1 cm of ampulla must remain intact but cannot be evaluated radiographically in the occluded tube. The hysterosalpingogram can demonstrate the length of the proximal segment to aid in planning the reversal procedure, but may be unable to distinguish interstitial from isthmic tube, which is of importance when only a very short segment of proximal tube remains. Visualization of the uterine cavity is an added benefit to diagnose unsuspected intrauterine pathology, such as polyps or submucous myomata.

Postoperative findings characteristic of fallopian tubal operations are many and variable (10) (Fig. 9.26). In the case of an ampullary-isthmic tubal reanastomosis, the tubal lumen will vary in size on either side of the anastomosis, with the degree of difference depending on where the anastomosis was done (Figs. 9.27 and 9.28). The cornual area may have lost its characteristic wedge-like appearance if a "cork borer" technique was used for the implantation of the tube into the uterine cavity at the area of the cornu (Fig. 9.29).

Reimplantation may also be accomplished in cornual occlusion by a technique in which the myometrium is "shaved" down until a patent interstitial lumen is found. Reanastomosis is then carried out, which likely leaves the cornual area with its usual anatomic configuration, or at least less distortion than when the area is opened with a uterine reamer (Fig. 9.30).

A technique occasionally encountered but now rarely performed implants the proximal portion of the tube into the uterine cavity below the level of the fundus (11). Al-

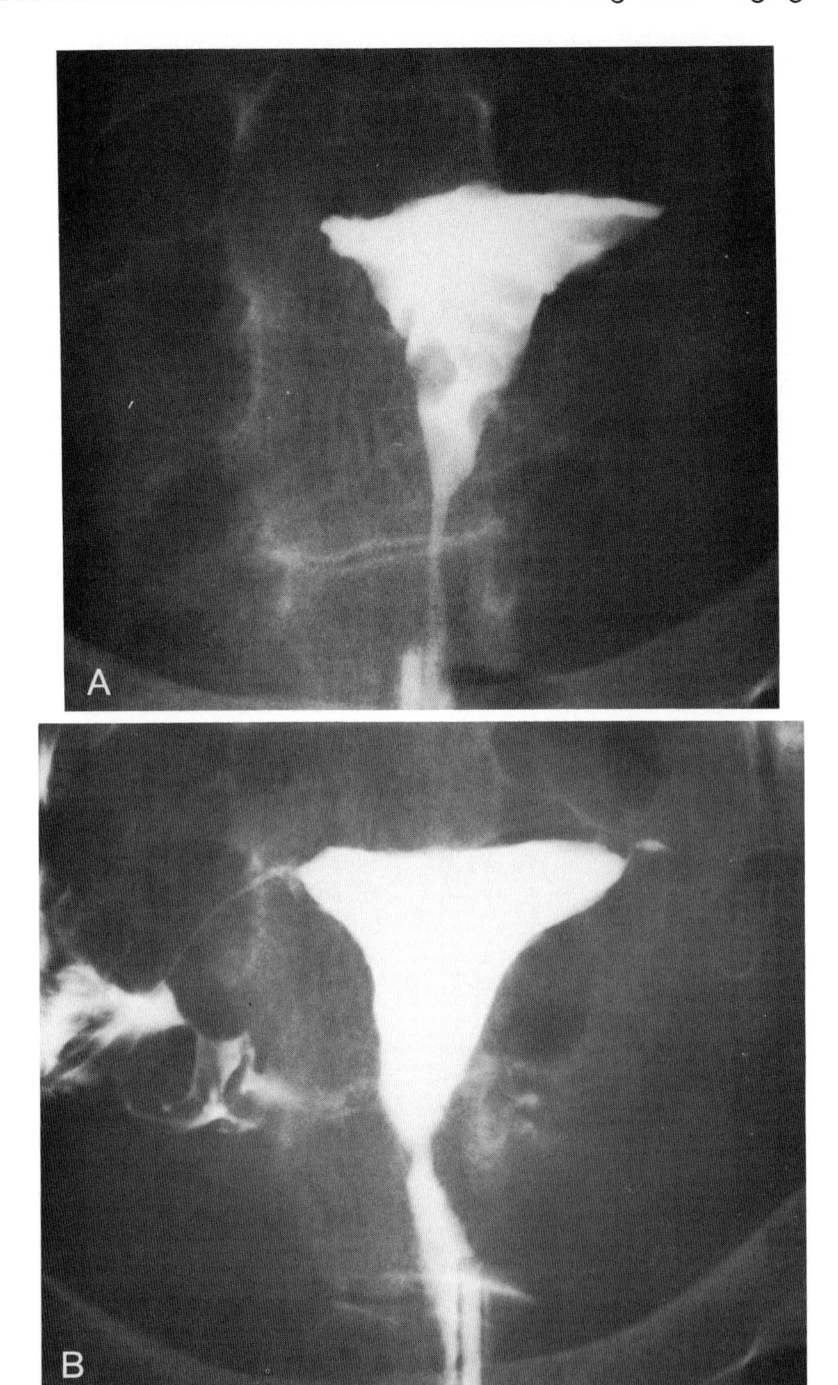

Figure 9.12. *A*: Multiple endometrial polyps throughout uterine cavity. *B*: Hysterosalpingogram following hysteroscopic resection and curettage. The cavity now has a normal appearance.

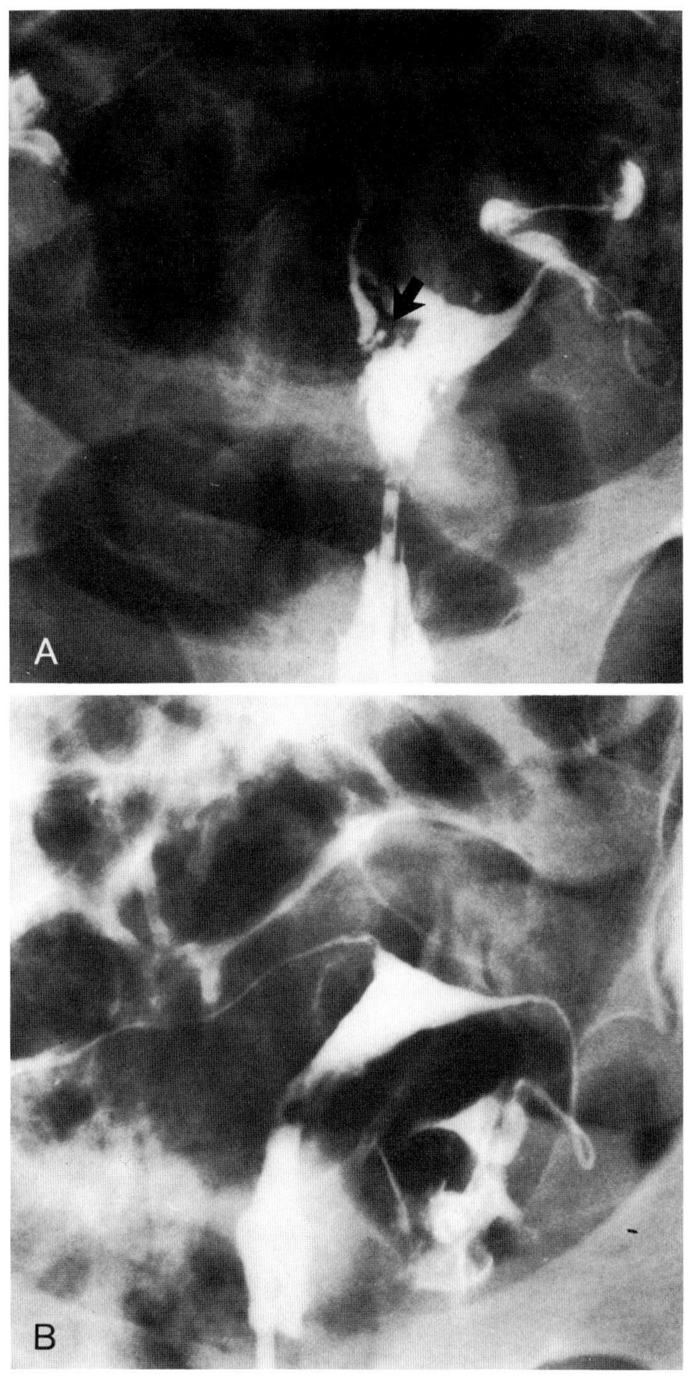

Figure 9.13. *A*: Synechia formation (*arrow*). *B*: Hysterosalpingogram following hysteroscopic resection and uterine cavity distension with Foley catheter for 2 weeks. Normal-appearing cavity results.

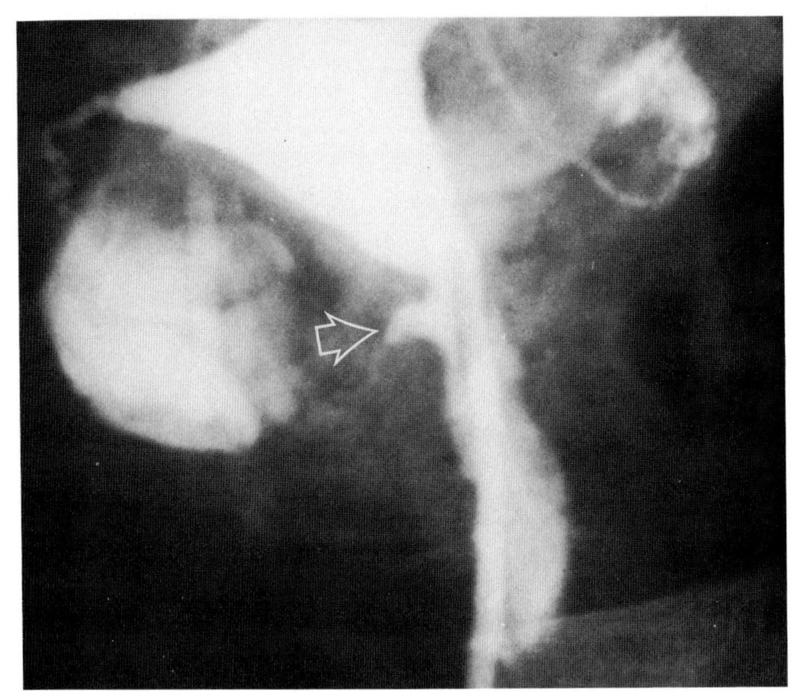

Figure 9.14. Typical appearance of deformity of the lower uterine segment (*arrow*) secondary to previous cesarean section.

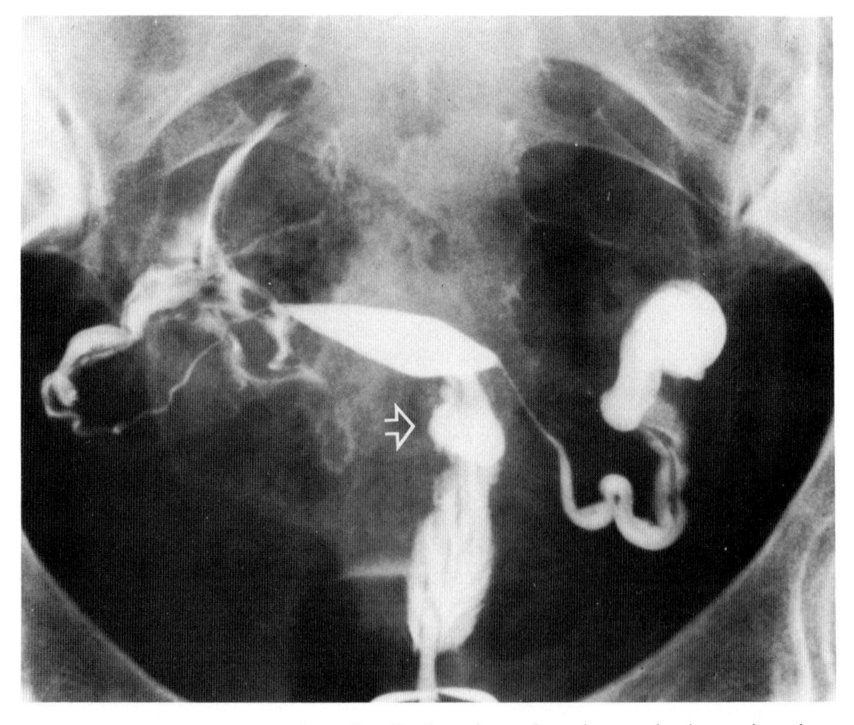

Figure 9.15. Postcesarean section diverticulum, broad and saccular in contour (*arrow*).

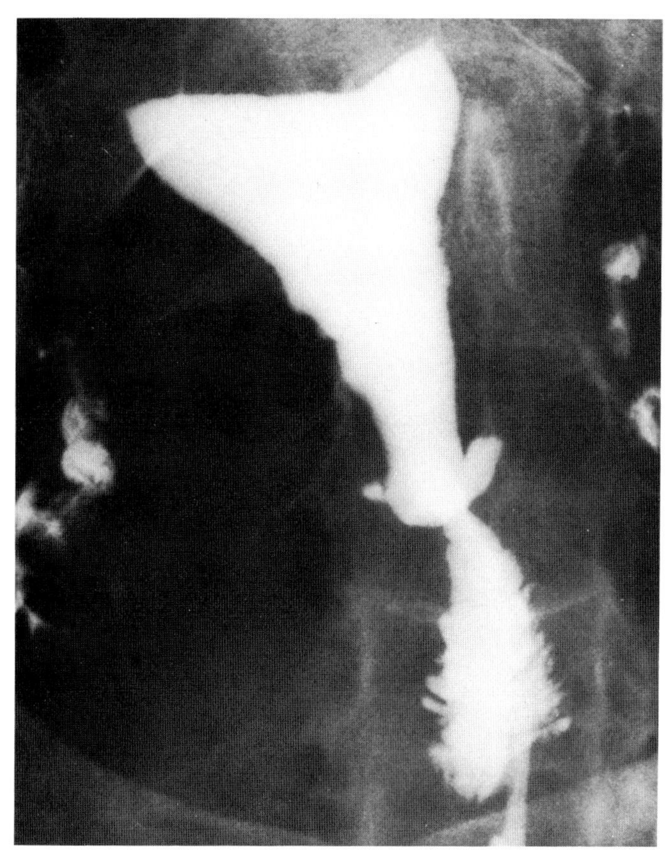

Figure 9.16. Bilateral diverticular outpouching resulting from cesarean section in the past.

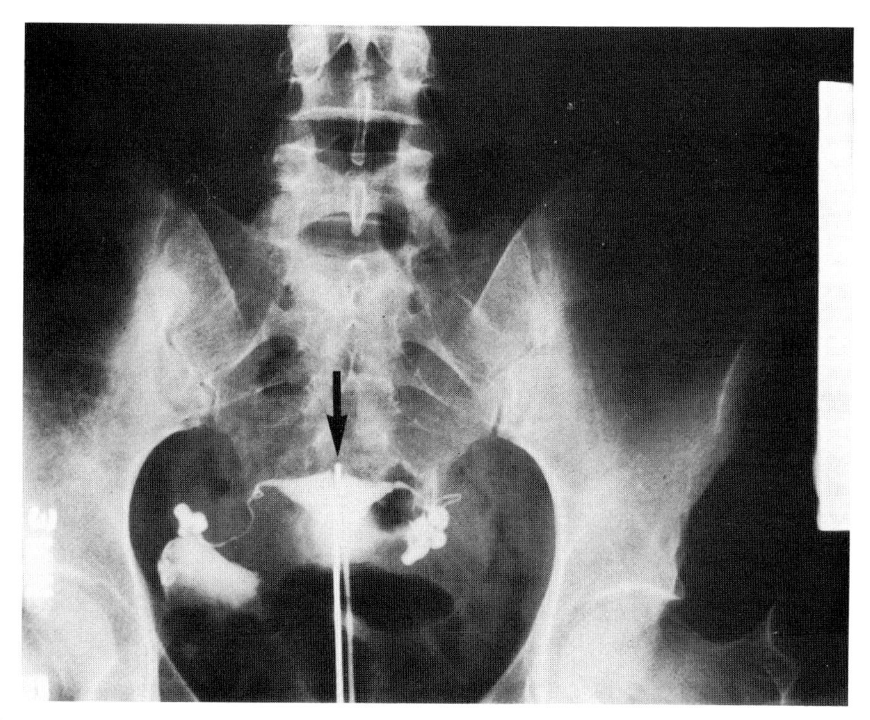

Figure 9.17. Probe inadvertently imbedded in the myometrium (*arrow*). Occasionally such instrumentation may traverse the muscular wall and enter the peritoneal cavity.

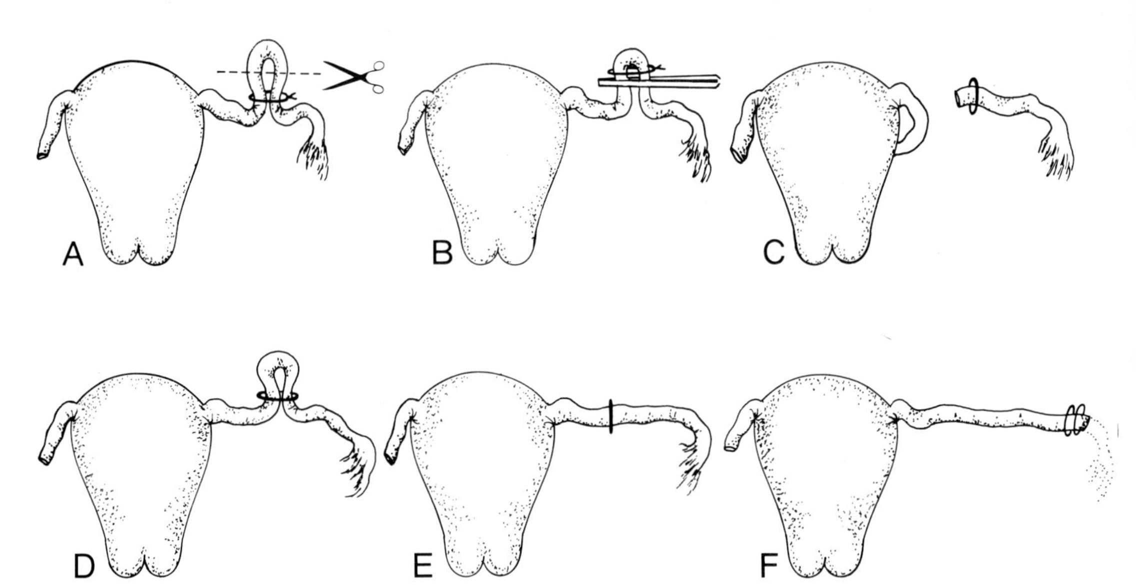

Figure 9.18. Various techniques of tubal ligation and interruption currently encountered. *A*: Pomeroy. *B*: Madlener. *C*: Irving. *D*: Falope ring. *E*: Hulka clip. *F*: Fimbriectomy.

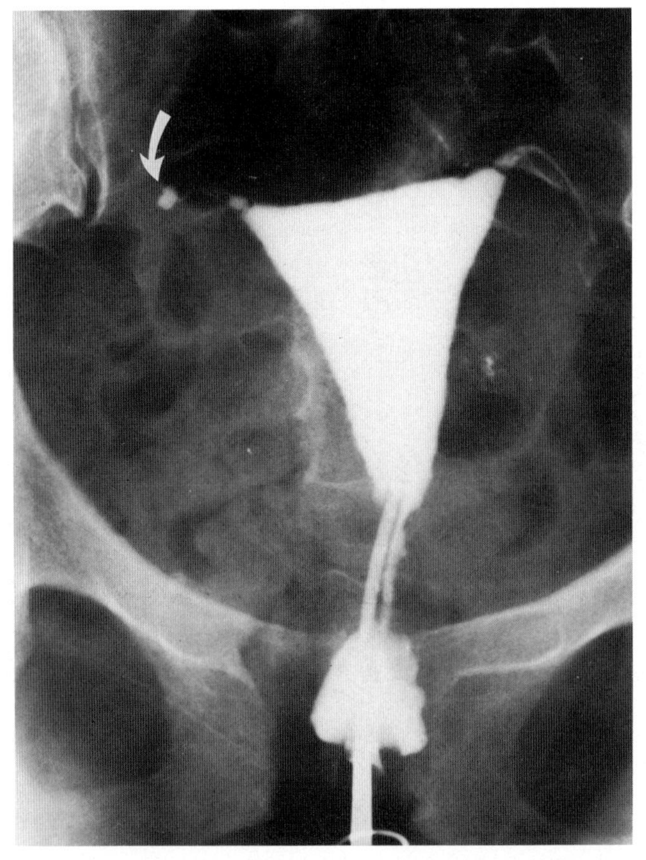

Figure 9.19. Posttubal occlusion (Pomeroy technique). Occlusions are characteristically in the isthmic segment and often show slight bulbous dilatation at the point of obstruction (*arrow*).

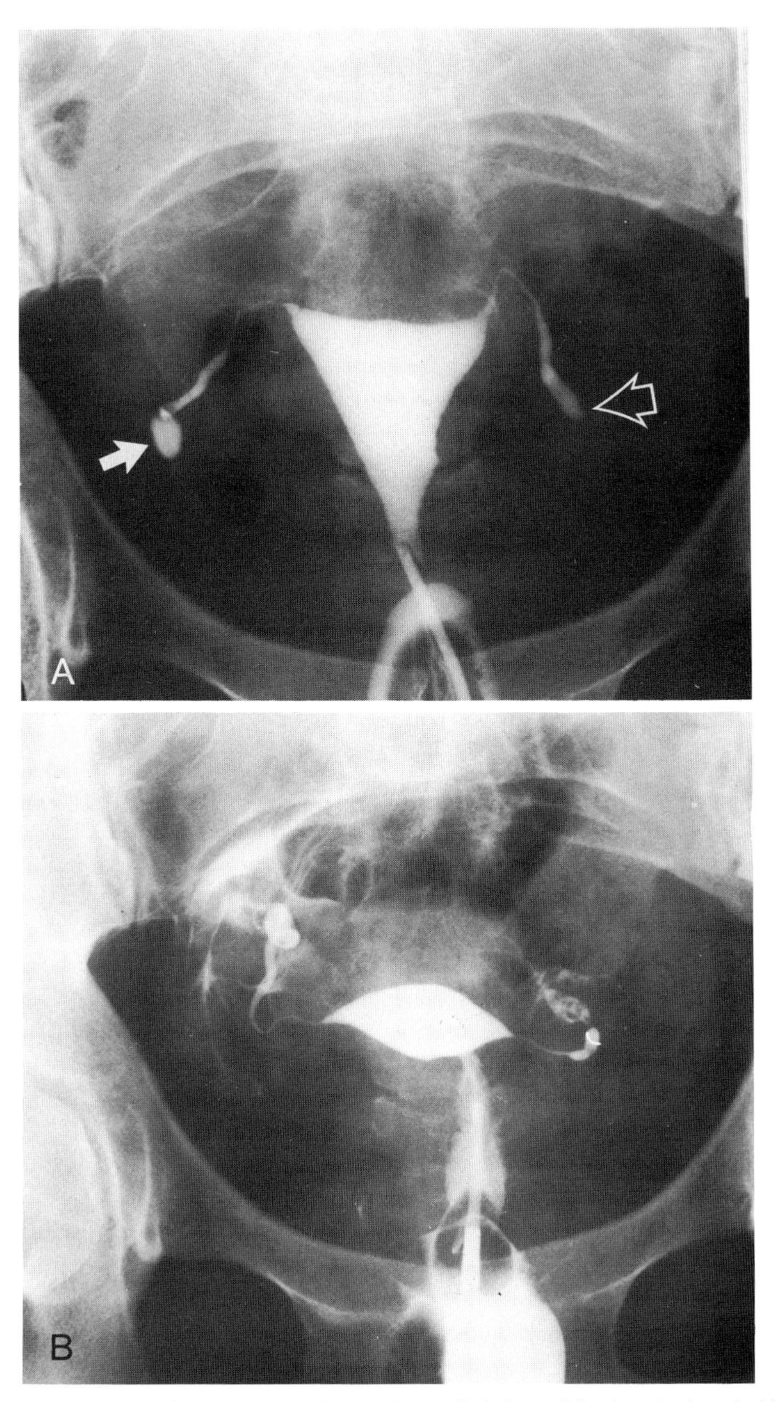

Figure 9.20. *A*: Bilateral tubal occlusions with prominent dilatation at the terminal end of the right tube (*arrow*), and a club-like termination of the left tube (*open arrow*). *B*: Same patient following successful reanastomosis. Both tubes are patent.

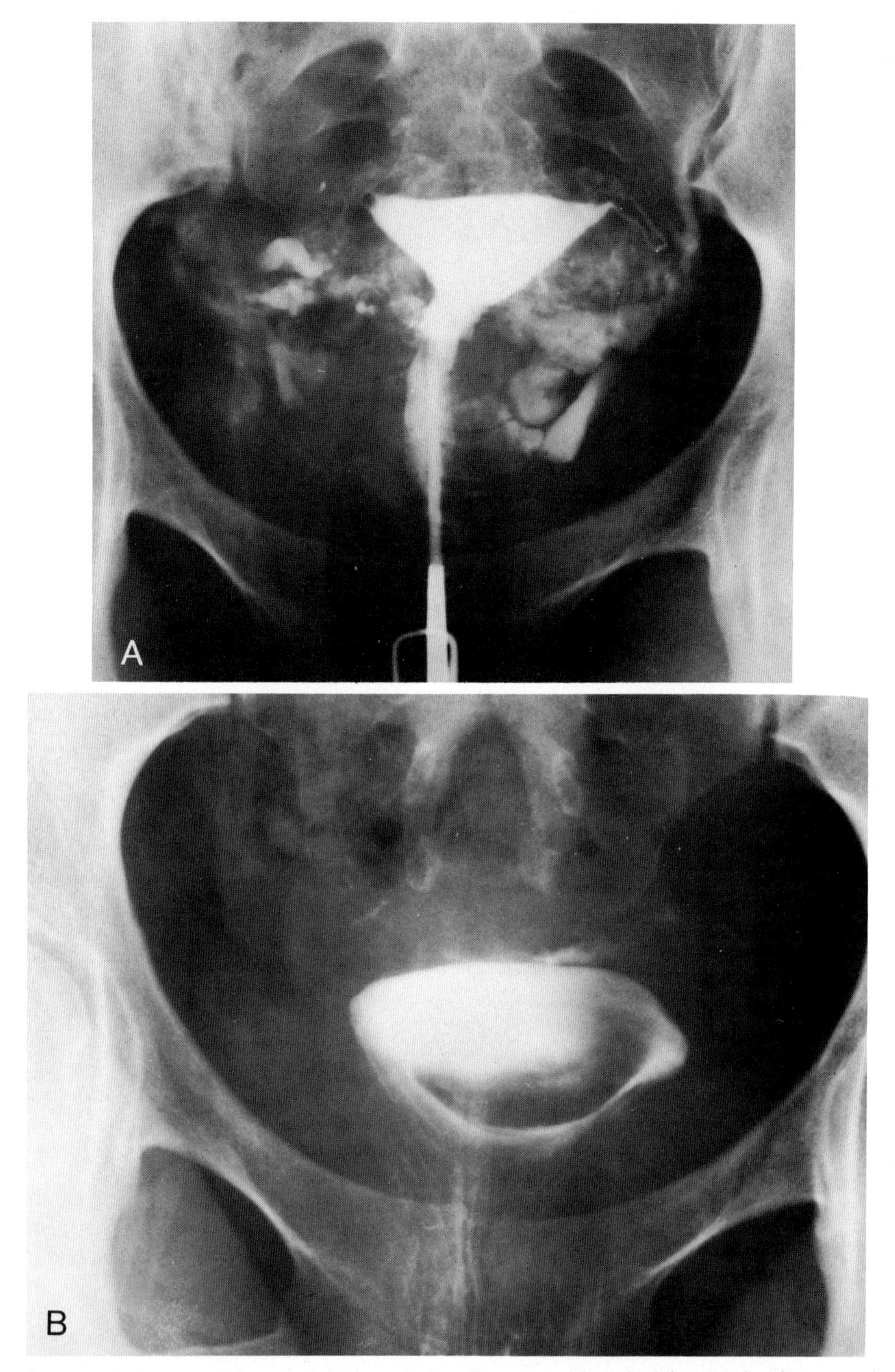

Figure 9.21. *A*: Irving procedure for tubal occlusion. The severed end of the proximal segment of the tube is embedded in the myometrium. The contrast material thus fills the venous channels draining the uterus. *B*: Delayed film (same study), within a few minutes, shows emptying of the vascular channels seen in A.

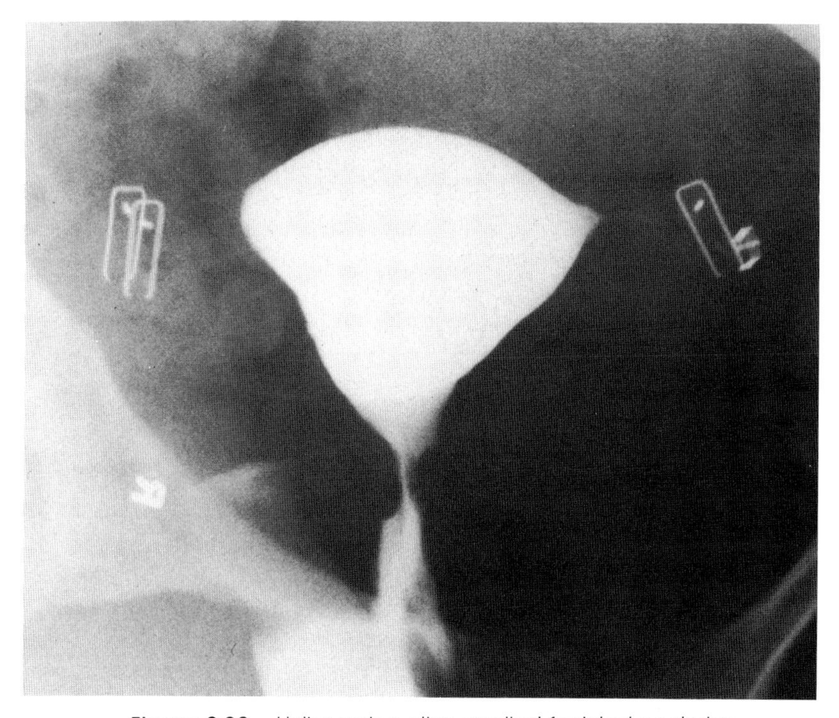

Figure 9.22. Hulka spring clips applied for tubal occlusion.

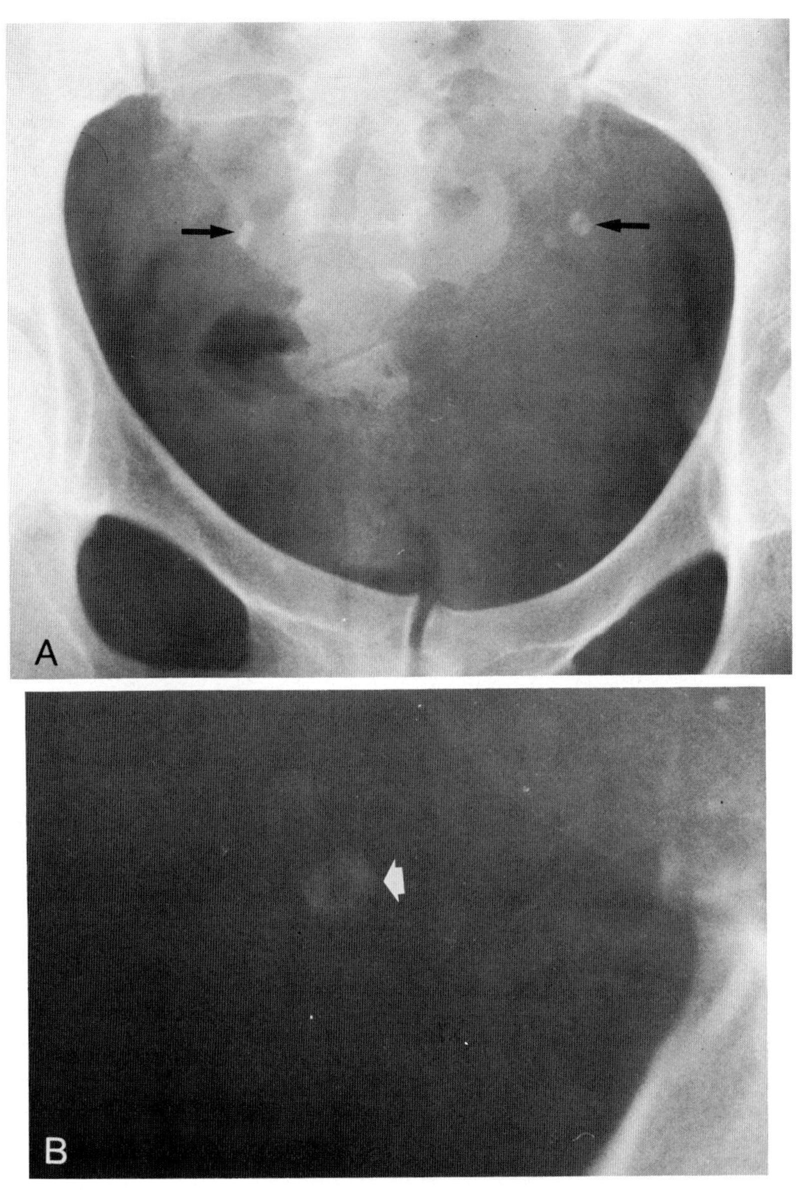

Figure 9.23. *A*: Falope rings (*arrows*) applied for tubal occlusion. This appearance may, on occasion, be confused with ureteral calculi in some symptomatic patients. *B*: Note characteristic ring-like appearance (*arrow*).

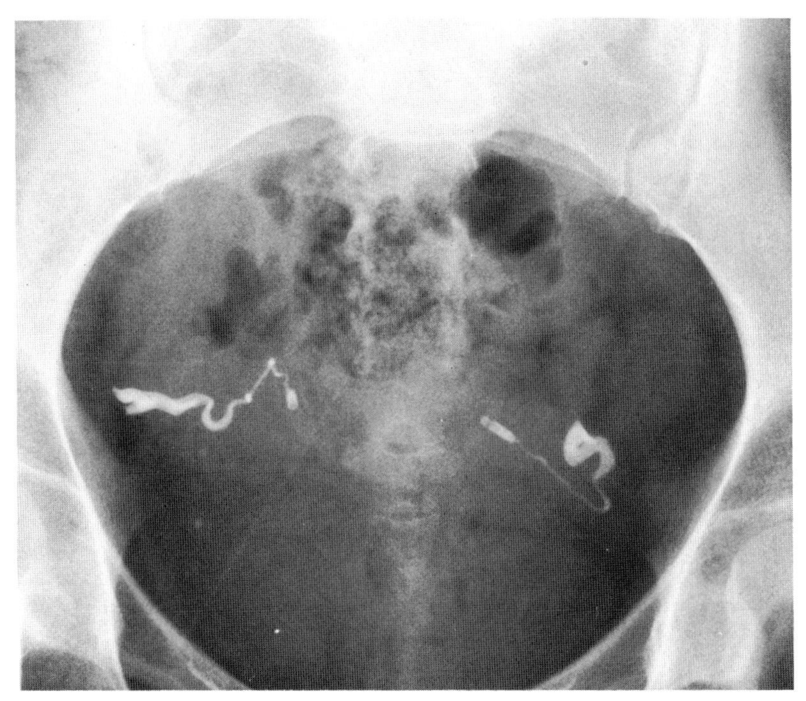

Figure 9.24. Characteristic appearance of fallopian tubes filled with silicone plugs for occlusion. Radiopaque appearance results from silver salts impregnating the silicone.

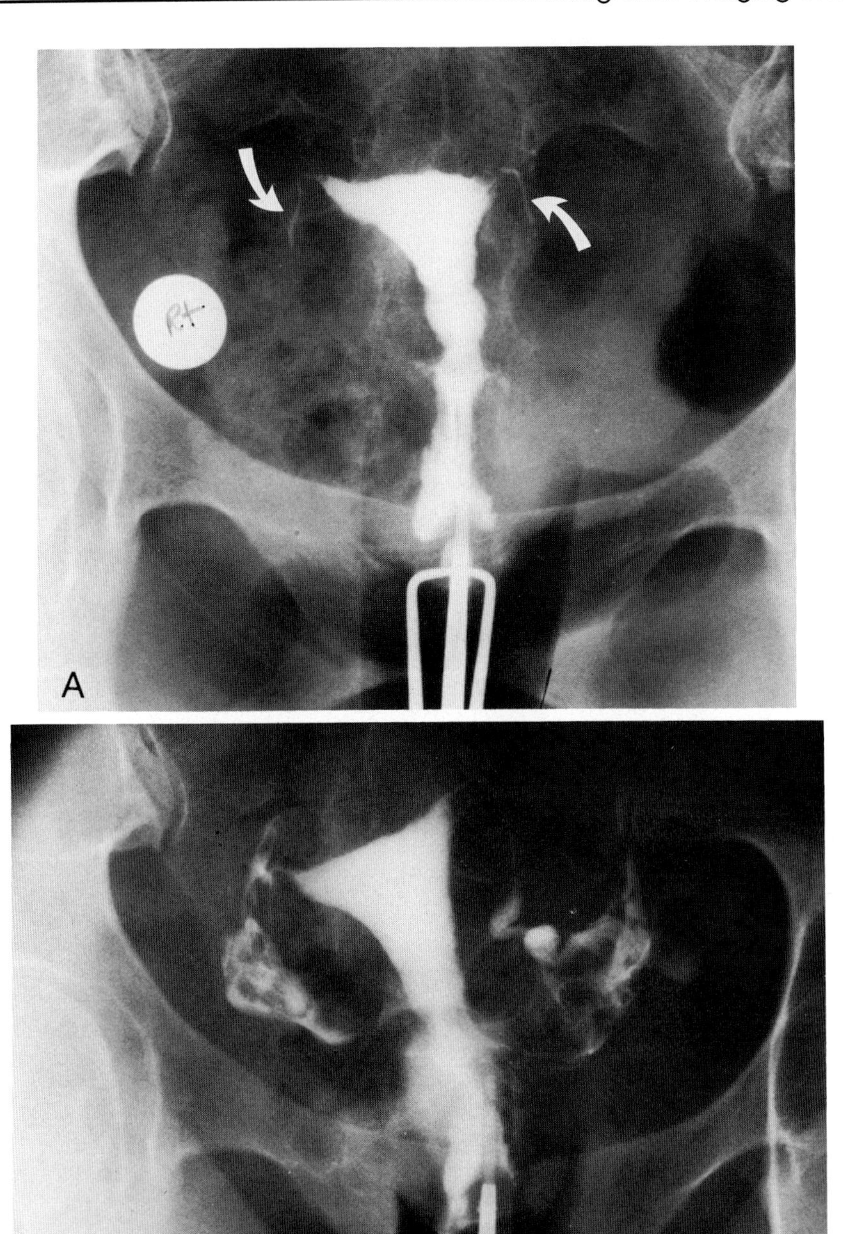

Figure 9.25. *A*: Bilateral tubal occlusions. Prior to a surgical attempt to reanastomose the occluded tube segments, the position of the separation must be assessed. Note that the isthmic segment of the tube is ample for plastic repair (*arrows*) and suggests that an adequate portion of the distal tube remains. *B*: Successful tubal reanastomoses. Both tubes are patent.

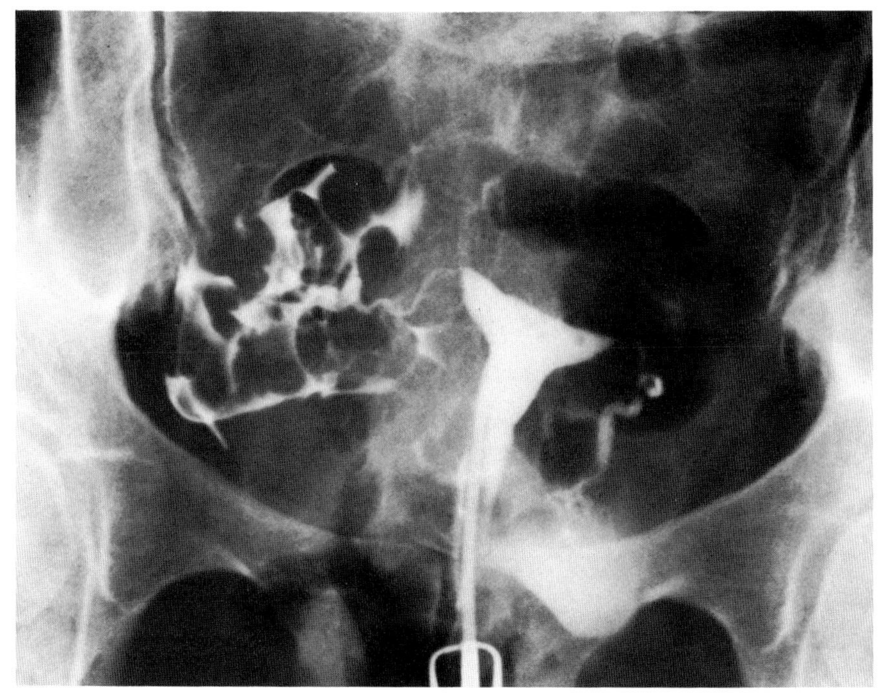

Figure 9.26. Successful tubal reanastomoses. Both tubes are patent. Location of repair is difficult to identify.

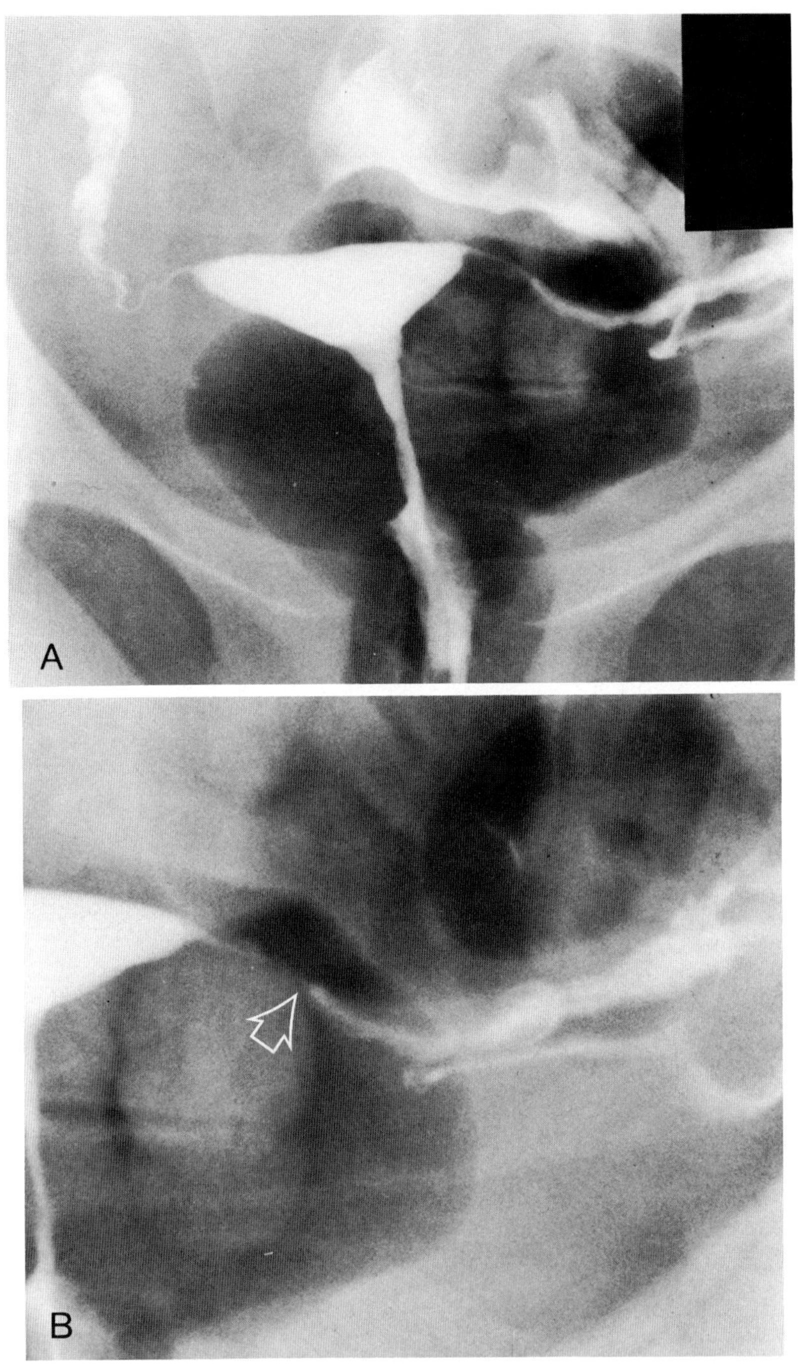

Figure 9.27. *A*: location of tubal anastomosis can be identified by sudden change in caliber of fallopian tube, as seen in enlarged view (*B*) (*arrow*).

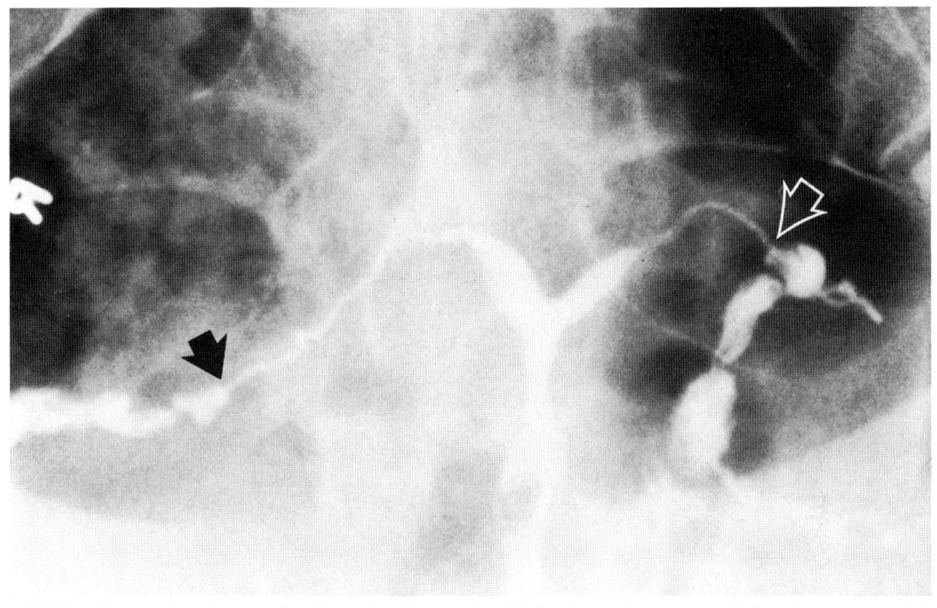

Figure 9.28. Dramatic change in diameter (*arrows*) of fallopian tube lumen following successful reanastomosis.

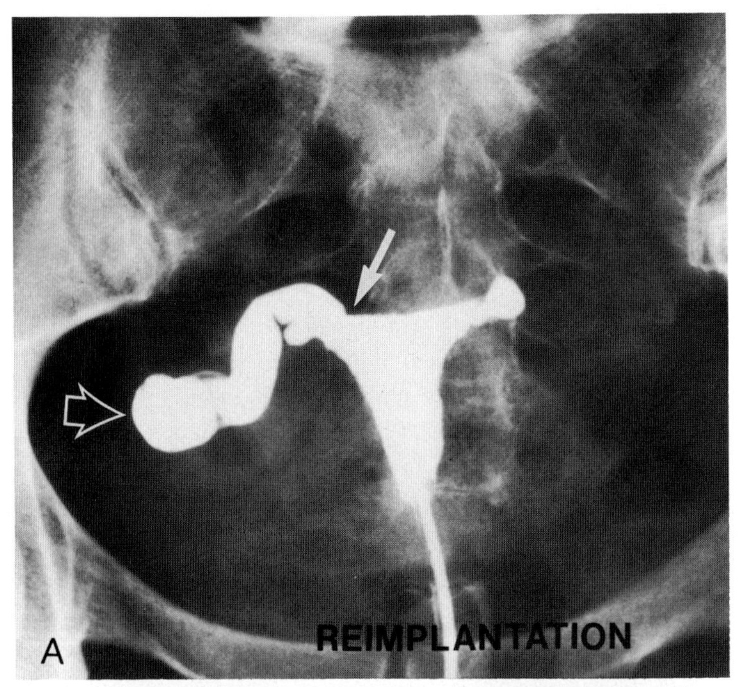

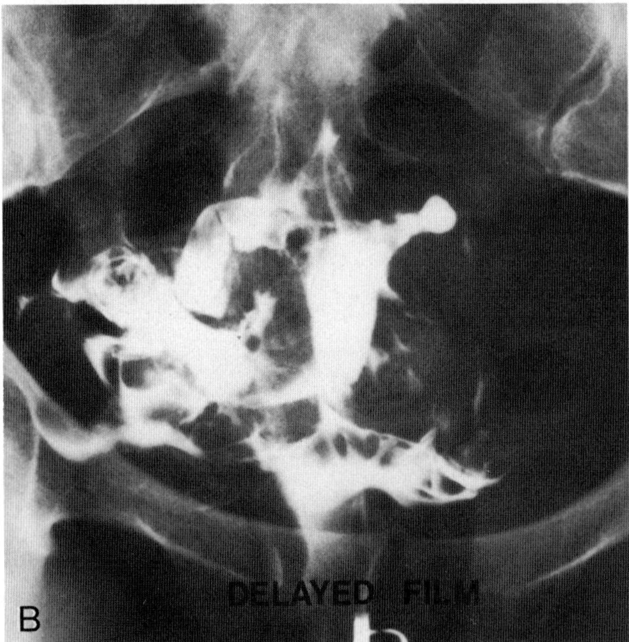

Figure 9.29. *A*: Reimplantation of right fallopian tube into cornua following "cork borer" technique to establish an adequate lumen (*arrow*). Dilated ampullary portion of the tube was opened at the same time (*open arrow*). Left fallopian tube is obstructed. *B*: Delayed film demonstrates that the right fallopian tube is patent, although deformity persists.

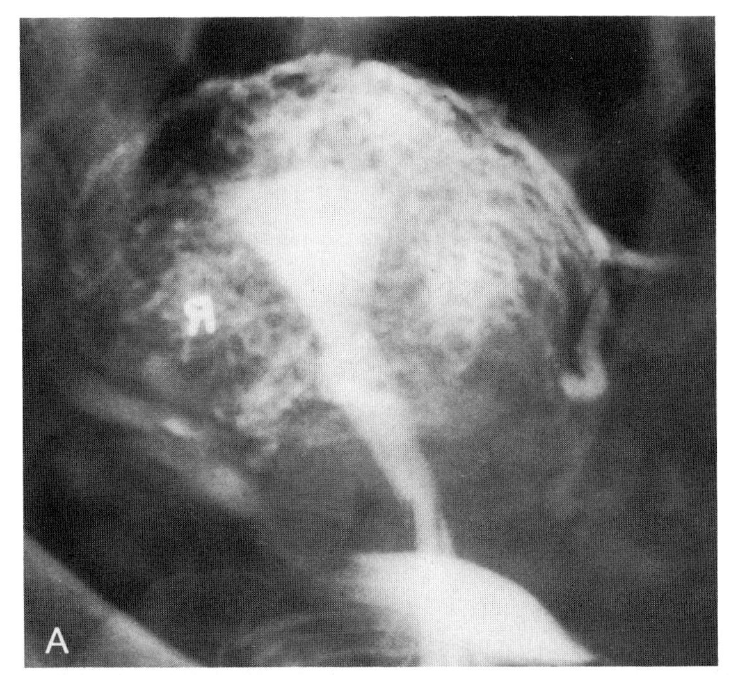

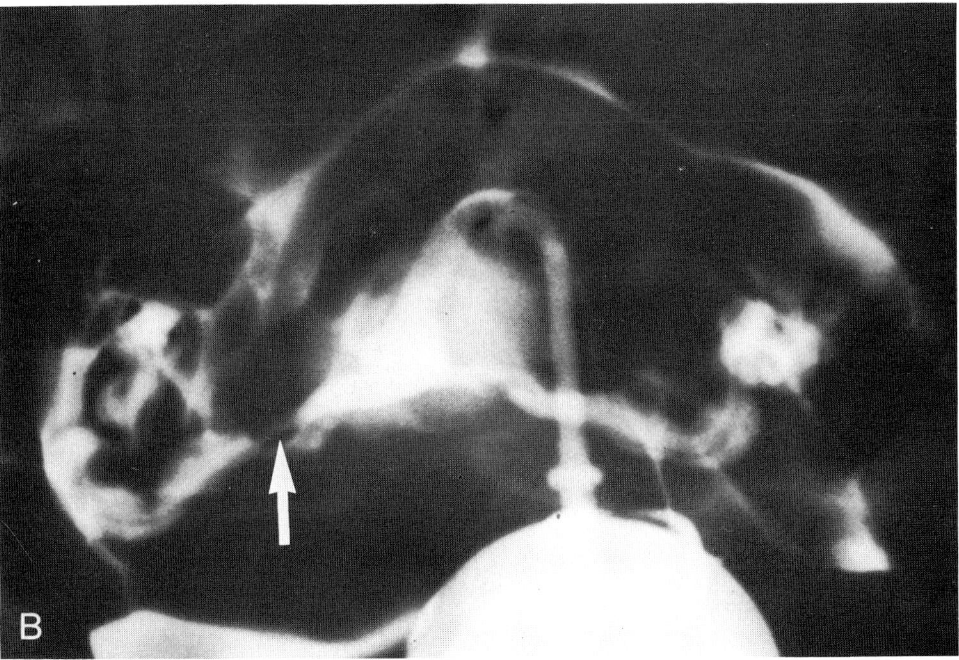

Figure 9.30. *A*: Bilateral cornual occlusion. Extravasation of contrast material in myometrium and vascular channels is related to the obstruction and increased pressure in the endometrial cavity. *B*: Reimplantation with very little deformity at the uterotubal junction (*arrow*).

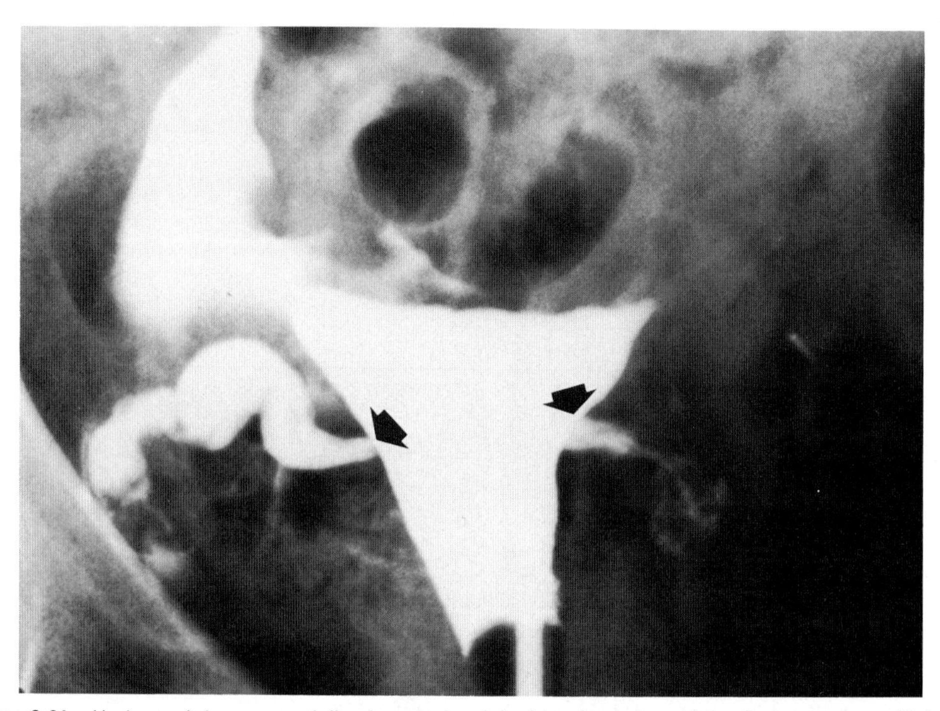

Figure 9.31. Hysterosalpingogram following a uterotubal implantation of the Peterson type. Note the low position of the tubal insertion (*arrows*) in reference to the fundus. (Courtesy of T.A. Baramki and F. Bottiglieri, Baltimore, Maryland.)

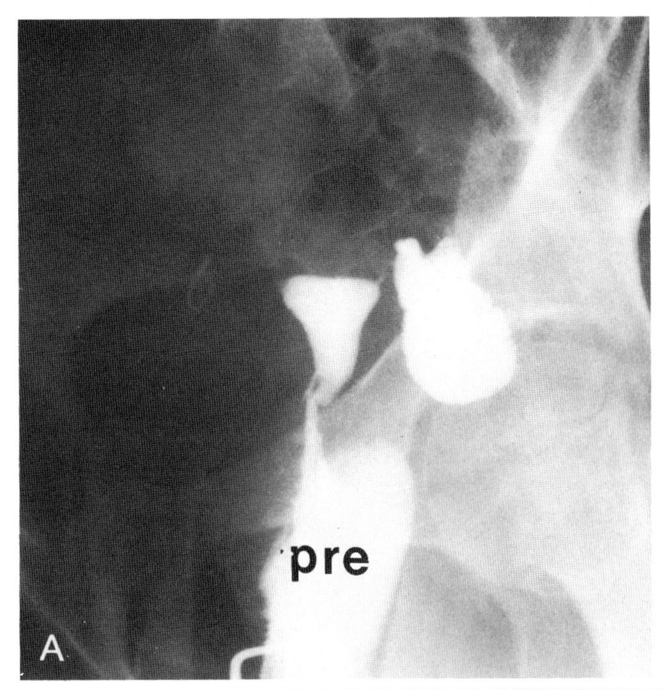

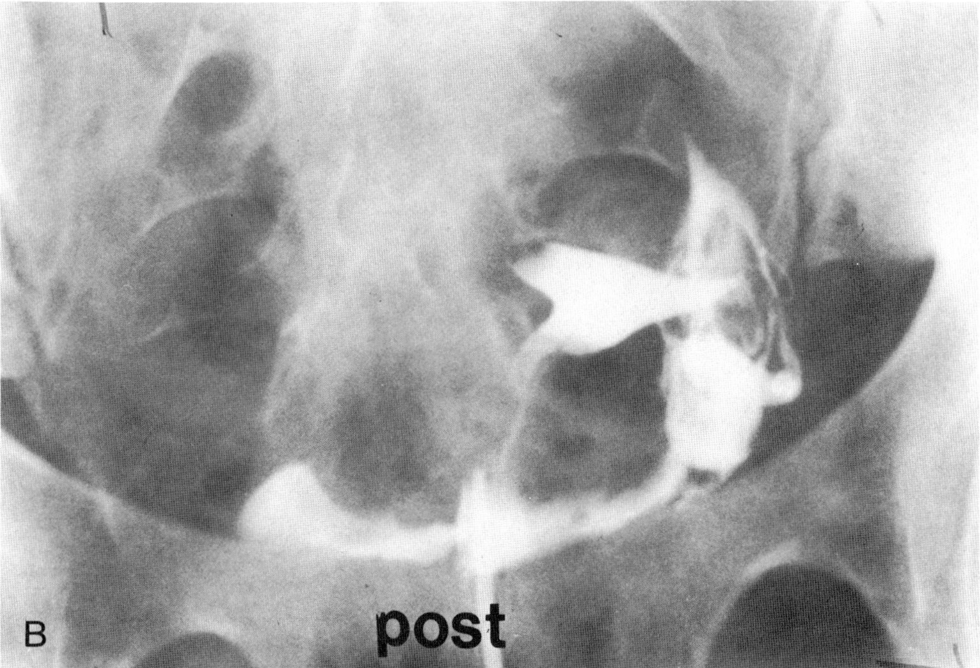

Figure 9.32. *A*: Marked hydrosalpinx on left. Right fallopian tube has been surgically removed. *B*: Following laser laparoscopic surgery, the dilatation persists but there is obvious tubal patency.

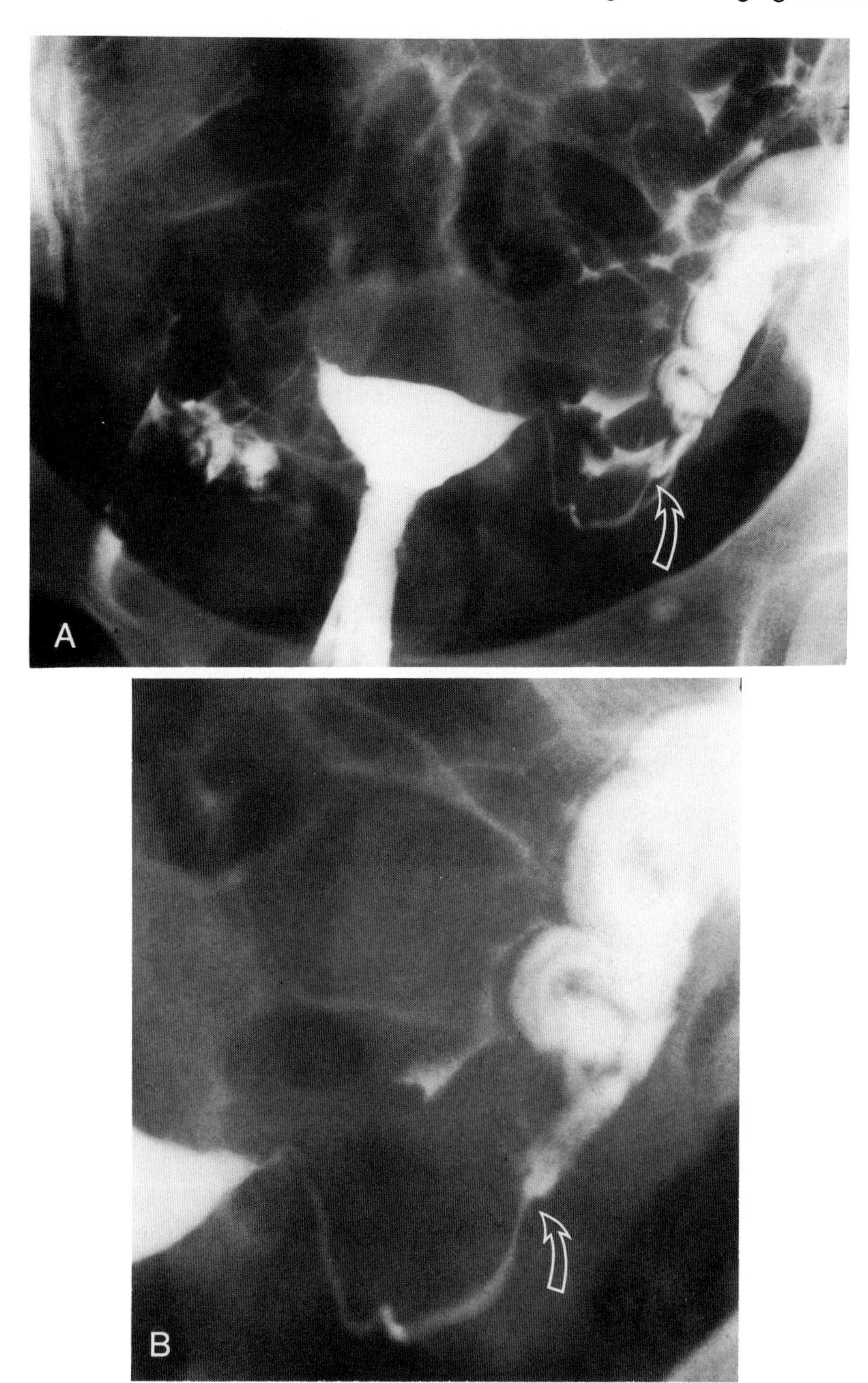

Figure 9.33. *A*: Bilateral laser salpingostomies. Ectopic pregnancies have necessitated surgery to both fallopian tubes. Site of surgery is identifiable (*arrow*), but only minimal deformity is recognized. *B*: Enlarged view of A.

though initial results were encouraging, this procedure seems to offer no significant advantage over other techniques (Fig. 9.31).

At follow-up hysterosalpingography, some tubes will demonstrate occlusion *distal* to the site of anastomosis.

Other Postoperative Findings

One of the most dramatic changes is the postoperative appearance of the hydrosalpinx. Surgical incision and opening can be done by laparoscopy in some cases, although usually an open laparotomy is required. The postoperative hysterosalpingogram may reveal decompression of the hydrosalpinx with or without evidence of damage to the endothelial surface, although often the tubal dilatation persists (Fig. 9.32). Frequently, the normal rugae may be obliterated, which suggests that the thin-walled hydrosalpinx may have no ciliated or nonciliated lining cells left. Rugae may sometimes be seen in hydrosalpinges, although these may have a thickened appearance suggesting intratubal edema or the presence of postinflammatory folds.

Site of Salpingostomy for Ectopic Pregnancy

Linear incisions over an ectopic pregnancy can be accomplished by laser laparoscopy with evacuation of the tubal products of conception or by more conventional methods at open laparotomy. The postoperative hysterosalpingographic picture will vary with the degree of damage, and may range from complete occlusion to fistula formation to a tubal lumen indistinguishable from normal (Fig. 9.33).

Ordinarily, there are no characteristic findings of postoperative tubal procedures, other than a lack of rugae and perhaps problems with tubal mobility due to peritubal adhesions.

REFERENCES

1. Rock JA, Jones HW Jr: The clinical management of the double uterus. *Fertil Steril* 28:798, 1977.
2. Wortman J: Tubal sterilization—review of methods. *Population Rep*, Series C, May 1976.
3. Ayers JWT, Johnson RS, Ansbacher R, Menon M, LaFerla JJ, Roberts JA: Sterilization failures with bipolar tubal cautery. *Fertil Steril* 42:526, 1984.
4. Jordan JA, Edwards RL, Pearson J, Maskery PJK: Laparoscopic sterilization and follow-up hysterosalpingogram. *J Obstet Gynaecol Brit Commonw* 78:460, 1971.
5. Sheikh HH: Hysterosalpingographic follow-up of laparoscopic sterilization. *Am J Obstet Gynecol* 126:181, 1976.
6. Sheikh HH: Hysterosalpingographic follow-up of the partial salpingectomy type of sterilization. *Am J Obstet Gynecol* 128:858, 1977.
7. Grunert GM: Late tubal patency following tubal ligation. *Fertil Steril* 35:406, 1981.
8. Dan SJ, Goldstein MS: Fallopian tube occlusion with silcone: Radiographic appearance. *Radiology* 151:603, 1984.
9. Silber SJ, Cohen R: Microsurgical reversal of female sterilization: The role of tubal length. *Fertil Steril* 33:598, 1980.
10. Schwimmer M, Heiken JP, McClennan BL: Postoperative hysterosalpingogram: Radiographic surgical correlation. *Radiology* 157:313, 1985.
11. Peterson EP, Musich JR, Behrman SJ: Uterotubal implantation and obstetrical outcome after previous sterilization. *Am J Obstet Gynecol* 128:662, 1977.

10 Sonography in Gynecologic Infertility

ARTHUR C. FLEISCHER, M.D.
WAYNE S. MAXSON, M.D.

The recent increased availability of real-time ultrasound scanners has had a major impact on the clinical management of infertile women. The improved resolution and decreasing price of most currently available real-time scanners relative to conventional static scanners has made this modality more extensively available to infertility specialists. Sonography now allows the infertility specialist a means to more accurately assess, in a noninvasive manner, follicular and endometrial development. In addition, sonography affords a means to evaluate more accurately a number of disorders that are related to gynecologic infertility. The most common of these are occlusive and/or inflammatory tubal disease and endometriosis.

In this chapter, the use of sonography in a variety of gynecologic disorders that are associated with infertility will be discussed and illustrated.

Instrumentation

The widespread availability of sector real-time scanners in the late 1970s and early 1980s dramatically enhanced the capability of sonography to monitor follicular development within the ovary compared to that which was possible with static scanners (Figs. 10.1, 10.2, and 10.3) (1). Today, the most widely used scanners for evaluation of the pelvic organs are mechanical sector real-time scanners. These scanners use transducers of 3.5 or 5.0 MHz that are selectively focused at a specific depth. The higher the frequency utilized, the better the detail that can be depicted. However, use of the higher frequency transducers (5.0 MHz) is limited in some large patients due to decreased penetration. Therefore, they cannot be used for every patient. Thus, for obese or large patients a 3.5-MHz transducer must be used.

Rotating wheel mechanical sector transducers are preferred over hysteresing varieties that utilize the single element that oscillates back and forth to create a sector field of view. Phased sector scanners also have excellent resolution of the uterus and ovaries, but at present are more expensive than mechanical sector scanners due to the computerization required to form and steer the beam. In general, the lateral resolution of transducers that have a rotating wheel configuration is superior to that of oscillating transducers. The axial resolution of a 3.5-MHz transducer is dependent on the wavelength of the ultrasound emitted, and therefore is 1–2 mm, whereas the lateral resolution or the ability to image two objects perpendicular to the beam is between 2 and 3 mm. Transducers for evaluation of the ovaries should be focused at a depth of 9–11 cm. The operator should be aware of the focal range of the transducer so that images can be optimized for a particular patient.

Urinary bladder distension is required for evaluation of the ovaries. With bladder distension, the ovaries are usually displaced laterally from the uterus, and immediately posterior to the bladder (Fig. 10.1B,C). This position optimizes their sonographic depiction, displacing the loops of bowel cephalad and the uterine fundus posteriorly. In most cases, the ovaries are located 9–11 cm from the anterior abdominal wall. In some patients who have ovaries that are low in the cul-de-sac or very anterior along the anterior peritoneal surface, sonographic delineation may be suboptimal using conven-

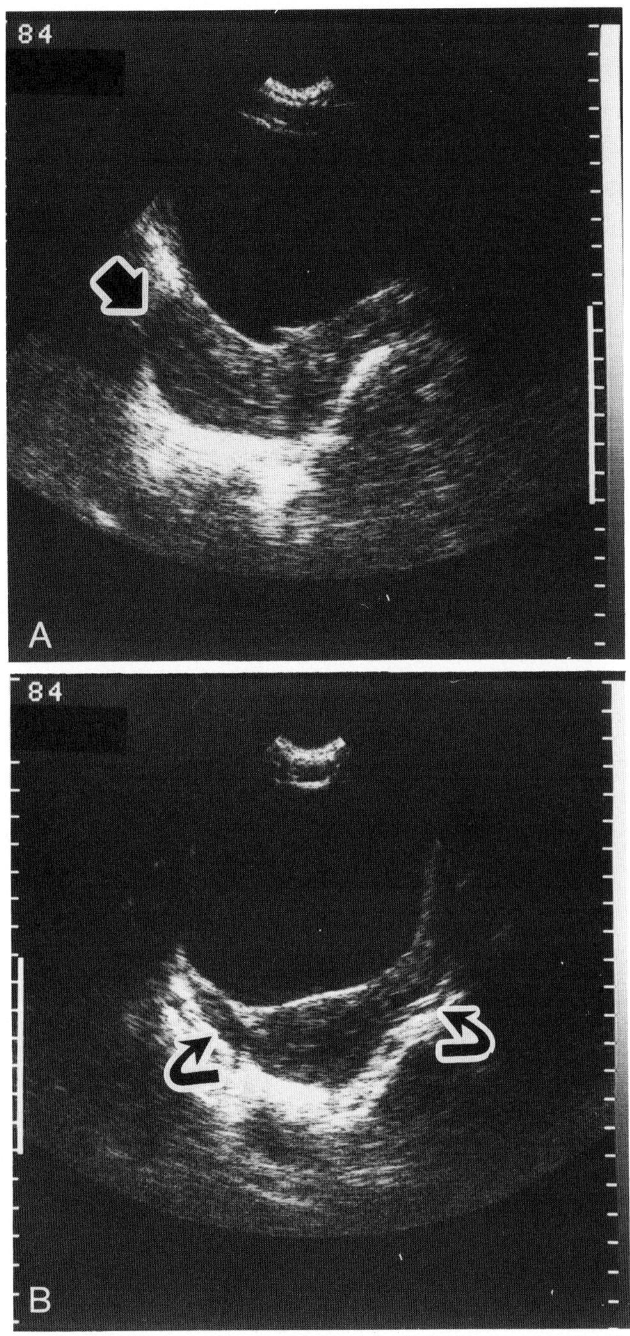

Figure 10.1. Sonographic appearance of normal ovaries. *A*: Before evaluating the ovaries, the uterus should be identified in its long axis (*arrow*). *B*: Once the uterus is identified, a 90° turn of the transducer reveals both ovaries (*arrows*) flanking the uterus.

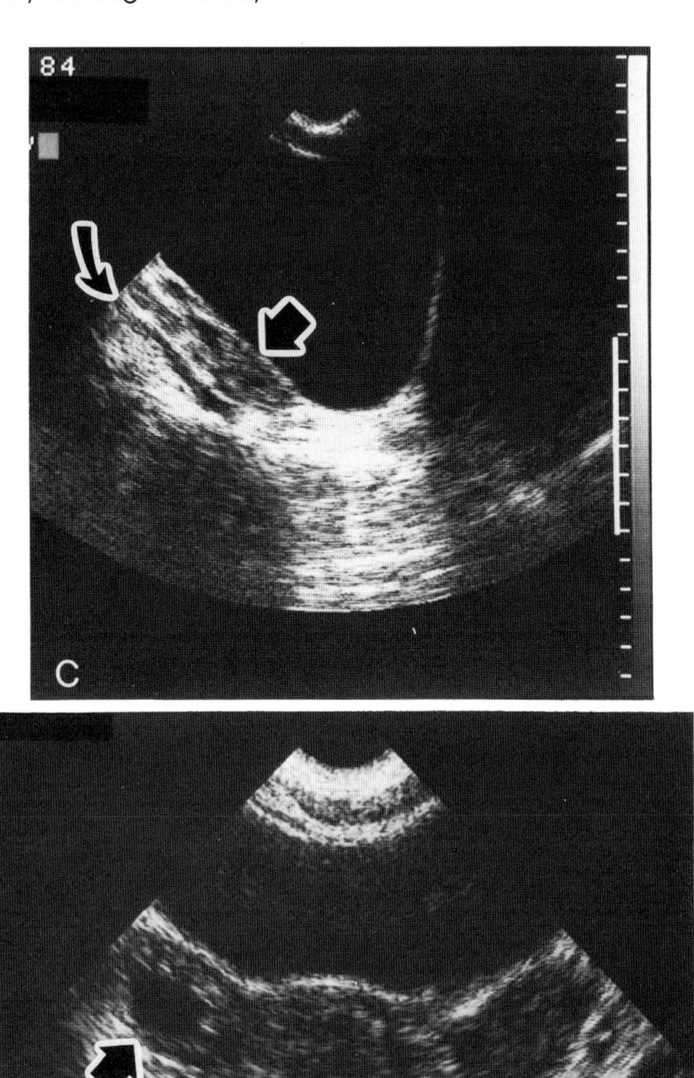

Figure 10.1. *C*: Long axis of normal left ovary (*arrow*). The distal ureter is seen (*curved arrow*). *D*: Same patient as in Figure 10.1A and B (magnified view) at cycle day 12 showing development of a single mature follicle (*arrow*) within right ovary.

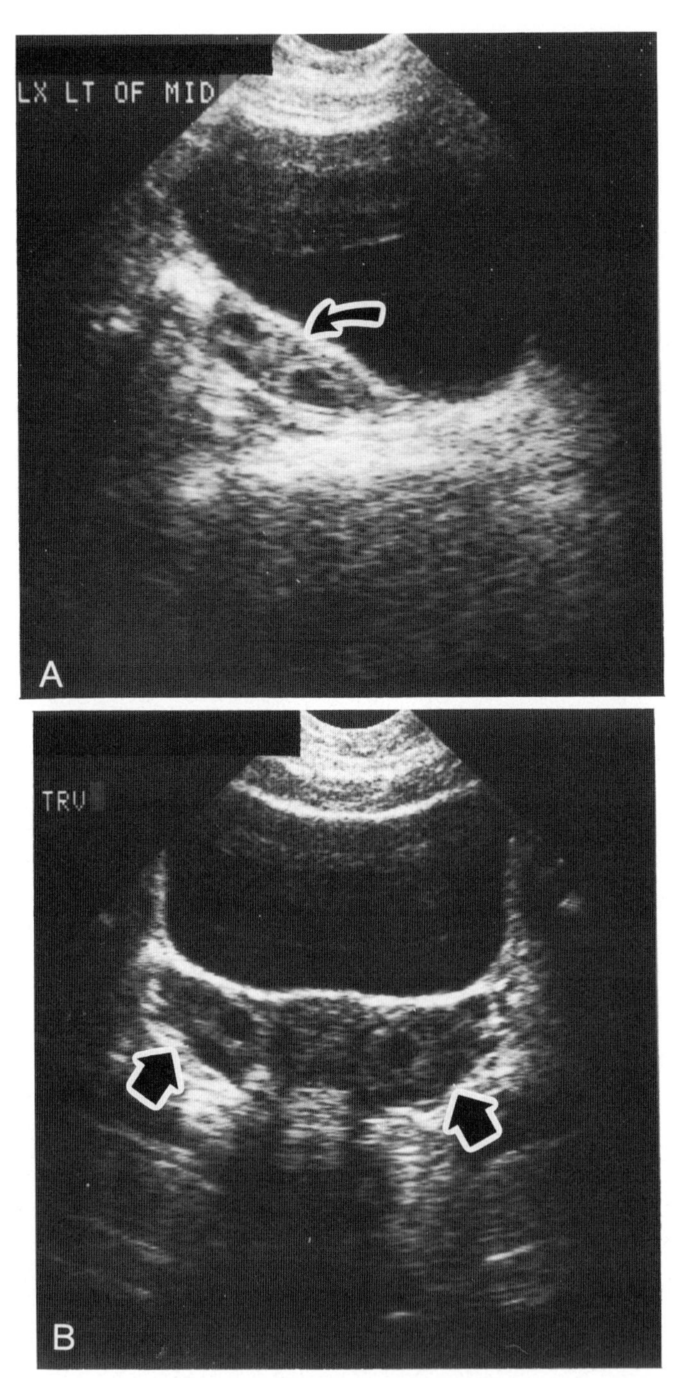

Figure 10.2. Sonographic appearance of anatomic variants of ovary. *A*: Long axis of left ovary (*curved arrow*) that measures 4.5 cm in long axis. As long as the ovary maintains a fusiform shape and is less than 4–5 cm in any one dimension, it probably is normal. *B*: Transverse sonogram demonstrating bilaterally rounded and enlarged ovaries (*arrows*) with several immature follicles consistent with polycystic ovary disease.

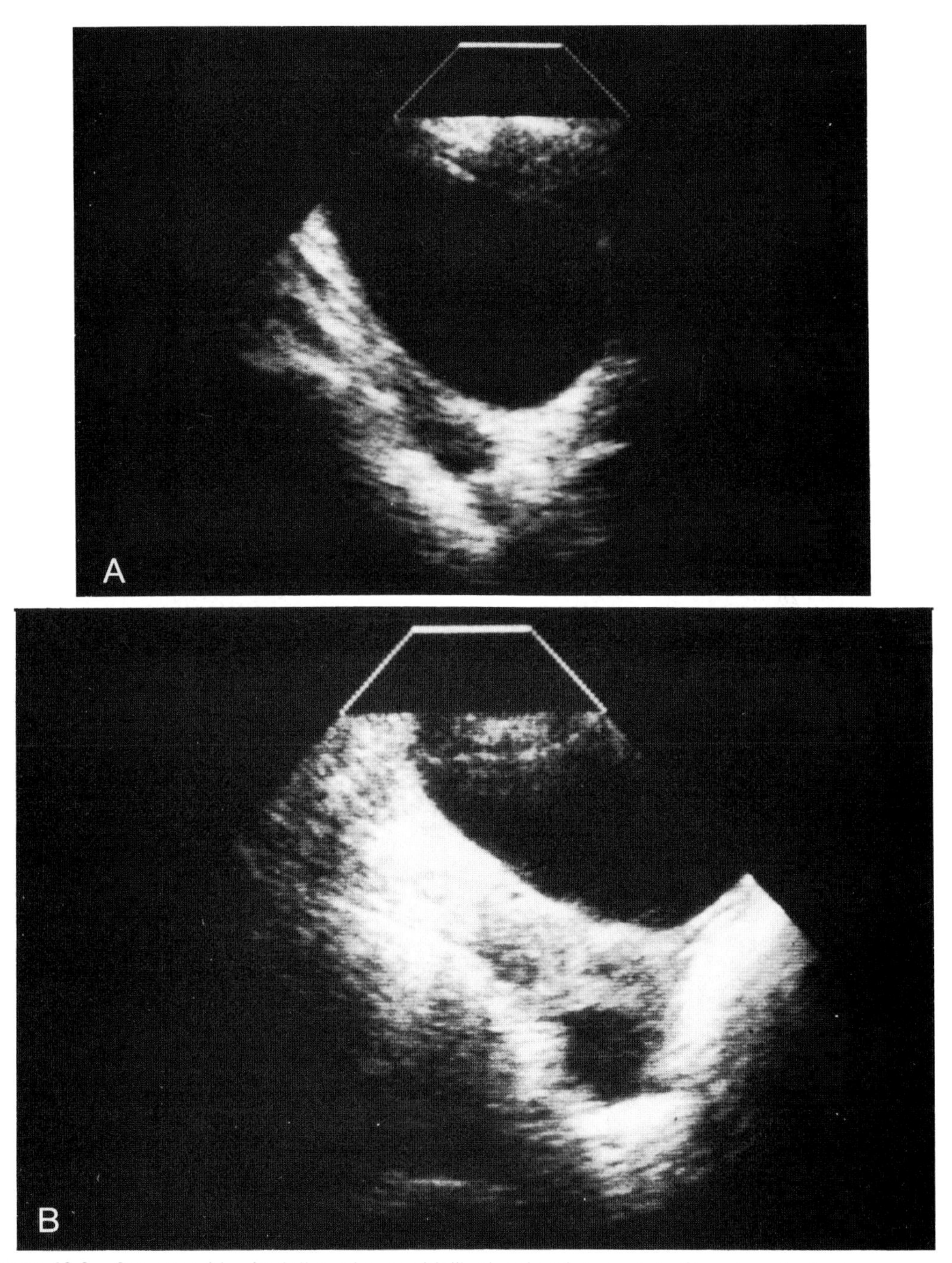

Figure 10.3. Sonographic depiction of normal follicular development. A–C are longitudinal sonograms of the right ovary. *A*: At day 6, mean dimension of 8 mm. *B*: At day 12, mean dimension of 12 mm.

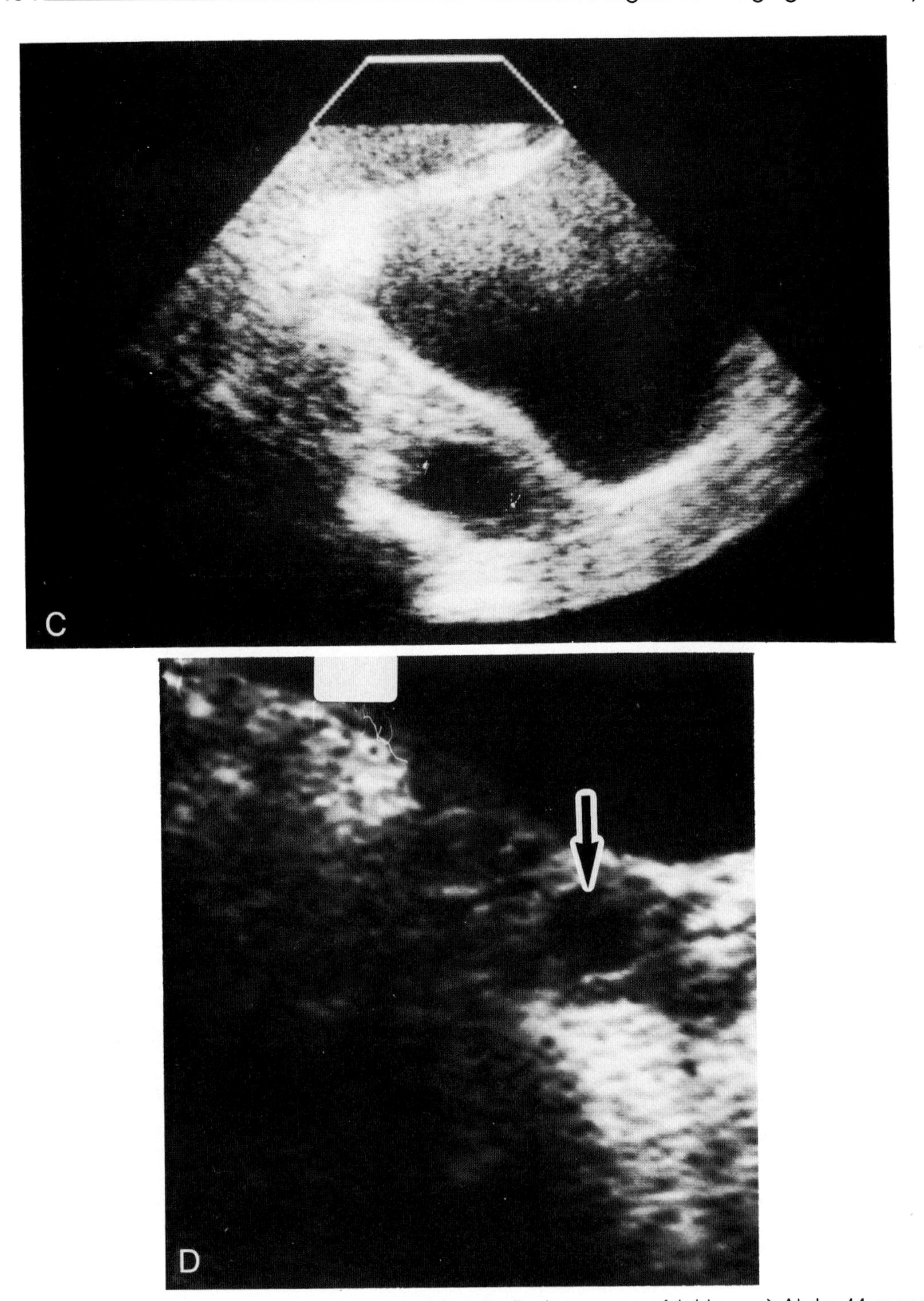

Figure 10.3. *C*: Normal follicular development (longitudinal sonogram of right ovary). At day 14, mean dimension of 20 mm. *D*: Long axis of mature follicle demonstrating cumulus oophorus (*arrow*).

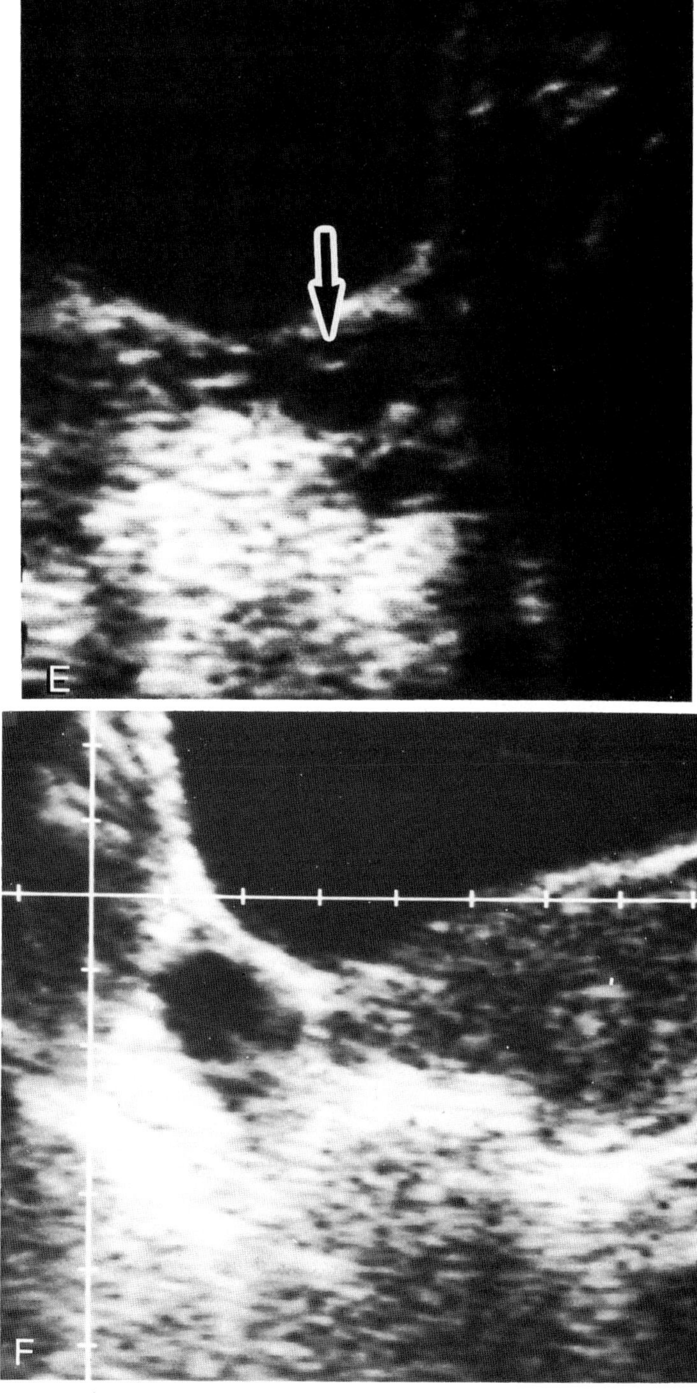

Figure 10.3. *E*: Transverse image of mature follicle demonstrating cumulus oophorus (*arrow*). *F*: Appearance of follicle within the right ovary one day prior to ovulation.

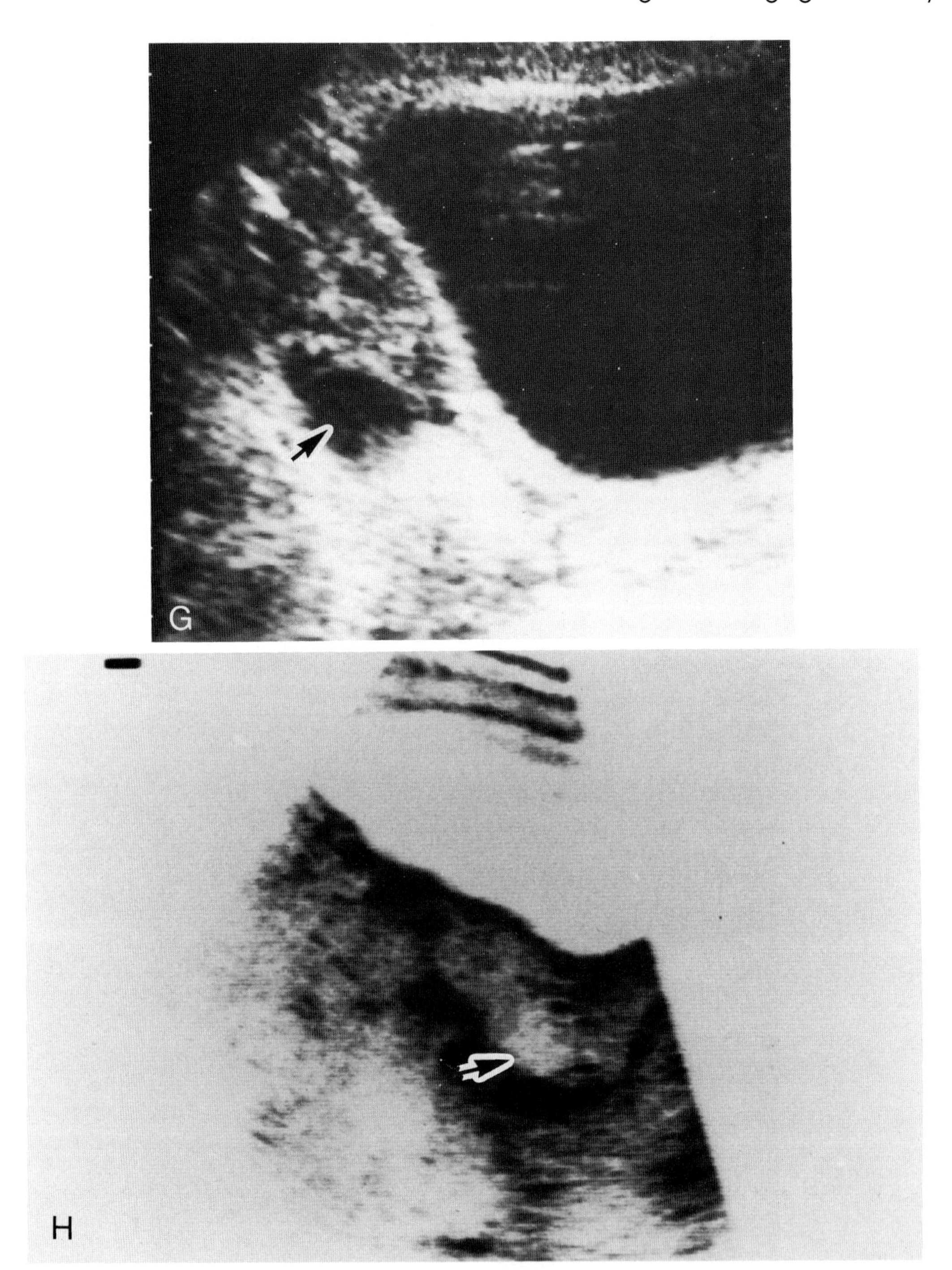

Figure 10.3. *G*: Same patient as in Figure 10.3F. Next day (at ovulation) sonogram demonstrates low level internal echoes. Note corpus luteum (*arrow*). *H*: Static scan of corpus luteum (*arrow*) at ovulation day +1.

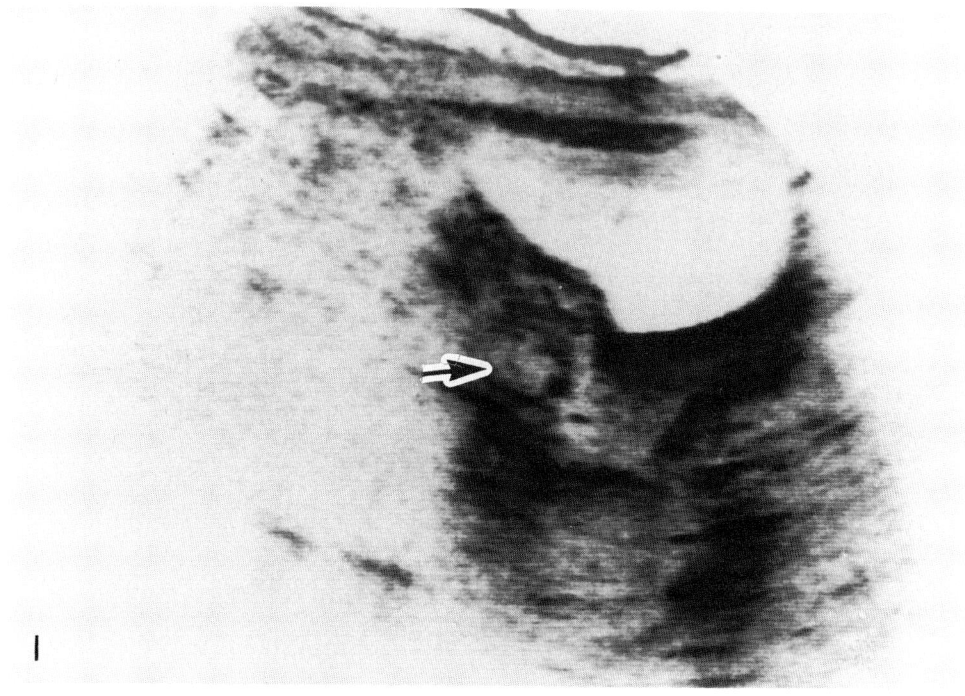

Figure 10.3. *I*: Same patient 3 days later demonstrating sonographic appearance of mature corpus luteum (*arrow*).

tional scanning techniques. For ovaries located deep in the cul-de-sac, the transducer can be placed within the vagina for more accurate delineation (2). An end-fired transducer covered by a condom is used for this purpose. For the ovary that is located near the anterior peritoneal surface, a small parts transducer with water path offset may be utilized to image the ovary in the most optimal focal zone.

Real-time scanners that allow image magnification are preferred for evaluation of the ovaries. Electronic magnification of the region of interest or the image memory improves the ability to measure accurately follicular and endometrial dimensions. Systems that actually utilize the entire image matrix on magnification rather than expanding an image stored in memory are preferred since theoretically greater detail can be obtained. When possible, the calipers on the display should be used for measurement of follicles since this is more accurate than measurement with hand-operated calipers.

Scanning Technique

Mechanical sector scanners allow the sonographer/sonologist to direct the beam in order to optimize depiction of the ovaries. Table 10.1 lists the images included in our routine series for follicular monitoring.

Initially, it is helpful to begin scanning in the longitudinal plane and identify the greatest long axis of the uterus. The first images obtained on film should be of the

Table 10.1
Routine Views for Follicular Monitoring

Image #	Feature Depicted
1–2	Long axis of uterus
3	Magnified, measured endometrium
4–5	Longitudinal right ovary, follicles
6–7	Longitudinal left ovary, follicles
8–10	Transverse of right ovary, follicles
11–12	Transverse of left ovary, follicles
13	Cul-de-sac for fluid

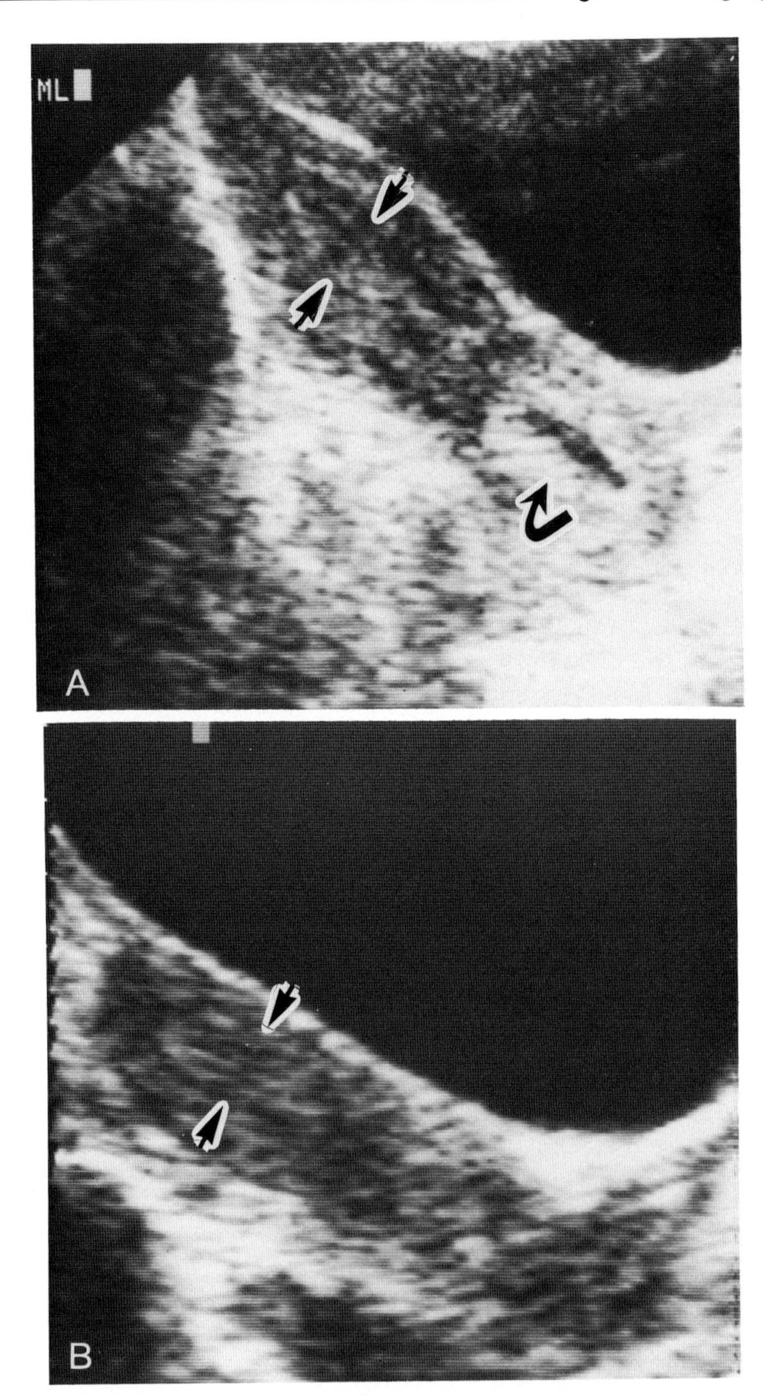

Figure 10.4. Uterine changes. *A*: Thick (5 mm single-layer) endometrium (*between arrowheads*) in a patient who achieved pregnancy with in vitro fertilization/embryo transfer. Fluid (*curved arrow*) within the cervical mucus in the periovulatory period is also present. *B*: Thick (5 mm single-layer) endometrium (*between arrowheads*) in a patient who did not achieve pregnancy after hMG/hCG stimulation.

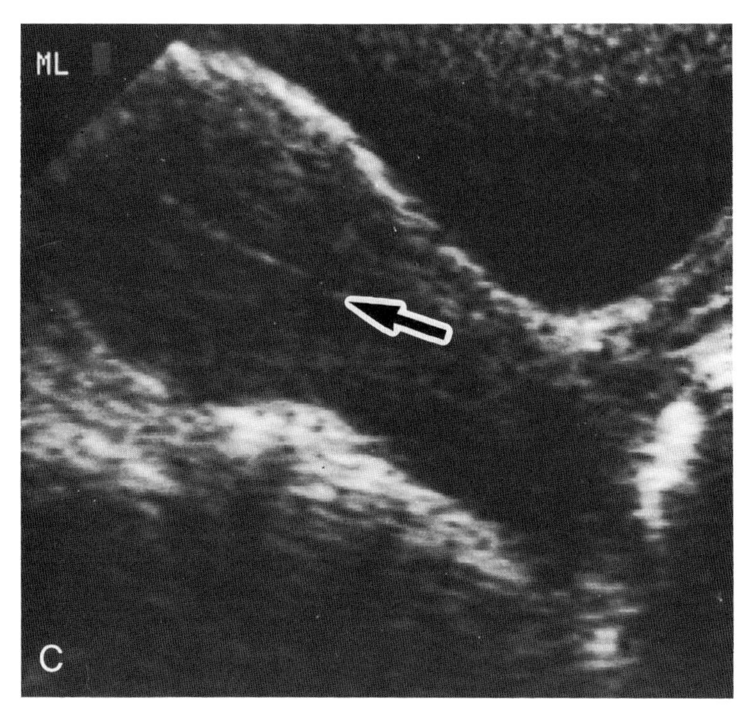

Figure 10.4. *C*: Thin (3 mm single-layer) endometrium (*arrow*) in a patient who did not become pregnant after in vitro fertilization/embryo transfer. This lack of normal development of the endometrium may have prognostic implications.

uterus (Fig. 10.1A). Its size, contour, internal texture, and flexion should be documented. The size of the uterus relates to the parity of the individual. Images of the endometrium may be documented after satisfactory images of the uterus are obtained. Images of the endometrium are obtained by magnifying the images of the uterus when the greatest portion of the central lumen of the uterus is depicted (Fig. 10.4). The endometrium can be measured with digital calipers from the echogenic endometrial-myometrial interface (Figs. 10.4A,B).

Once images of the uterus are obtained, the ovaries should be scanned (Figs. 10.1B,C). One can begin sonographic evaluation of the ovaries in an angled sagittal scan with the transducer placed in the midline over the bladder and angled toward the pelvic side wall. The ovaries can be identified as almond-shaped structures, which usually lie anterior to the internal iliac vein and artery and distal ureter. Once a follicle is found in the longitudinal plane, its longitudinal and anteroposterior dimensions should be measured.

The ovaries and uterus can also be imaged in the transverse plane. Oblique scans can be obtained empirically to optimize sonographic depiction of the ovaries, follicles, and uterus.

The cul-de-sac and lower abdominal peritoneal cavity should be examined for the presence of intraperitoneal fluid (Fig. 10.5G). Fluid resulting from ovulation can be loculated and located outside of the posterior cul-de-sac in the anterior cul-de-sac, near the uterine fundus, or surrounding the bowel loops (Fig. 10.5C).

Safety Considerations

There have been no proven adverse bioeffects of ultrasound used at diagnostic intensities (3). One report has suggested that the use of ultrasound could have contributed to premature ovulation in patients undergoing ovulation induction (4). How-

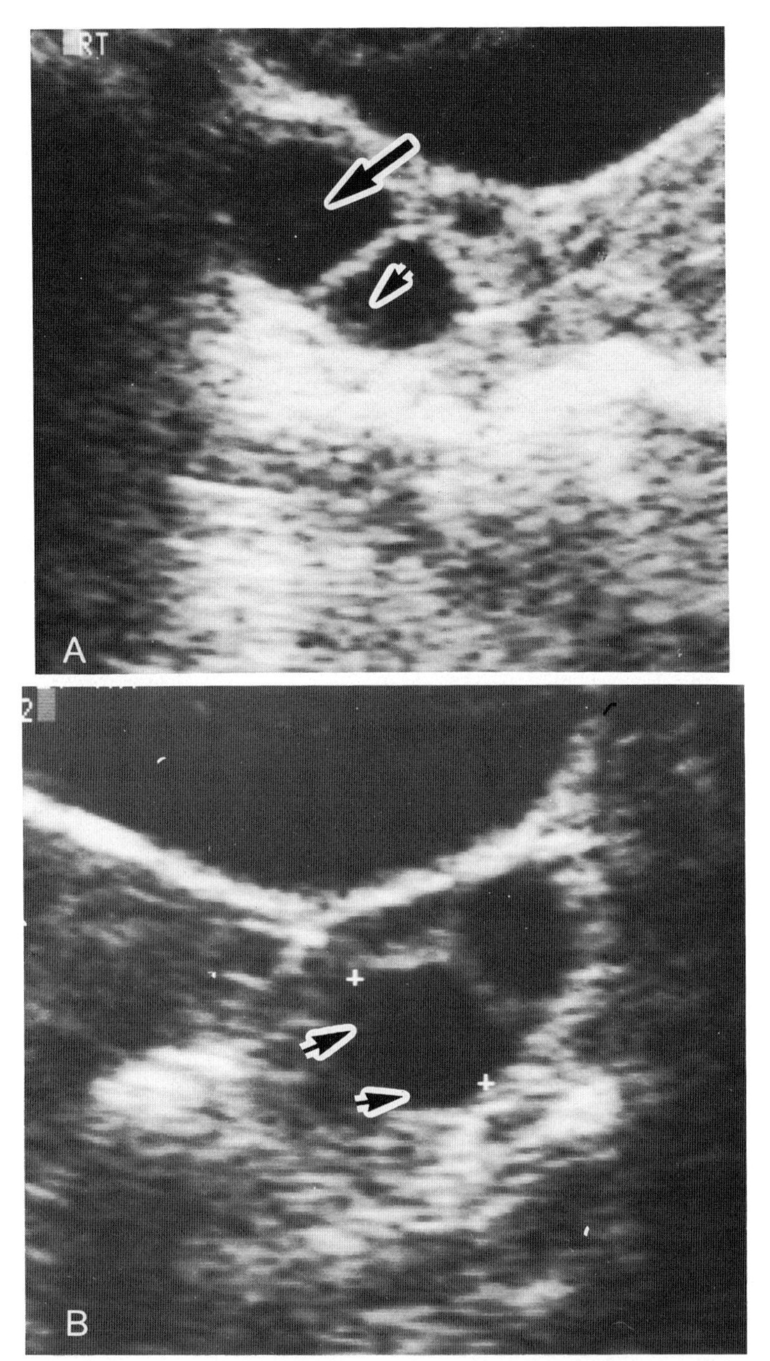

Figure 10.5. Sonographic depiction of normal follicular anatomy. *A*: Bulge along inner wall of the lower follicle (*arrowhead*) that represents the cumulus oophorus. The cumulus is floating within the upper follicle (*arrow*). This occurs immediately prior to ovulation. *B*: Intrafollicular membrane (*arrowheads*) probably representing a layer of granulosa cells that has separated from the theca cell layer.

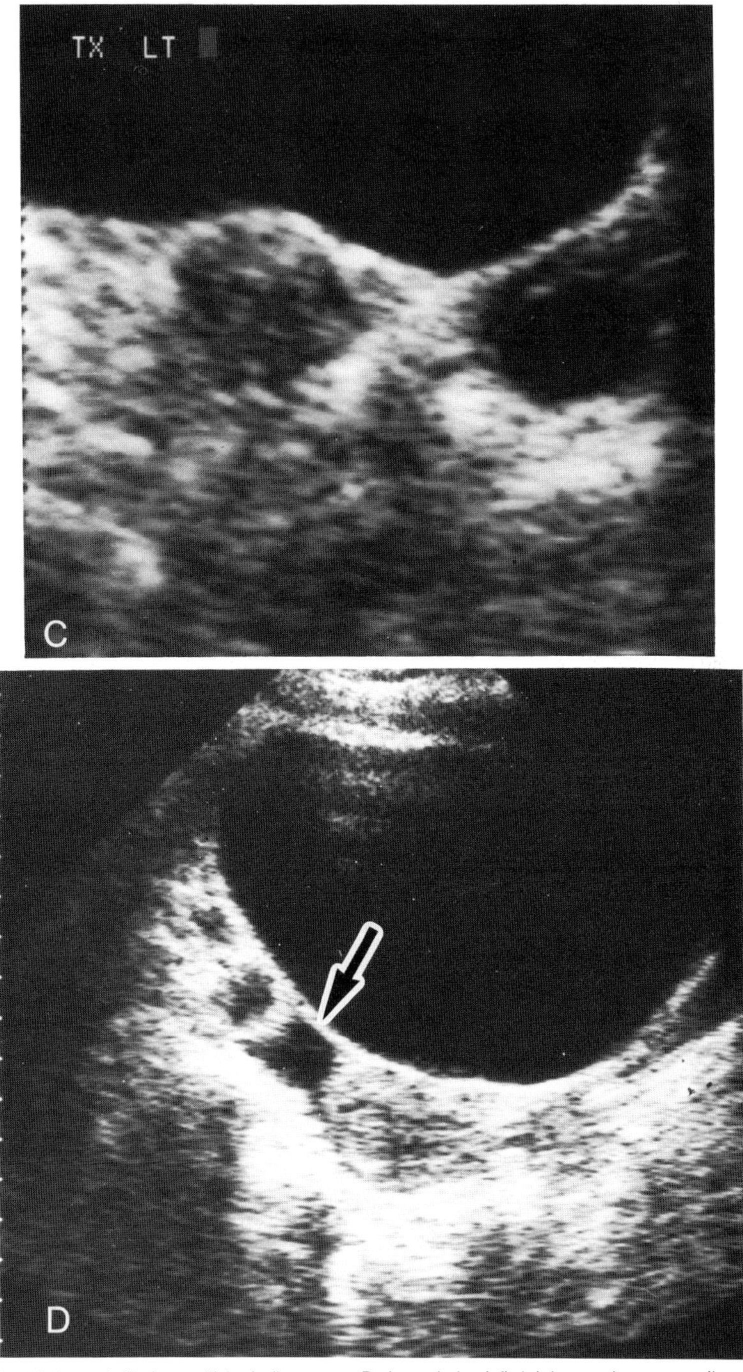

Figure 10.5. *C*: Mature follicles within left ovary. *D*: Loculated fluid (*arrow*) surrounding a bowel loop superior to the uterine fundus.

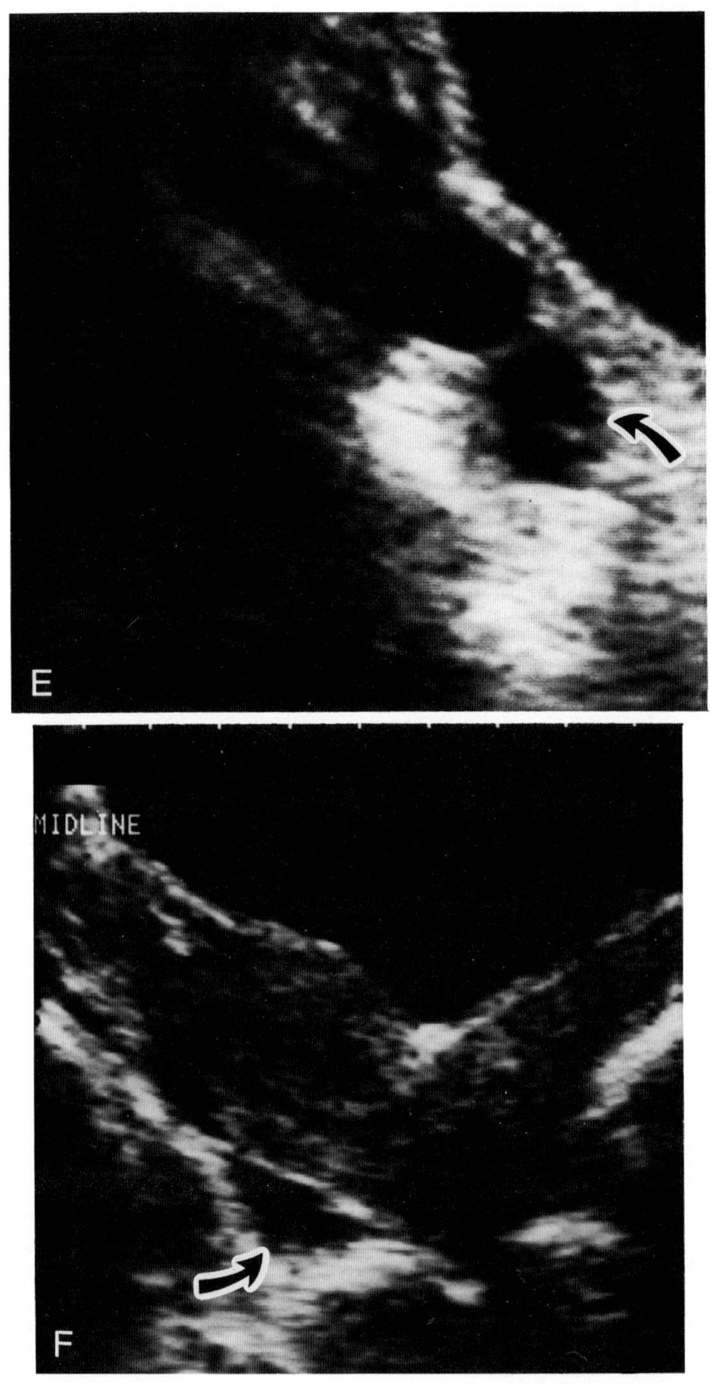

Figure 10.5. *E*: Follicle with a crenated border (*curved arrow*) indicating ovulation has just occurred. *F*: Intraperitoneal fluid (*curved arrow*) in a portion of the cul-de-sac immediately postovulation.

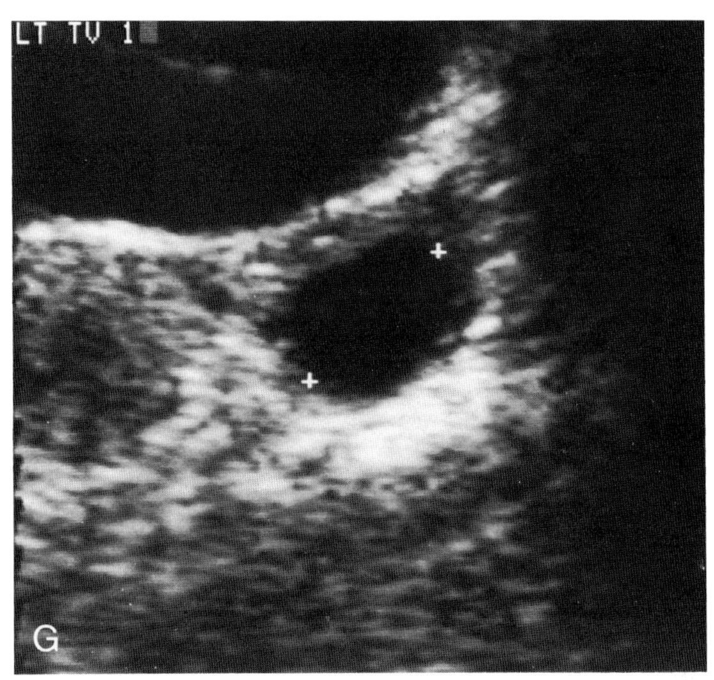

Figure 10.5. *G*: Same patient as in Figure 10.5D the next day; the follicle (*between* +'s) has slightly enlarged and now contains low level internal echoes. The presence of low level echoes seems to indicate a periovulatory, mature follicle.

ever, this may have been caused by excessive bladder distension with resultant follicular rupture before, during, or after bladder decompression rather than a direct effect of the ultrasound study. There is continuing research concerning the possible bioeffects of various intensities of ultrasound on the rat ovary (5). All of the bioeffects observed thus far have been at intensities of 10 to 1000 times greater than used for diagnosis. In general, ultrasound exerts its major bioeffect on the membrane of the cell, and not on the nucleus. It is fair to state that all currently available data indicate that ultrasound has no significant bioeffects to the patient or operator.

Normal Menstrual Cycle

In order to discuss and illustrate the uses of sonography in patients undergoing treatment for gynecologic infertility, it is important to present the sonographic appearances of the uterus, ovary, and endometrium during the normal menstrual cycle.

Ovaries are best delineated when the patient has a fully distended bladder. Bladder distension improves sonographic delineation of the pelvis by displacement of gas-containing small bowel loops cephalad from the pelvis, shifting of the ovaries to a lateral position between the urinary bladder and pelvic floor, and placement of the anteflexed uterus in a more horizontal position relative to the incident beam. This situation affords optimal depiction of the uterine internal anatomy since the axial rather than lateral resolution capabilities are operant. In most cases, the distended bladder places the ovaries at between 9 and 11 cm deep to the anterior abdominal wall. Thus, depiction of the ovaries with transducers that are long internally focused are best for most patients. In some patients, marked bladder distension can displace ovaries outside of this focal range. In these individuals, partial emptying of the overly distended urinary bladder may be helpful for better sonographic depiction of the ovaries (Fig. 10.6A,B). As stated previously, some ovaries in the cul-de-sac that are only moder-

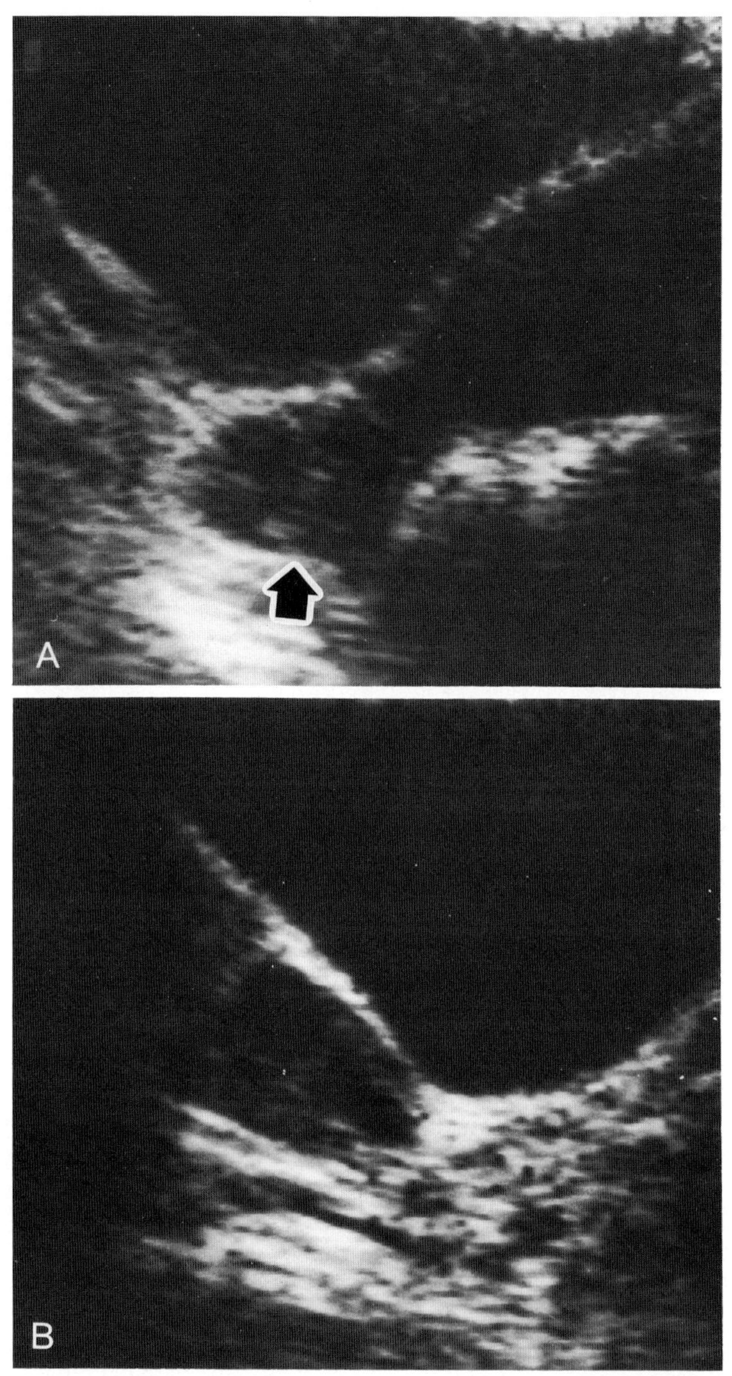

Figure 10.6. Technical aspects of sonographic follicular monitoring. *A*: Transverse image of right ovary (*arrow*) is difficult to depict in this plane with an overly distended urinary bladder. *B*: Same patient as in Figure 10.6A after partial voiding. The follicles are much better depicted because the ovary is now in the focal plane of the transducer.

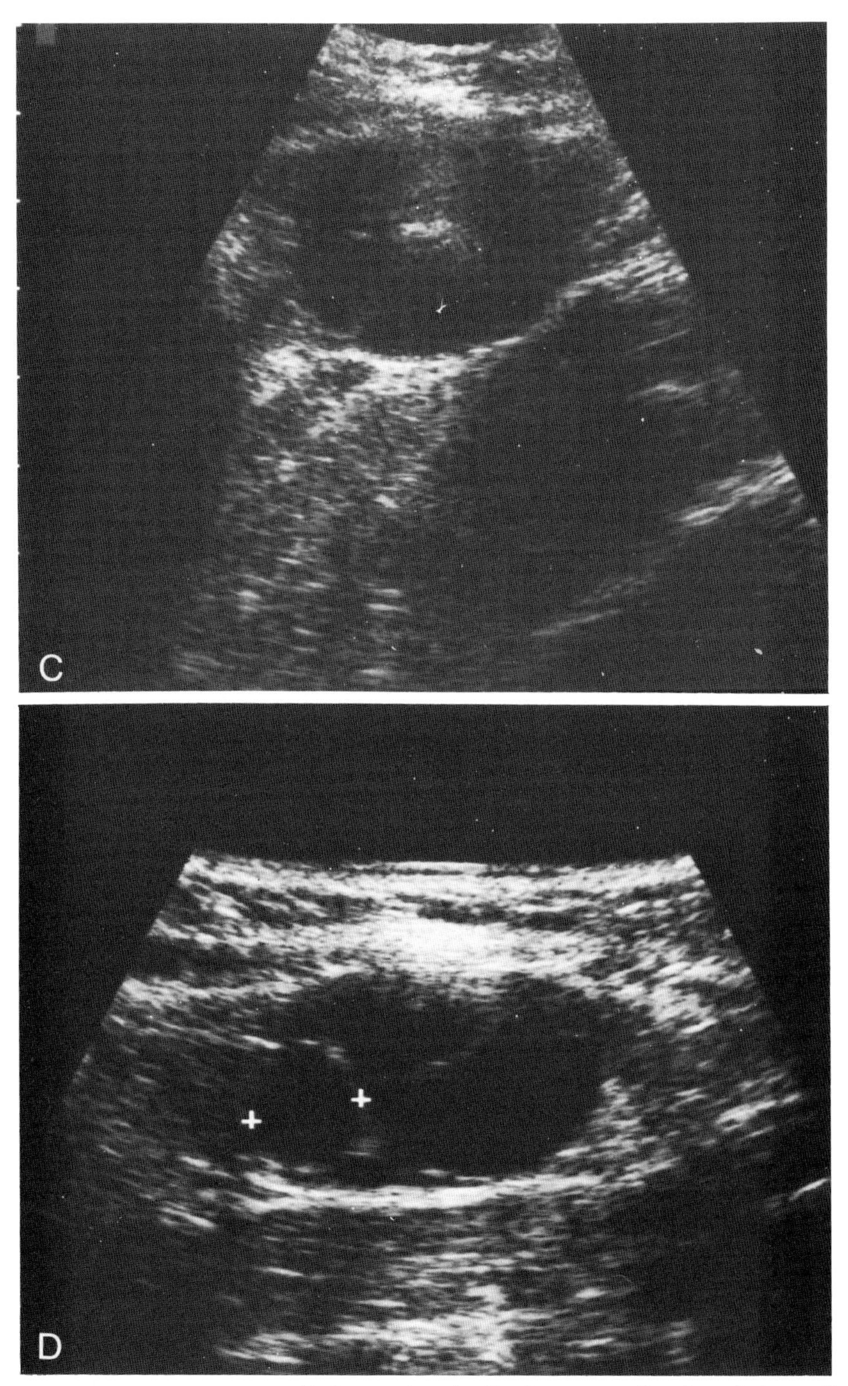

Figure 10.6. *C*: Real-time sonogram of left ovary located in close apposition to anterior peritoneal wall. *D*: Same patient as in Figure 10.6C using small parts scanner with water path offset. The follicular dimensions (*between* +'s) are much better depicted.

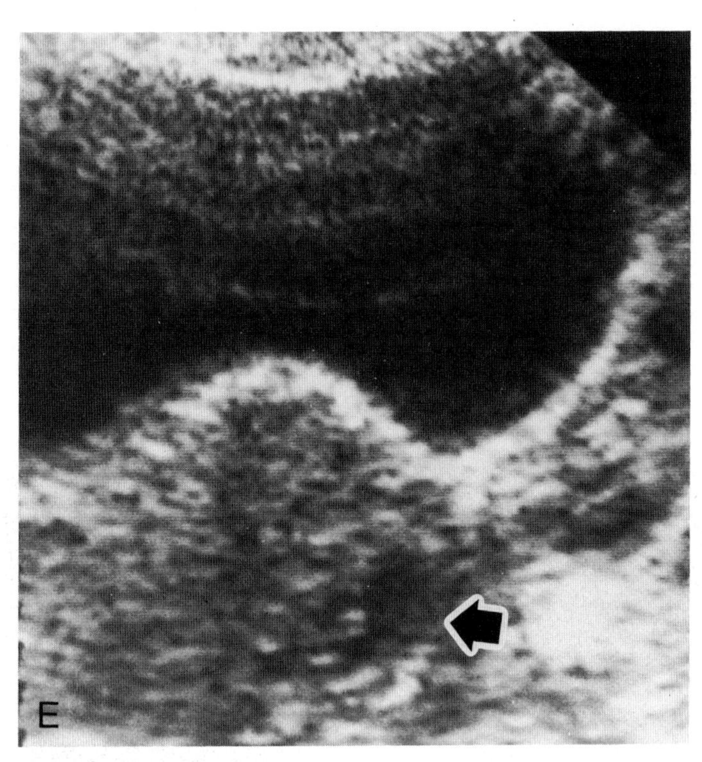

Figure 10.6. *E*: Transverse sonogram of patient who had undergone ovarian suspension. Difficult to depict the ovary (*arrow*), which now lies along the posterior aspect of the uterine fundus.

ately well delineated transabdominally can be better imaged using a transducer placed within the vagina. Similarly, ovaries that are unusually anterior in location may be depicted better by superficial organ scanners or imaging through a water bath such as a bag containing saline placed on the patient's abdomen, which accomplishes placement of the structure of interest within the focal place of the transducer (Figs. 10.6C,D).

In general, ovaries in postpubescent, premenopausal women appear as almond-shaped structures measuring approximately 3 cm in greatest axis by 2 × 2 cm (Figs. 10.1B,C). Anechoic areas as small as 2–3 mm can be seen within the normal ovary, most likely representing immature or atretric follicles. Occasionally, the vessels supplying the ovaries coursing in the infundibulopelvic ligament can be imaged and recognized by their typical course and pulsation.

During a normal cycle, sonography can begin to detect a maturing follicle in the 5–6 mm range (Fig. 10.3). Follicles tend to increase in average dimension 2 mm per day and at full maturity usually measure between 18 and 25 mm in average dimension (Figs. 10.3A,B,C) (6). In most individuals, even though more than one follicle may begin to enlarge, only one follicle will become fully mature and dominant. The size of the mature follicle seems to be constant within a particular individual, but varies from 18 to 25 mm in average dimension in different women (6).

The oocyte, which measures only 0.1 mm, cannot be readily depicted by sonography. However, occasionally the currently used scanners can resolve the cumulus oophorus, the cluster of granulosa cells that contains the oocyte, as a 1-mm mural protrusion (8).

By depicting the size of the follicles, sonography is indirectly helpful in assessing the maturity of the oocyte contained within them. However, sonography is often unable to differentiate a simple cyst with an atretic or absent oocyte from a developing follicle.

There do not appear to be any reliable sonographic signs of imminent ovulation. In one series, a subtle hypoechoic band occurring between the granulosa and theca cell layers was seen as an indicator of imminent ovulation (Fig. 10.5B) (9). In addition, the granulosa cell layer becomes folded upon itself or crenated immediately prior to ovulation. Intrafollicular echoes have also been observed immediately prior to ovulation (Figs. 10.5C,G) (8). However, one should be careful not to mistake spurious internal echoes arising from excessive gain settings for true intrafollicular echoes.

Immediately following ovulation, crenation of the follicular wall becomes even more pronounced (Fig. 10.5G). Over the next few days after ovulation, the follicle itself decreases in size and internal echoes can usually be observed within the center of the follicle (Figs. 10.3H,I). As the corpus luteum forms, the internal area of the follicle becomes isoechoic to the surrounding ovarian parenchyma (Figs. 10.3H,I). In some patients, the corpus luteum will be similar in sonographic appearance to the preovulatory follicle, demonstrating a smooth border (10). In others, the preovulatory follicle may totally collapse within a few days after ovulation to reappear in the late secretory phase as a smooth-walled cyst.

Follicular cysts are typically greater than 2.5 cm in average dimension. Follicular cysts can sometimes be differentiated from luteal physiologic cysts by the presence of a thicker and more irregular border in luteal cysts (Fig. 10.7E,I,N).

With the improved resolution afforded by real-time scanners, the endometrium of the uterus can consistently be depicted throughout the cycle. During menses, the endometrium appears as a thin, broken, echogenic, linear interface. During the proliferative phase the endometrium thickens and appears slightly hypoechoic compared to the surrounding myometrium. During the late proliferative and early secretory phases, the endometrium becomes more echogenic and typically measures between 3 and 5 mm average single-layer thickness. A hypoechoic halo can be seen surrounding the more echogenic endometrium that represents the vascular and compact inner layer of the myometrium. Immediately following ovulation, an internal hypoechoic layer within the echogenic endometrium can be observed. This probably results from edema in the compactum stratum that occurs immediately after ovulation (11). The endometrium achieves its thickest and most echogenic texture in the midsecretory phase and measures up to 6–7 mm in average thickness.

Small amounts of intraperitoneal fluid may be present throughout the menstrual cycle. The fluid collecting in the cul-de-sac region is most frequently seen immediately after ovulation (Fig. 10.5F) (12, 13). In patients with multiple adhesions or previous surgery, the fluid can be loculated around bowel loops or superior to the uterine fundus in the anterior cul-de-sac and therefore may be difficult to image by sonography (Fig. 10.5F) (12).

Infertility Evaluation

Sonography has an important role in the evaluation of patients with gynecologic infertility. Besides monitoring follicle development during normal or induced ovulation, sonography can be used to detect and evaluate uterine malformations, tubal disease, endometriosis, and polycystic ovary disease, and to suggest the possibility of abnormal folliculogenesis, specifically the luteinized unruptured follicle syndrome (14–17). Other major applications include prospective detection of those patients who are at risk for ovarian hyperstimulation and the guidance of transvesicle, follicle, or transvaginal aspiration.

Sonography has a role secondary to hysterosalpingography in the evaluation and detection of uterine malformations (14). However, unsuspected uterine malformations may be seen on routine sonography.

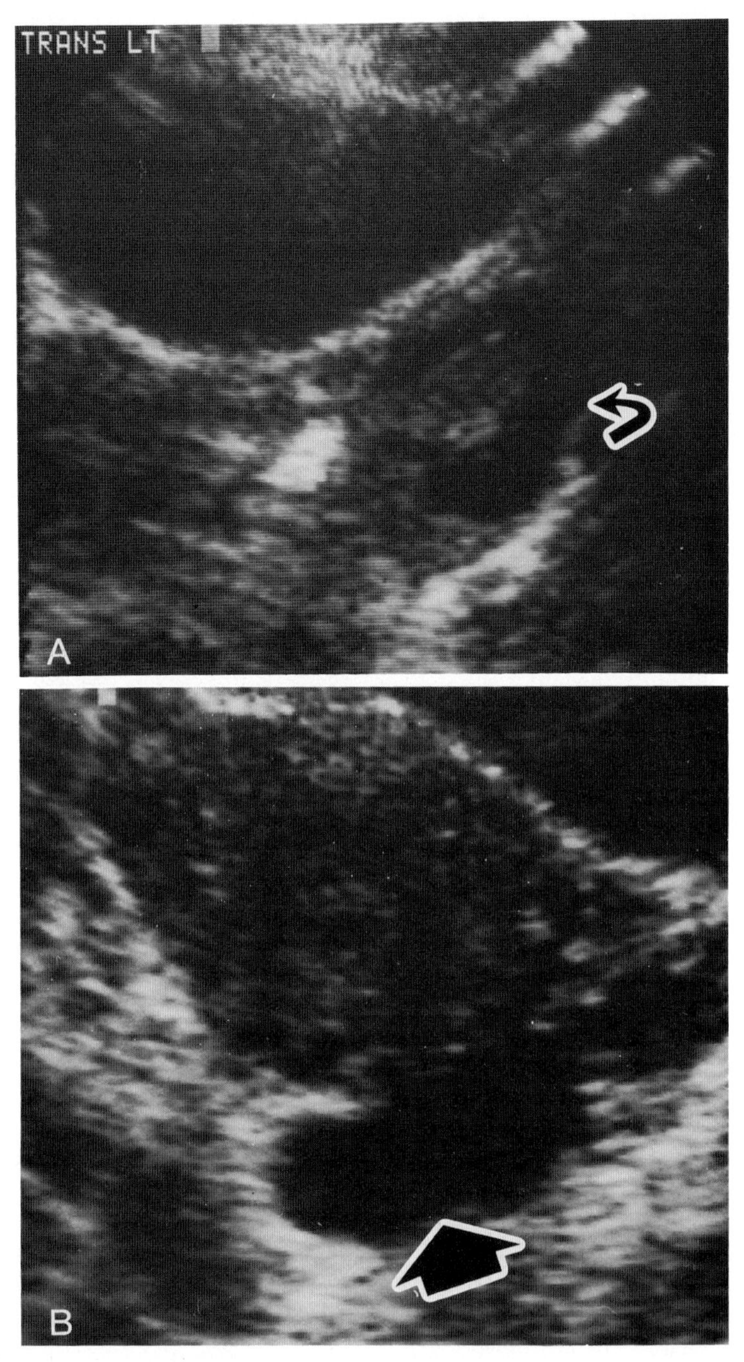

Figure 10.7. Sonographic appearance of adnexal masses that may be encountered in infertility patients. *A*: Magnified transverse sonogram of small hydrosalpinx (*arrow*) that surrounds the left ovary. *B*: Magnified transverse sonogram of larger hydrosalpinx (*arrow*) having a fusiform configuration.

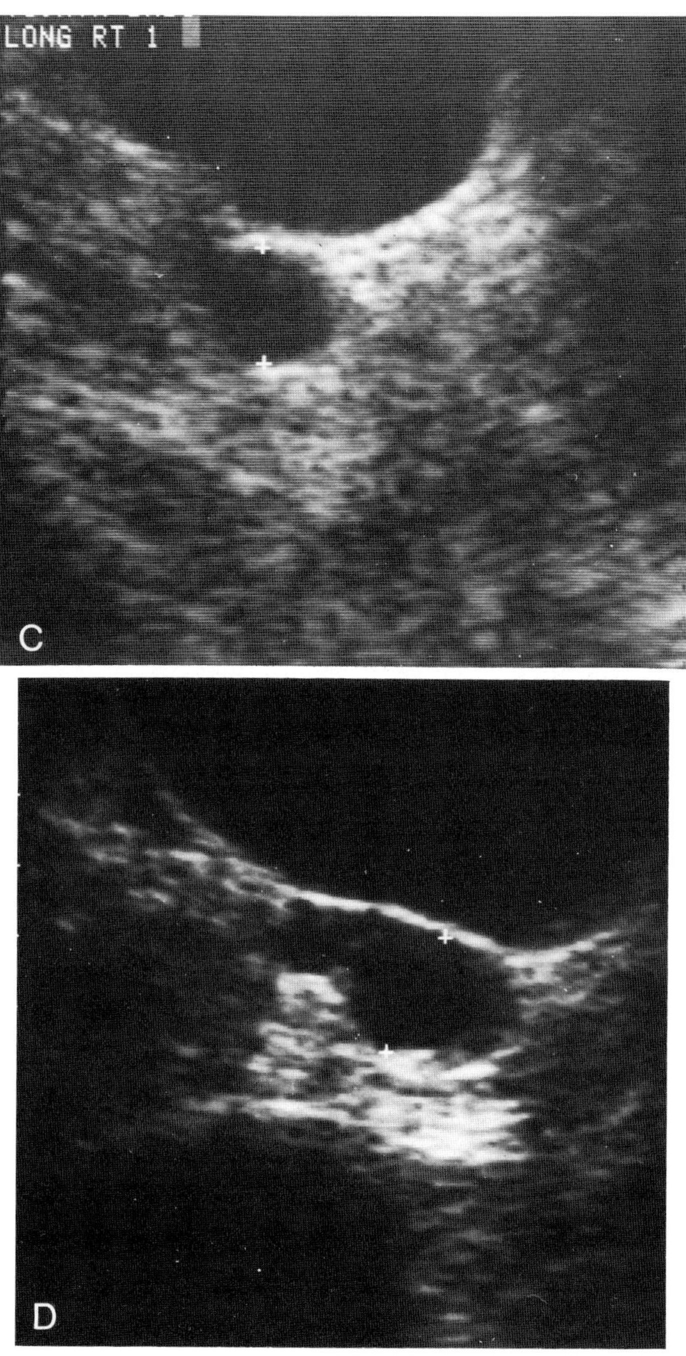

Figure 10.7. *C*: Longitudinal sonogram showing round cystic adnexal structure (*between* +'s). *D*: Same patient as in Figure 10.7C, 5 days later showing apparent enlargement of the cystic structure. At laparoscopy this was found to represent a hydrosalpinx.

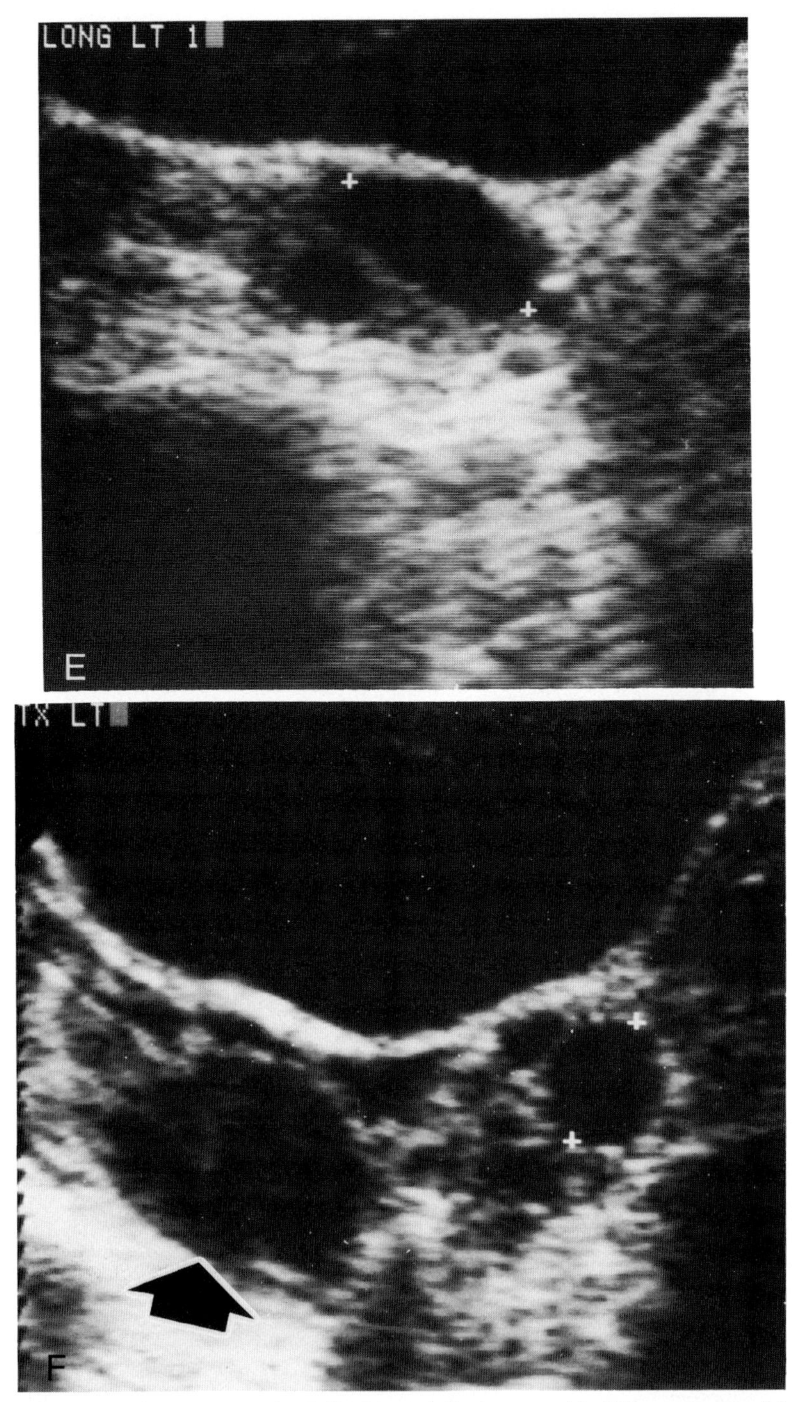

Figure 10.7. *E*: Magnified sonogram of a follicular cyst (*between* + 's). At laparoscopy this cyst did not contain an oocyte. *F*: Transverse magnified sonogram of peritoneal inclusion cyst (*arrow*) that was adjacent to the follicle-containing (*between* + 's) left ovary.

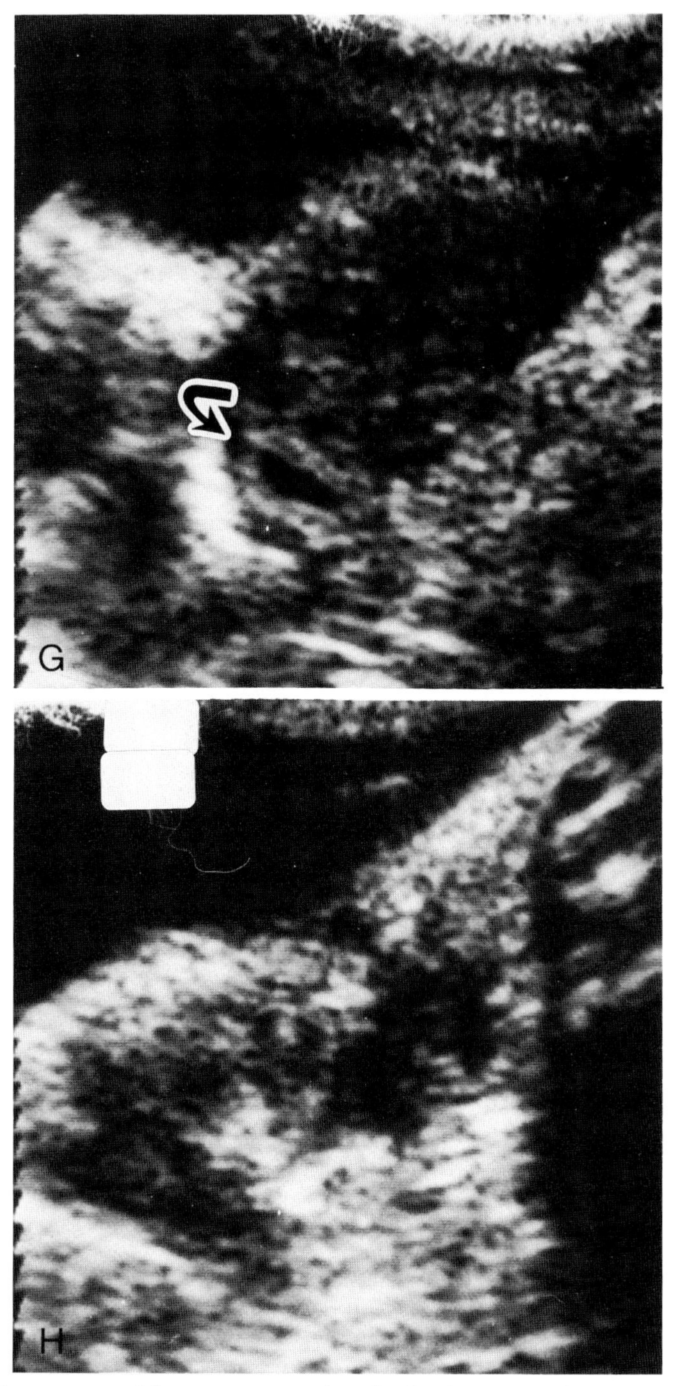

Figure 10.7. *G*: Transverse real-time sonogram of cystic structure (*curved arrow*) posterior to uterus. *H*: Same patient as in Figure 10.7F 1 sec later showing peristalsis of bowel.

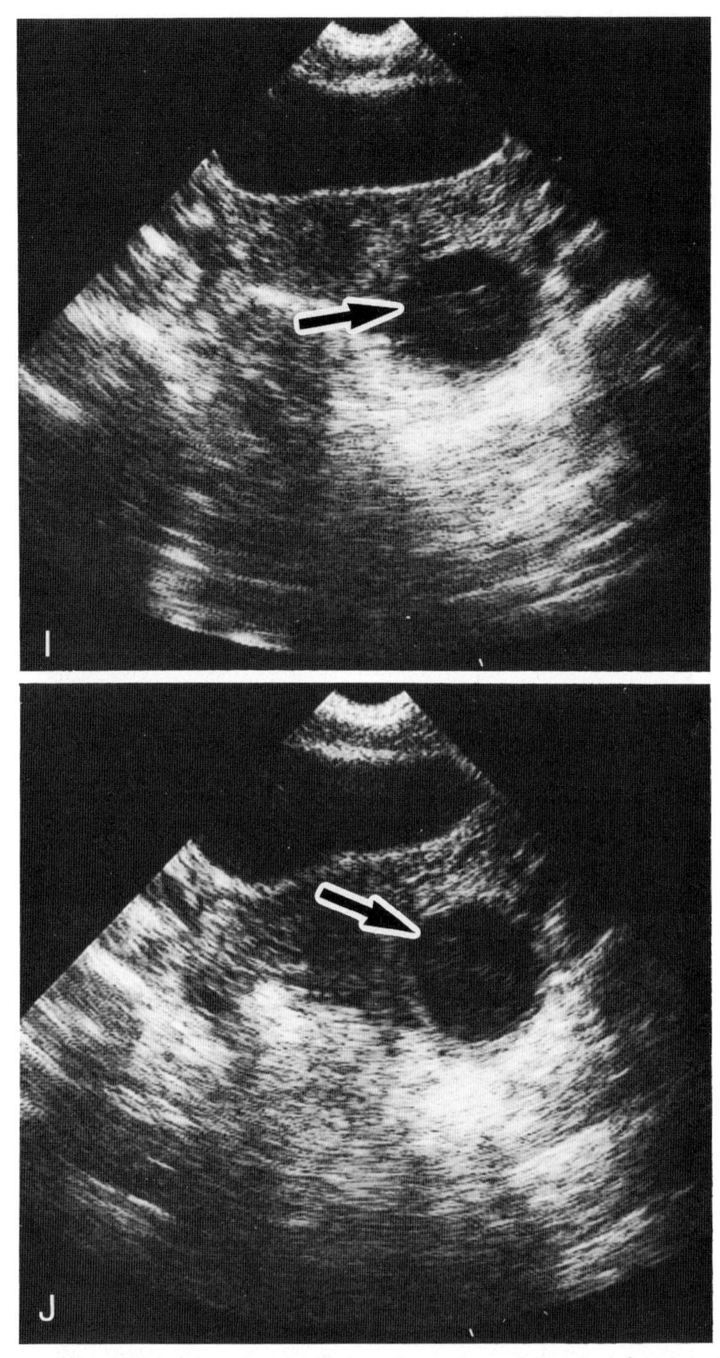

Figure 10.7. *I*: Transverse sonogram of cyst with layering echoes (*arrow*). *J*: Same patient as in Figure 10.7I in the left-side-up position showing change in orientation of internal echoes (*arrow*). This represented a hemorrhagic corpus luteum cyst that contained a layer of clotted blood.

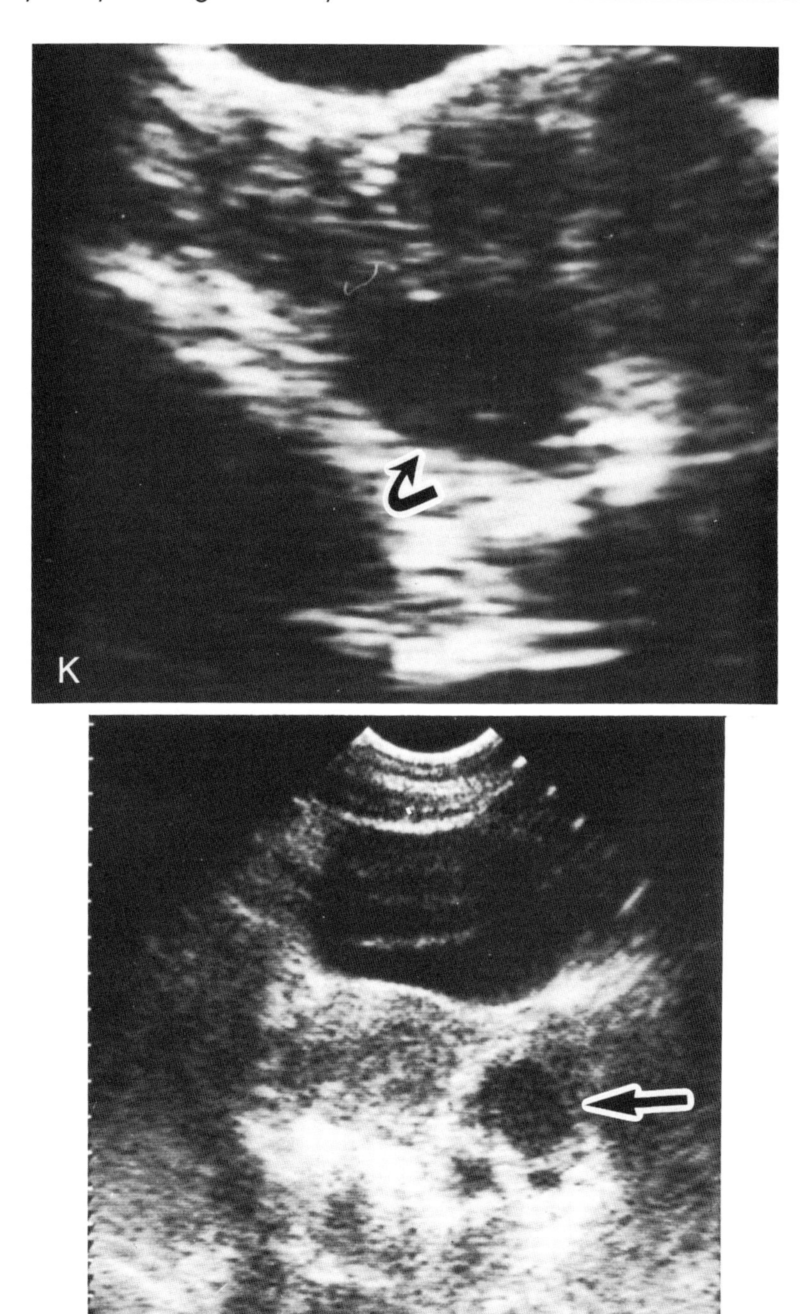

Figure 10.7. *K*: Tuboovarian abscess (*curved arrow*) simulating appearance of a follicle-containing ovary. *L*: Large endometrioma (*arrow*) with smaller endometriomas posteriorly. Endometriomas typically have irregular borders and internal echoes.

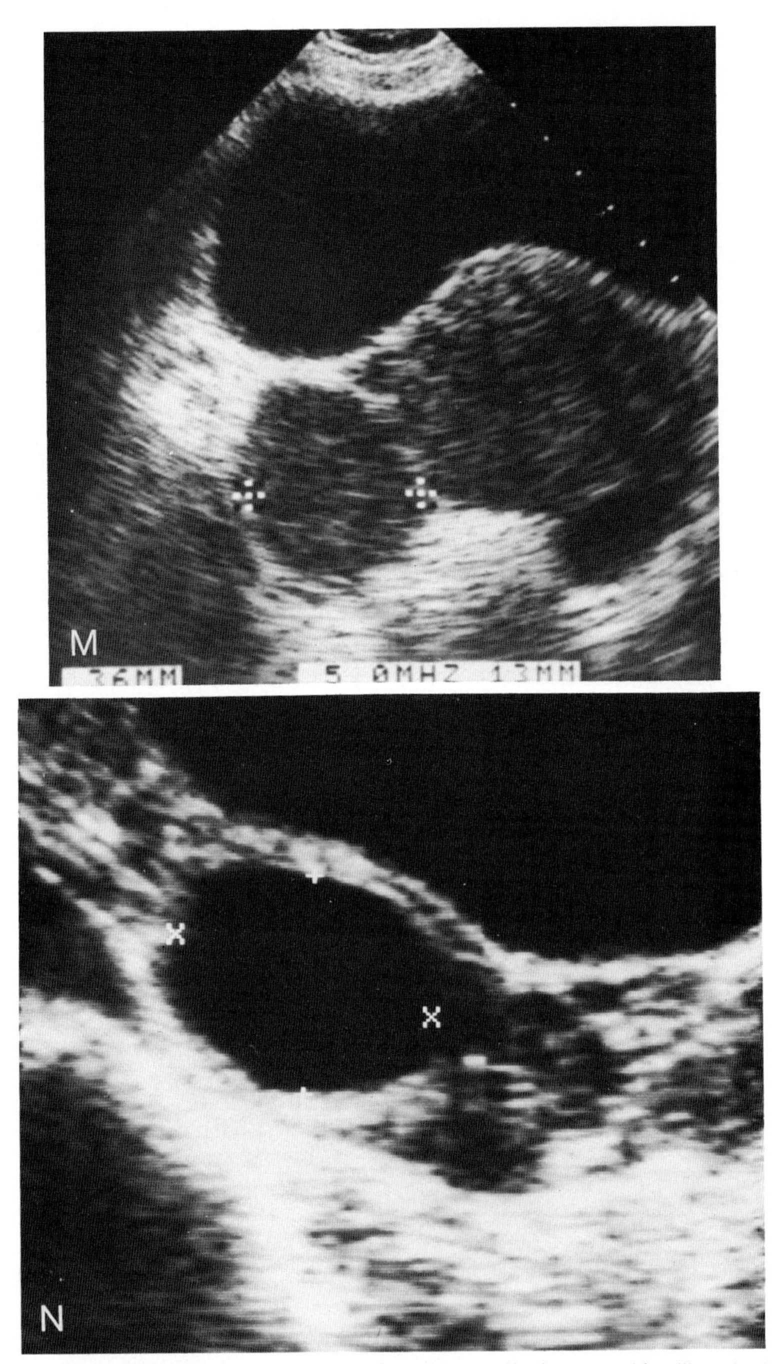

Figure 10.7. *M*: Patient with a large (3.6-cm) endometrioma (*between* +'s) with a texture that was similar to that arising from the uterus. *N*: Luteal cyst (*between* +'s) with relatively thick walls.

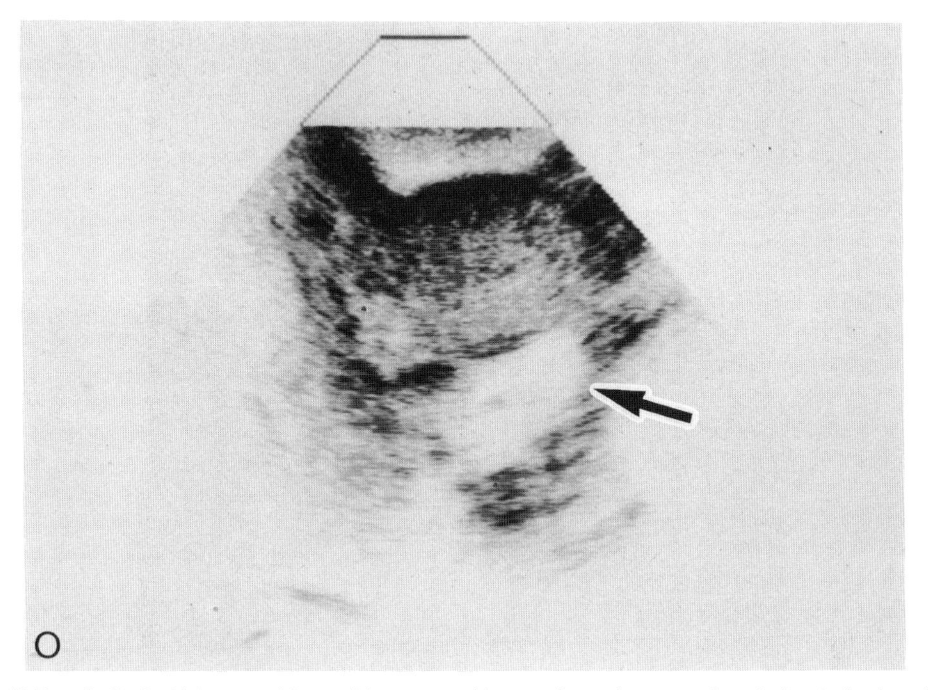

Figure 10.7. *O*: Cyst of Morgagni (*arrow*) in reversed image format, appearing similar to hydrosalpinx.

Bicornuate uteri appear as a binodular outline of the uterus when imaged in a transverse plane (Fig. 10.8B). Patients whose mothers took diethylstilbestrol (DES) during their pregnancies may have a shortened uterus that is wider in the transverse dimension, a so-called T-shaped uterus, named after the shape of the lumen in hysterosalpingography. Detailed evaluation of uterine malformations, however, must be obtained using hysterosalpingography and direct intraperitoneal visualization of the uterine fundus (18). Clinically unsuspected or nonpalpable fibroids can also be detected with sonography (Fig. 10.8C).

Sonography is helpful in detecting a hydrosalpinx and differentiating this condition from tuboovarian abscesses (Figs. 10.7A,J). Occasionally, a hydrosalpinx will have an appearance similar to a mature follicle and can actually enlarge during ovulation induction (Fig. 10.7C,D) (19). This may be related to fluid secreted or transudated into the lumen under the influence of ovarian hormones (20).

Sonography has been recently utilized in a role similar to hysterosalpingography for evaluation of tubal patency (15). In this procedure, a highly viscous fluid such as 32% dextran (Hyskon) is injected in a manner similar to that used for hysterosalpingography. High molecular weight dextran causes mechanical lavage of the tube, which may dislodge mucus plugs or lyse peritubal adhesions. Saline may be used instead of Hyskon. This material can be injected into the uterus using the same uterine injector used for hysterosalpingograms. In general, tubal patency can be implied by the appearance of intraperitoneal fluid after injection. In patients with both tubes, it may be difficult to ascertain whether one or both tubes are patent. In those patients who have had one of their tubes removed, sonosalpingography may have more predictive value than when both tubes are present. Tubal occlusion produces no intraperitoneal spillage, only distension of the uterine lumen with fluid.

Although sonosalpingography seems to be sensitive for detection of bilateral occlusion or tubal patency, it lacks the anatomic definition provided by hysterosalpingography. The advantages of sonosalpingography include lack of irradiation and detection of masses within the pelvis that may

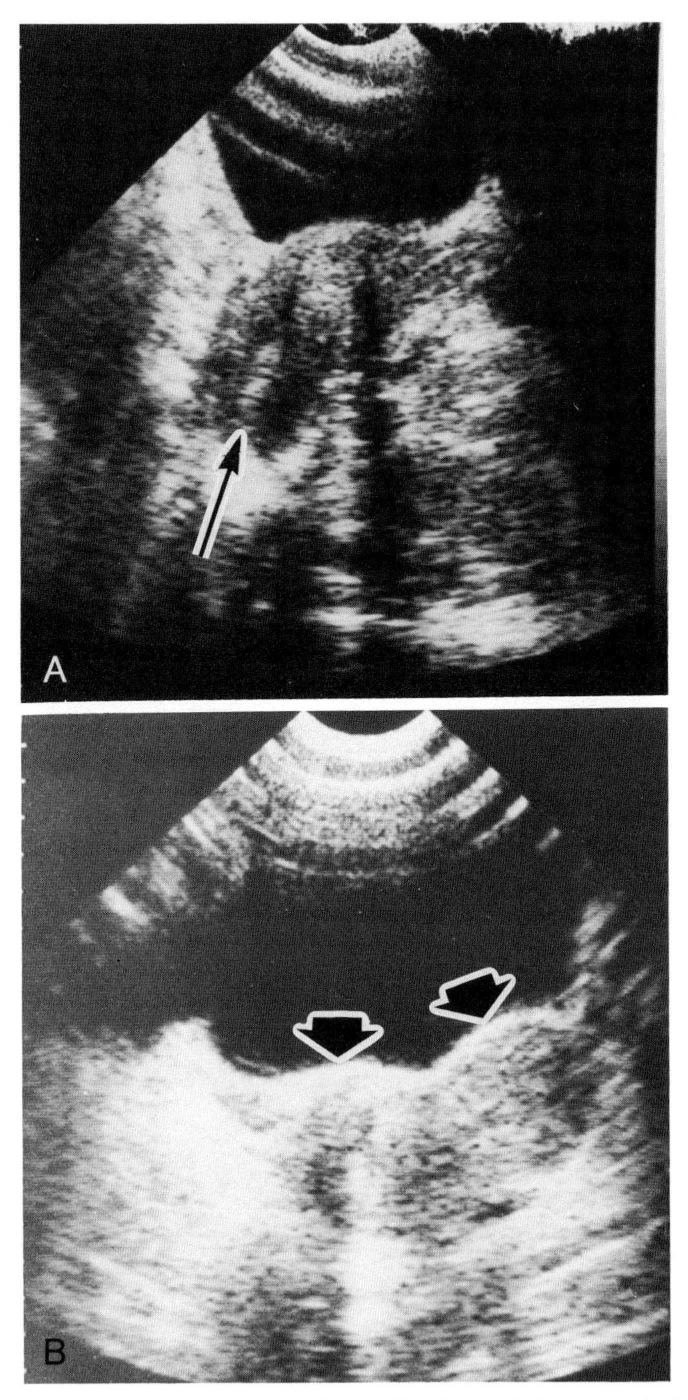

Figure 10.8. Anatomic variants and other disorders of the uterus. *A*: Severely retroflexed uterus within the fundus (*arrow*) directed posteriorly. *B*: Bicornuate uterus showing binodular outline (*arrows*).

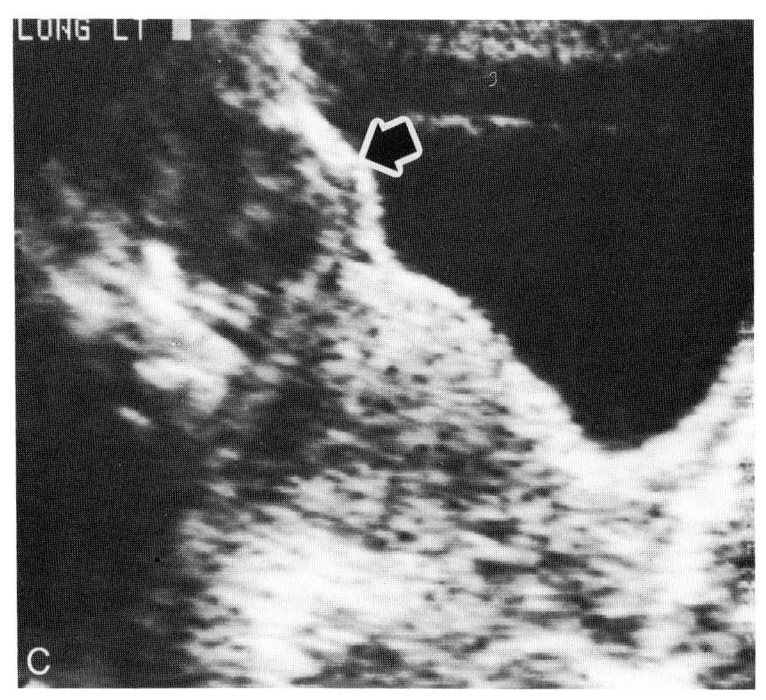

Figure 10.8. *C*: Subserosal pedunculated fibroid (*arrow*) arising from the uterine fundus in an infertile patient. This mass could not be palpated.

not be detected radiographically (15). Sonosalpingography can be performed in the gynecologist's office, whereas hysterosalpingography needs to be performed in a radiologic facility with fluoroscopy.

Sonography can be helpful for the evaluation and follow-up of patients with endometriosis. In general, small (<1 cm) endometriotic implants cannot be detected on sonography. However, when they are 1 cm or greater, they appear as hypoechoic to isoechoic masses, usually with an irregular border related to the fibrotic reaction that is incited around these masses (Fig. 10.7K). Sonography is helpful in follow-up of patients on medical treatment for endometriosis in that enlargement or regression of endometriomas can be documented.

Enlarged, rounded ovaries are typically encountered in patients with polycystic ovary (PCO) disease (Fig. 10.2B) (16). However, up to one-third of patients with PCO will have normal-size ovaries on sonography. In polycystic ovaries, numerous immature follicles can be seen along the periphery. Ovarian volumes in PCO are usually over 14 cm^3, whereas normal ovaries average 10 cm^3 (16).

One subtle cause of infertility that has been evaluated by sonography is the luteinized unruptured follicle syndrome (17). In this disorder, it is thought that the oocyte is trapped within the follicle and that actual ovulation of the follicle does not occur. Sonographically, this disorder is suggested by a failure to show the typical signs associated with ovulation, such as a decreased follicular volume, crenation, and the interval presence of intraperitoneal fluid after the expected time of ovulation (Fig. 10.9D,E). Although there is controversy concerning the association of this syndrome with infertility, it is clear that sonography is helpful in detecting abnormal folliculogenesis (18, 19). Abnormal folliculogenesis can be detected on serial sonograms as a failure to demonstrate mature follicles. This can be correlated with a lack of estradiol production and may be a sign of premature ovarian failure.

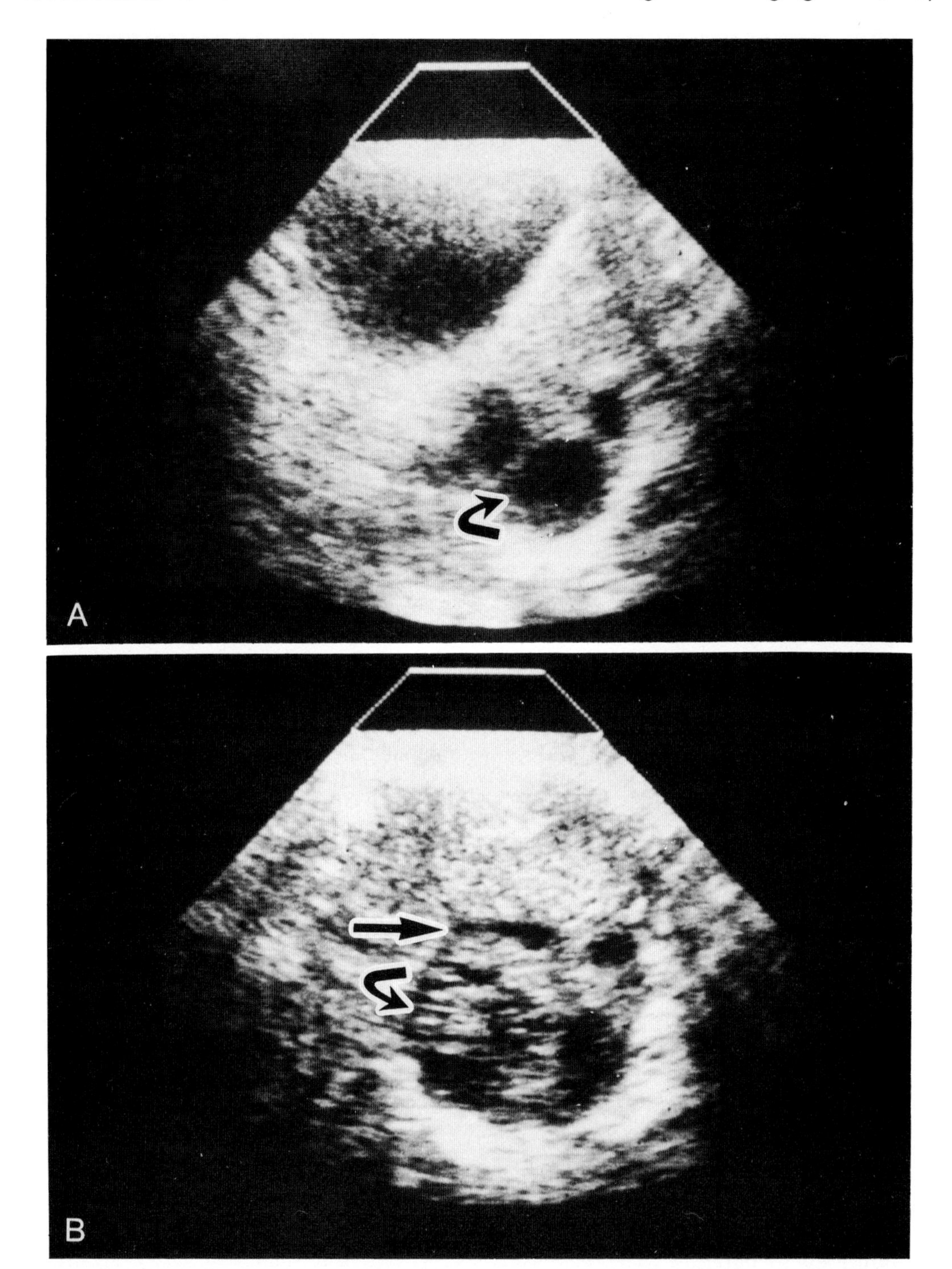

Figure 10.9. Sonographic signs of ovulation. *A*: Transverse real-time sonogram demonstrating a mature follicle (*curved arrow*) inferior to an immature one. *B*: Next day, prior to laparoscopy, the dominant follicle shows internal echoes (*curved arrow*). In addition, there is a sliver of fluid (*straight arrow*) between the ovary and the uterus.

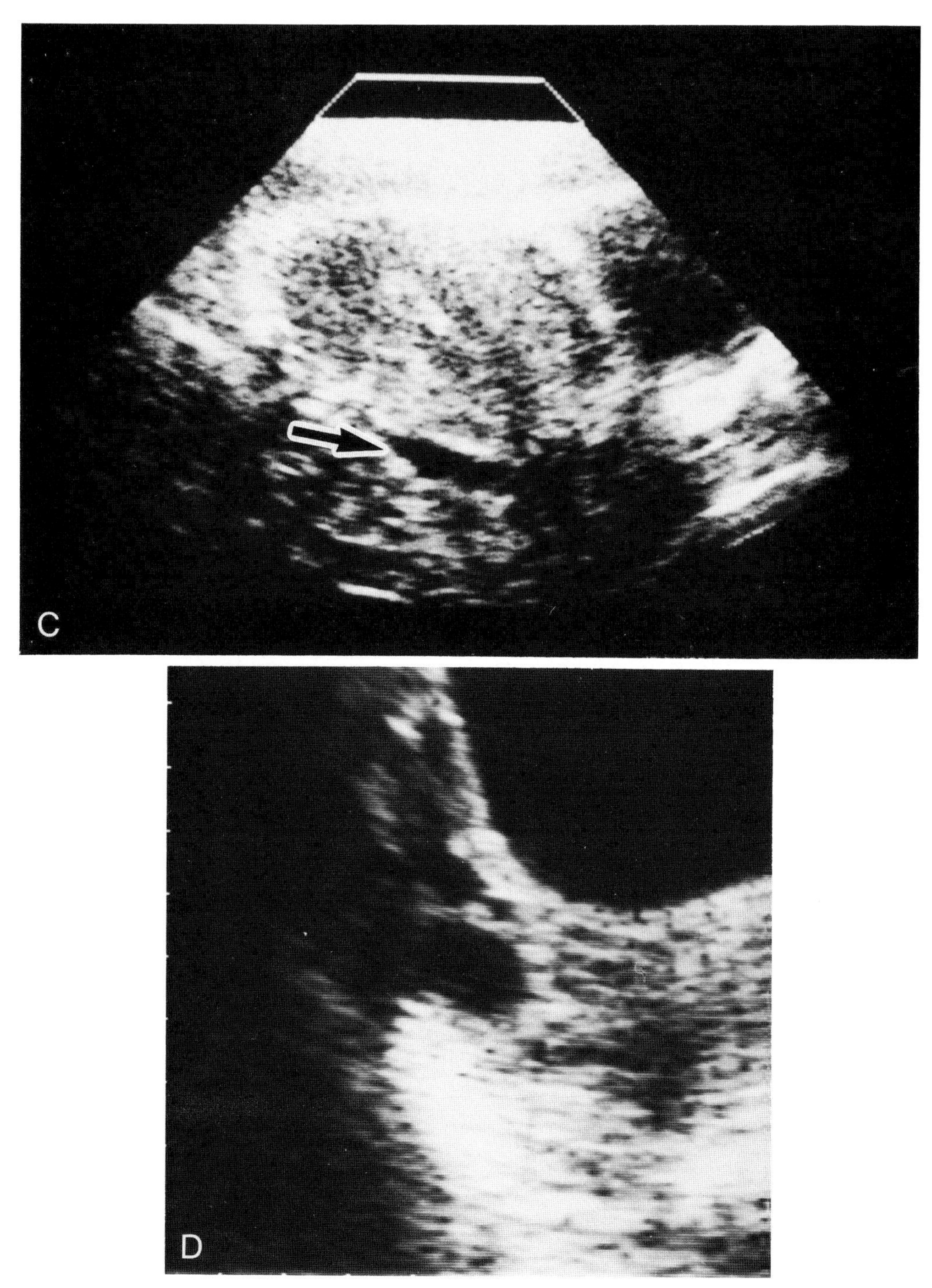

Figure 10.9. *C*: Long axis view again demonstrating sliver of fluid between the anterior surface of ovary and uterus resulting from ovulation (*arrow*). *D*: Appearance of mature follicle in spontaneous cycle 1 day prior to expected time of ovulation.

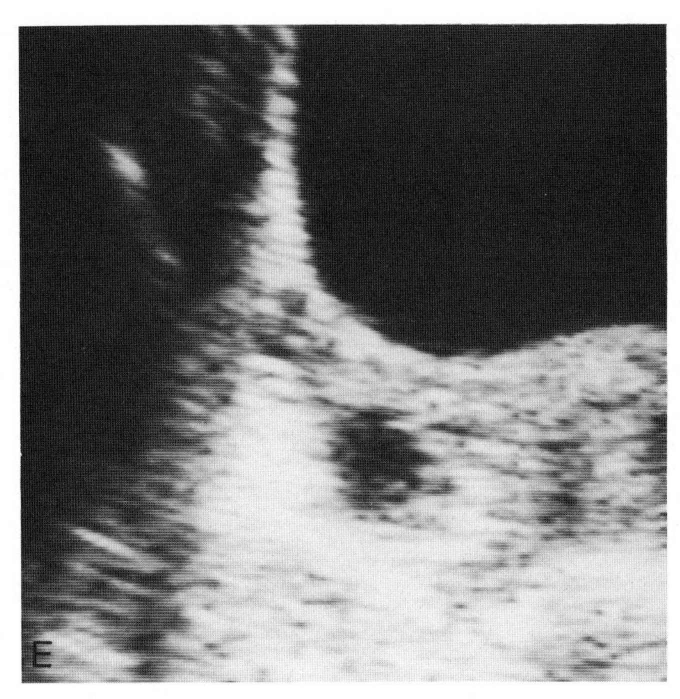

Figure 10.9. *E*: Two days later, the follicle has decreased in size, developed crenated borders but no intraperitoneal fluid. This may represent an example of the luteinized unruptured follicle syndrome.

Follicular Monitoring

Sonographic monitoring of follicular size is helpful in ovulation induction, timed artificial insemination, and in vitro fertilization/embryo transfer protocols. The size, number, and location of maturing follicles can be ascertained (Fig. 10.9A). The anatomic information obtained on sonography can be correlated with other physiologic parameters such as estradiol (E_2) and luteinizing hormone (LH) levels.

Disorders of ovulation account for 15–25% of infertility cases (19). Depending on the cause, patients can be treated with a variety of ovulation induction medications. The aim of this therapy is to induce follicular maturation of one or more follicles while avoiding hyperstimulation.

The two major medications used are clomiphene citrate (CC) and human menopausal gonadotropin (hMG). Clomiphene citrate stimulates endogenous follicle-stimulating hormone (FSH) and LH production and, when administered early in the cycle, induces a cohort of follicles to develop (Figs. 10.10A,B). The process of selection and dominance of follicles may be overriden, and multiple follicles often develop. Human chorionic gonadotropin (hCG) is given with CC to induce final meiosis and ovulation. Treatment with exogenous gonadotropins does not require an intact hypothalamus or pituitary and stimulates the ovaries directly, usually overriding the process of selection and inducing the development of multiple follicles (Figs. 10.10C,D). Pure FSH is occasionally used when conventional regimens are unsuccessful. Typically FSH is associated with production of multiple follicles (Fig. 10.10G,H).

The sonographic measure of follicle size must be correlated with estradiol values and other factors such as cervical mucus score to assess the timing of ovulation (21–25). The correlation of estradiol value with follicle size is imprecise at best since serum (E_2) concentration represents the combined estrogen output from follicles in both ovaries as well as a small amount from peripheral conversion of androgen and estrogen precursors. In general, an E_2 value of 400

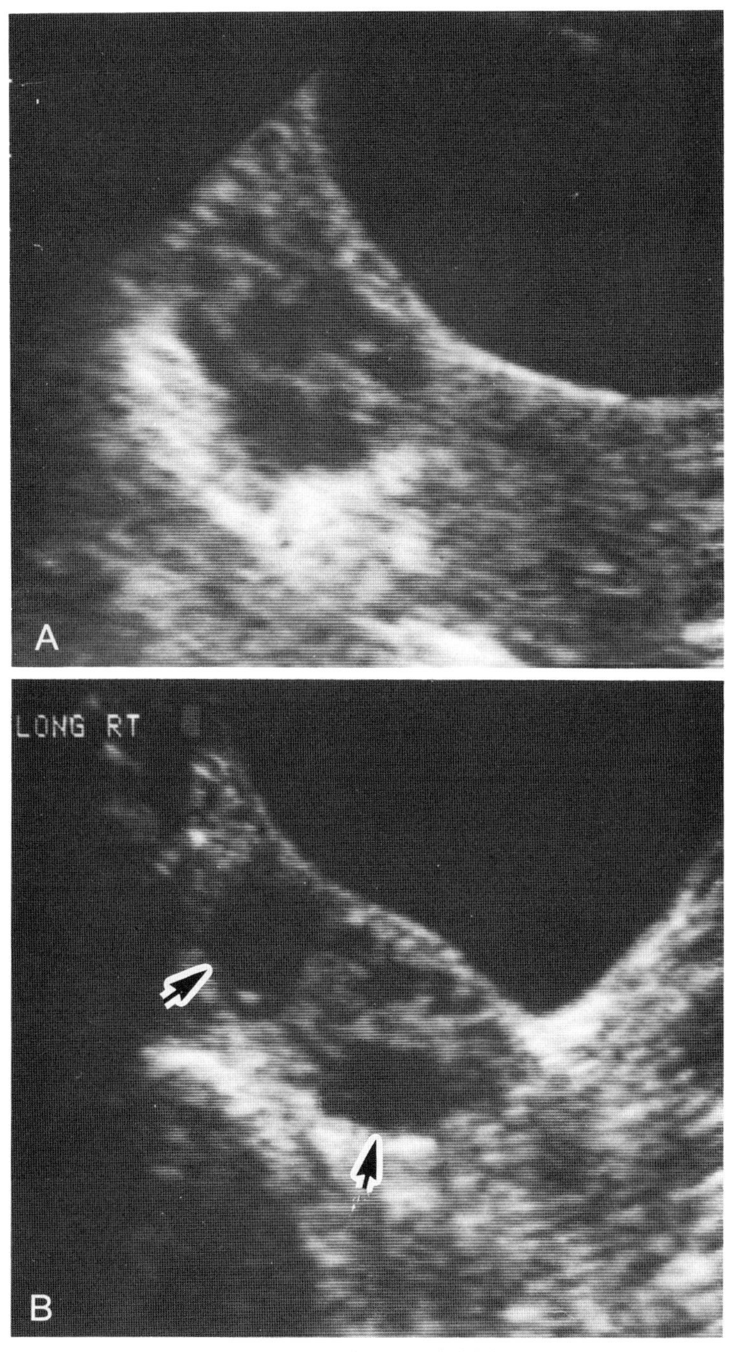

Figure 10.10. Follicular monitoring in ovulation induction. *A*: Multiple medium-size follicles on day 11 of Clomid cycle. *B*: Same patient 3 days later demonstrating maturation of two follicles (*arrowheads*).

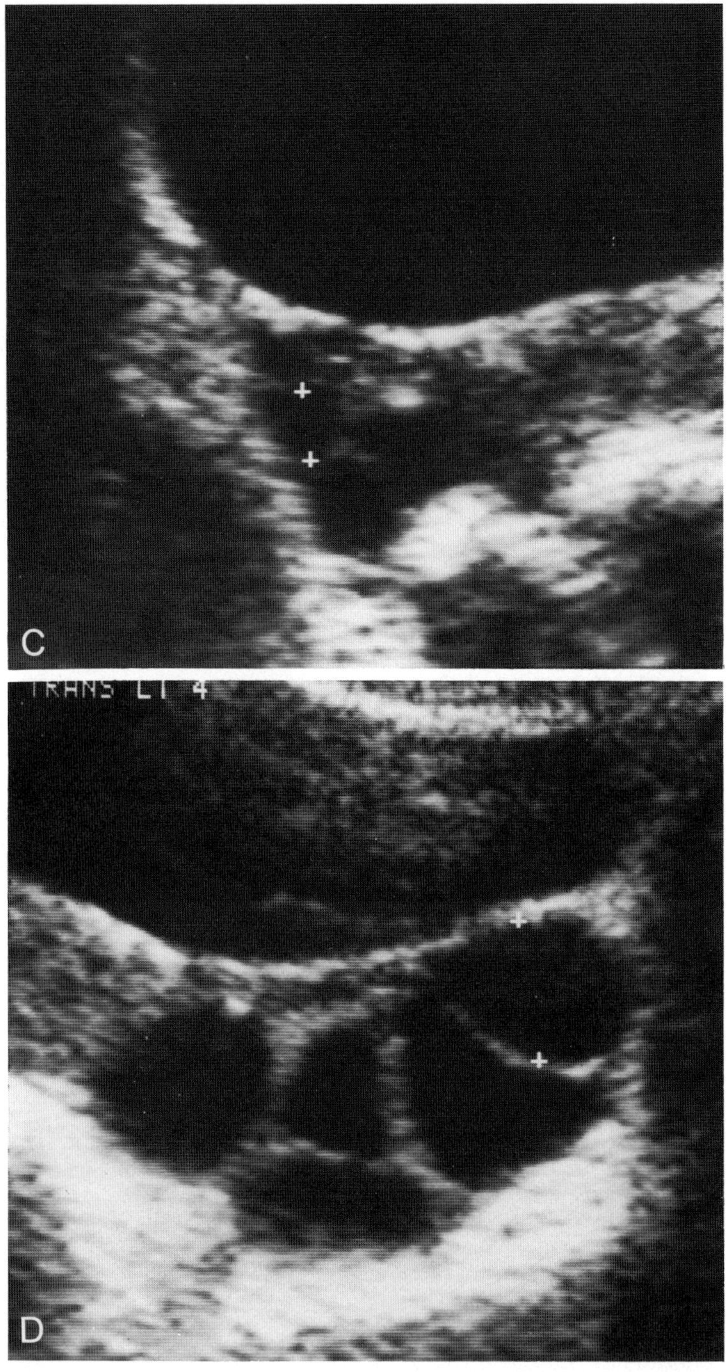

Figure 10.10. *C*: hMG stimulation producing multiple immature follicles. *D*: Same patient as in Figure 10.10C 4 days later showing development of several mature follicles.

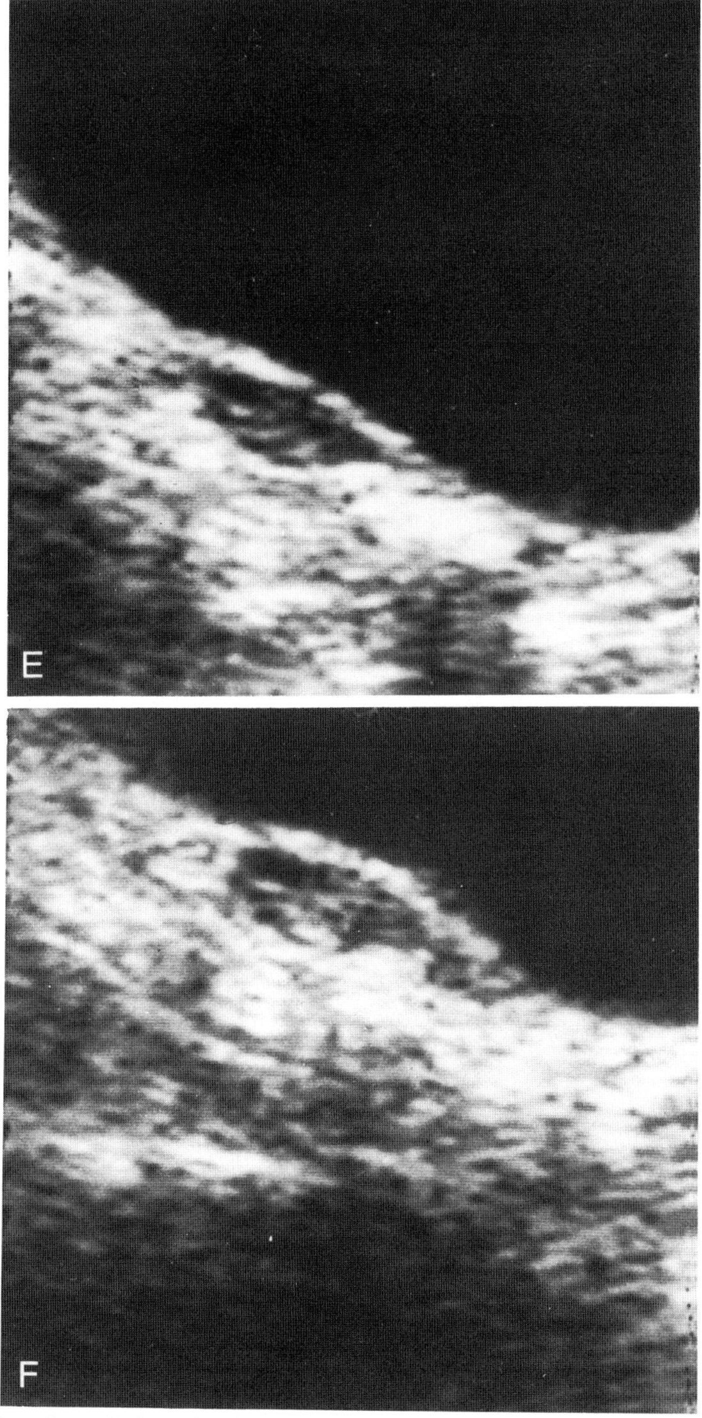

Figure 10.10. *E*: Despite multiple doses of hMG, this patient failed to develop mature follicles. *F*: Same patient as in Figure 10.10C 2 weeks later.

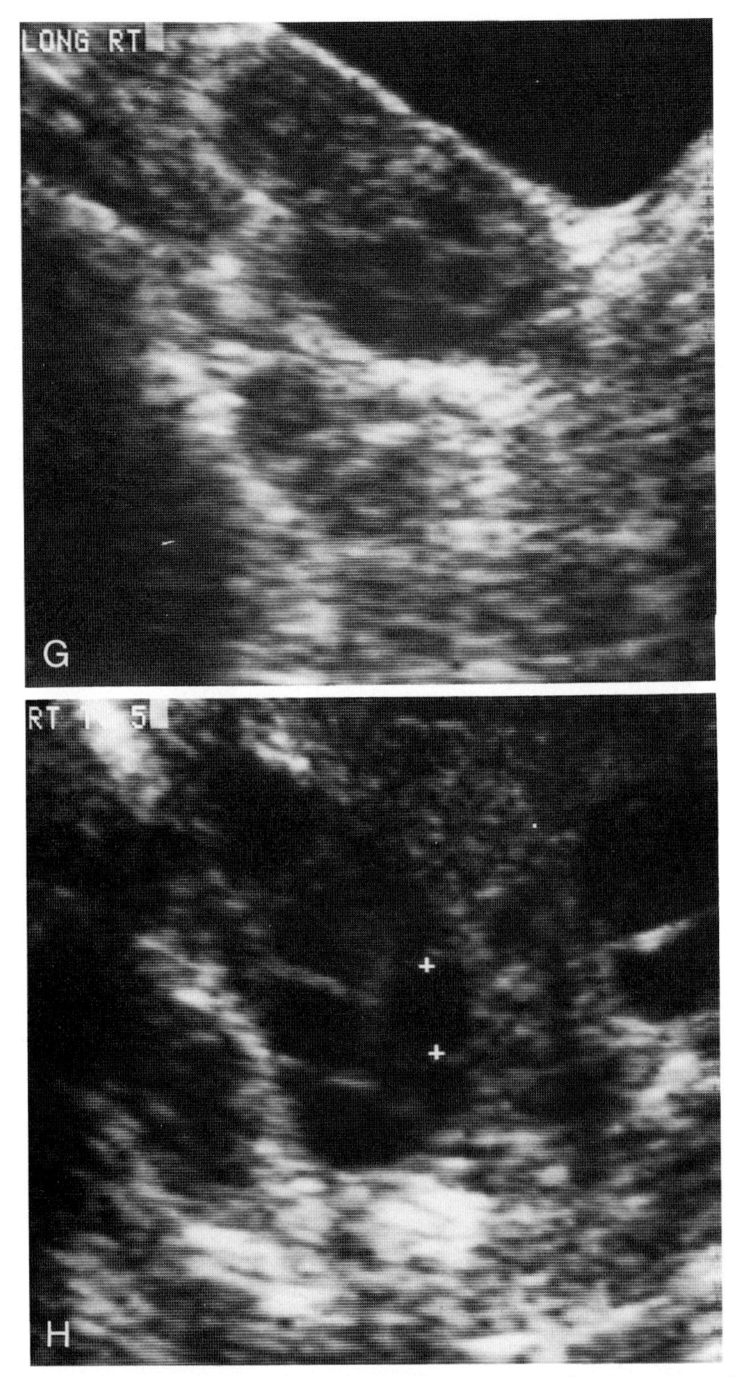

Figure 10.10. *G*: Multiple mature follicles in a patient undergoing ovulation induction with FSH. *H*: Same patient as in Figure 10.10G, 4 days later showing enlargement of several follicles.

pg/ml can be associated with production from one dominant follicle (26).

Sonography is particularly useful in the timing of hCG administration in CC or hMG cycles. Administration of hCG too early can cause follicular atresia and inhibition of ovulation. O'Herlihy et al. reported a 92% ovulation rate and a 70% pregnancy rate in patients on CC who received hCG when the largest follicle reached 17–18 mm maximum diameter by sonography (27).

Measurement errors can also result from improper scanning techniques (28). The follicle should be measured on at least two axes. Marked bladder distension can compress the follicles into assuming an oblong shape. It is important to image the follicle in the focal field of the transducer utilized. If the bladder is overly distended or if the fluid is located in the near field, measurement errors of 1–2 mm can be expected to occur (28). Because of the variable position of the ovary and its follicles, dependent upon several factors such as the amount of bladder distension and lack of a reliable anatomic landmark, some of the difference in measured follicle size can be artifactual rather than real. Accordingly, it is important to measure the follicle in at least two dimensions, its long and short axis. We report these two dimensions and a mean; others use volumetric measurements that seem to correlate more closely with E_2 values than linear dimensions.

The internal echogenicity of the follicle should be examined closely with proper gain settings since the presence of intrafollicular echoes may have diagnostic importance. In one study, the mature follicles that developed internal echoes were associated with a greater likelihood of achieving pregnancies. Although it is not fully known what these internal echoes represent, they may arise from granulosa cells that detach from the follicle wall immediately prior to ovulation (8).

Recent research has indicated that certain findings of the endometrium may be an indication that it is more favorable for implantation (29, 30). In general, a favorable endometrium appears isoechoic or echoic relative to the myometrium and thick, with an average dimension of approximately 4–5 mm in single-layer thickness (Fig. 10.4A). The endometrium, since it is sensitive to circulating estrogens, may be a useful parameter for assessment of timing of hCG injection. The endometrium that is out of phase or poorly developed may be detected sonographically (Fig. 10.4C) (31). However, a recent study has shown that there is no significant difference in the thickness of the endometrium in conception or nonconception hMG/hCG in vitro fertilization cycles (32).

Sonography is helpful in detection of patients who may develop ovarian hyperstimulation syndrome. This disorder is important to detect since it may be life-threatening or lead to multiple pregnancies (33). Hyperstimulation syndrome is rare with CC treatment but occurs more often with hMG/hCG treatment. It usually becomes manifest 3–7 days after hCG intramuscular administration. There are several degrees of ovarian hyperstimulation, ranging from mild enlargement of the ovary and mild weight gain to massive enlargement of the ovary (greater than 10 cm) with weight gain, electrolyte imbalance, and intraperitoneal effusions (Fig. 10.11D). In general, patients who are susceptible to ovarian hyperstimulation syndrome are those with enlarged ovaries that contain more than five immature follicles. The detection of one or two larger and presumably dominant follicles is encouraging, since hyperstimulation rarely occurs in these patients despite the presence of multiple smaller follicles.

Sonography might be utilized for guidance of embryo transfer. However, performance of sonography during embryo transfer may be difficult in these patients since they may be in the knee–chest position and do not routinely have maximal bladder distension.

Sonographic Guidance for Follicular Aspiration

Sonography has been recently utilized for directing follicular aspiration (34, 35). Sonographic guidance, with or without a

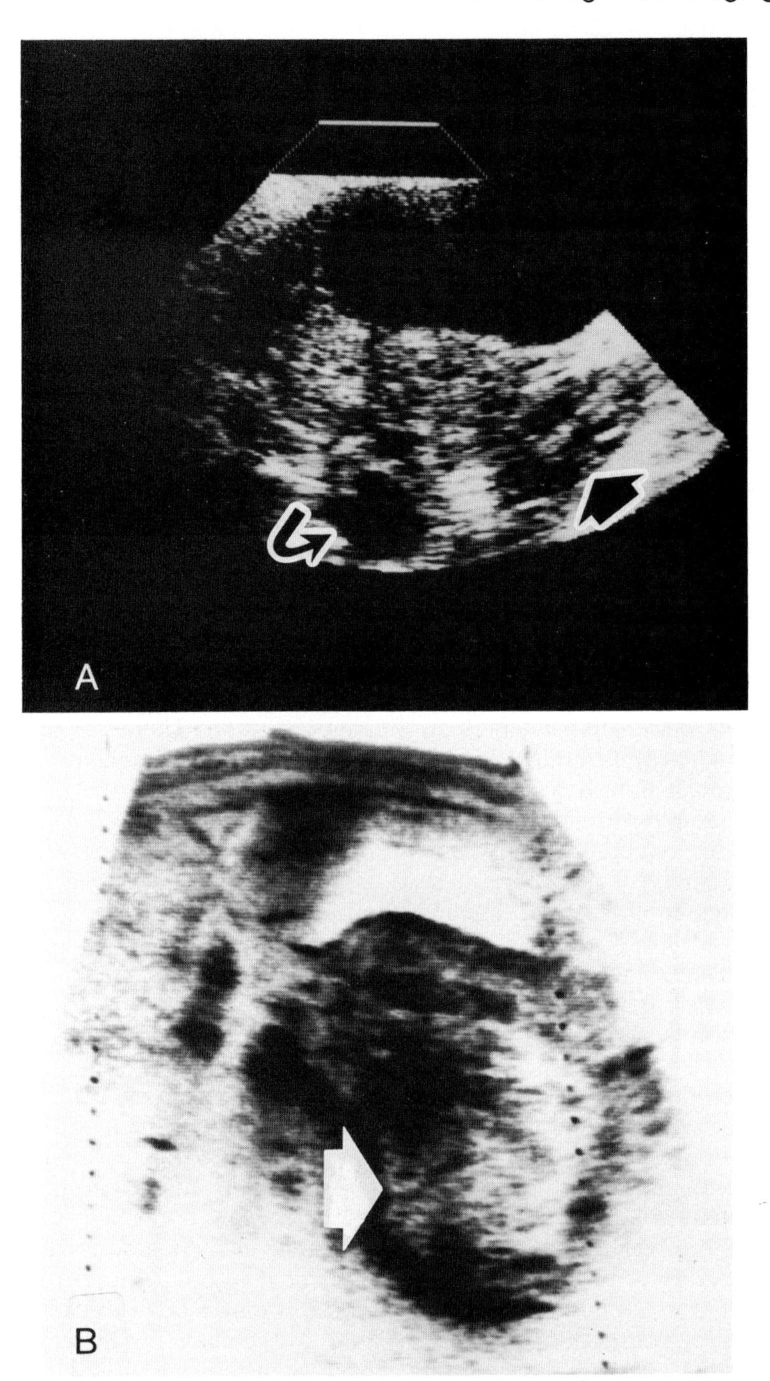

Figure 10.11. Sonographic appearance in ovarian hyperstimulation syndrome. *A*: Transverse sonogram showing dominant follicle in right ovary (*curved arrow*). Both ovaries (left ovary, *arrow*) are enlarged and rounded. Clinically the patient had polycystic ovarian disease. *B*: Static longitudinal sonogram in the midline demonstrating massively enlarged right ovary (*arrow*).

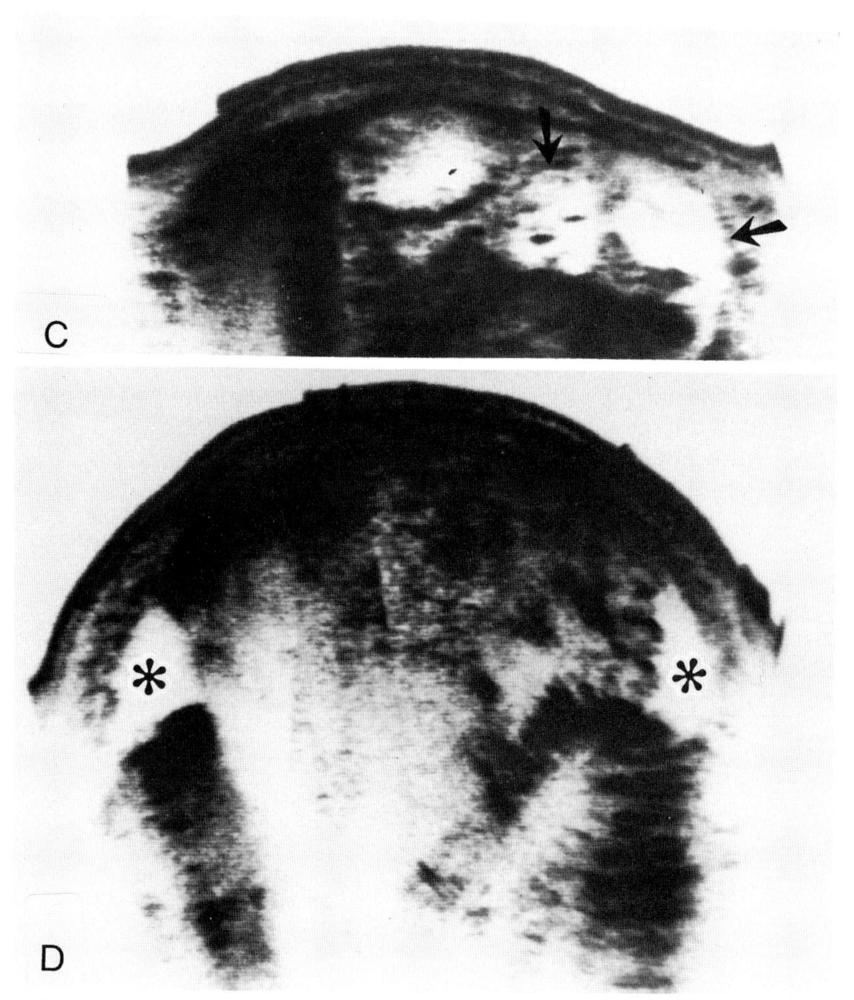

Figure 10.11. *C*: Transverse sonogram showing cystic areas within massively enlarged left ovary (*arrows*). *D*: Bilateral peritoneal effusions (*).

needle guide, is utilized for delineation of the path of the needle as it transverses the bladder either with the transabdominal or transurethral approach and locates within the follicle itself. The transurethral approach appears promising since it does not involve abdominal wall puncture and may afford a more optimal angle with which to aspirate a follicle (36).

Some authors have found the transvaginal approach to be most desirable for follicular aspiration (37). In this technique, transabdominal sonography is used to guide follicular aspiration through the fornix.

The rates for fertilization and cleavage of follicles that were obtained in this manner are similar to those obtained laparoscopically (34, 35). The advantage of sonographically directed follicular aspiration is that the procedure can be performed without general anesthesia. Follicles within ovaries that are surgically "inaccessible" can be aspirated using the sonographically guided percutaneous approach. Some limitations of this technique include the requirement that the ovaries be located immediately adjacent to the urinary bladder and the possibility of bladder or bowel damage. However, except for limited hematuria, there have been no adverse effects reported resulting from this technique (34, 35). In fact, sonographically guided aspiration is the preferred method in many centers, with laparoscopy used only in those cases where sonographically guided aspiration is unsuccessful (34, 35).

Future Developments

Improved resolution of real-time scanners will increase the accuracy of sonographic evaluation of follicular development. In particular, the oocyte itself might be more consistently imaged with scanners of better resolution, thereby distinguishing a follicle that does not contain immature oocytes from those that do.

REFERENCES

1. Fleischer AC, Pittaway DE, Wentz AC, et al: The uses of sonography for monitoring ovarian follicular development, in Sanders R, Hill M (eds), *Ultrasound Annual 1983*. New York, Raven Press 1983.
2. Schwimer SR, Lebovic J: Transvaginal pelvic ultrasonography: Accuracy in follicle and cyst size determination. *J Ultrasound Med* 4:61, 1985.
3. Biological effects of ultrasound, in Ziskin M (ed): *Clinics in Ultrasound*. New York, Churchill-Livingston, 1985.
4. Testart J, Thebault A, Frydman R: Premature ovulation after ovarian ultrasonography. *Br J Obstet Gynecol* 89:694, 1982.
5. Bailey KI, O'Brien WD Jr, Dunn F: Ultrasonically induced morphological damage to mouse ovaries. *Ultrasound Med Biol* 9:25, 1983.
6. Fleischer AC, Daniell JR, Rodier J, et al: Sonographic monitoring of ovarian follicular development. *J Clin Ultrasound* 9:275, 1981.
7. Davies J: Microscopic anatomy of the female reproductive tract, in Danforth O (ed): *Textbook of Obstetrics and Gynecology*. New York, Harper & Row, 1971, p 83.
8. Mendelson EB, Friedman H, Neiman HL, et al: The role of imaging in infertility management. *AJR* 144:415, 1985.
9. Picker R, Smith D, Tucker M, et al: Ultrasonic signs of imminent ovulation. *J Clin Ultrasound* 11:1, 1983.
10. Queenan JT, O'Brien GD, Bains LM, et al: Ultrasound scanning of ovaries to detect ovulation in women. *Fertil Steril* 34:99, 1980.
11. Hackloer B: The role of ultrasound in female infertility management. *Ultrasound Med Biol* 10:35, 1984.
12. Nyberg D, Laing P, Jeffrey R: Sonographic detection of subtle pelvic fluid collections. *AJR* 143:261, 1984.
13. Davis J, Gosink B: Fluid in the posterior cul-de-sac: Cyclical patterns. *J Ultrasound Med* 3:102, 1984.
14. Malini S, Valdes C, Malinak R: Sonographic diagnosis and classification of anomalies of the female genital tract. *J Ultrasound Med* 3:397, 1984.
15. Richman TS, Viscomi GN, deCherney A, et al: Fallopian tubal patency assessed by ultrasound following fluid injection. *Radiology* 152:507, 1984.
16. Hann L, Hall D, McArdle C, et al: Polycystic ovarian disease: Sonographic spectrum. *Radiology* 150:571, 1984.
17. Liukkenon S, Koskimies AI, Tenhunen A, Ylostalo P: Diagnosis of luteinized unruptured follicle (LUF) syndrome by ultrasound. *Fertil Steril* 41:26, 1984.
18. Ritchie WGM: Ultrasound in the evaluation of normal and induced ovulation. *Fertil Steril* 43:167, 1985.
19. Geisthovel F, Skubsch U, Zabel G, et al: Ultrasonographic and hormonal studies in physiologic and insufficient menstrual cycles. *Fertil Steril* 39:277, 1983.
20. Hill G, Fleischer AC: Enlarging hydrosalpinges. *Fertil Steril* (in press).
21. Mantzavinos T, Garcia JE, Jones HW Jr: Ultrasound measurement of ovarian follicles stimulated by human gonadotropins for oocyte recovery and in vitro fertilization. *Fertil Steril* 40:461, 1983.
22. Sallam HN, Marinho AO, Collins WO, et al: Monitoring gonadotropin therapy by real-time ultrasonic scanning of ovarian follicles. *Br J Obstet Gynecol* 89:155, 1982.
23. Hackeloer BJ, Fleming R, Robinson HP, et al: Correlation of ultrasonic and endocrinologic assessment of human follicular development. *Am J Obstet Gynecol* 135:122, 1983.
24. McArdle CR, Seibel M, Weinstein F, et al: Induction of ovulation monitored by ultrasound. *Radiology* 148:809, 1983.
25. Dornbluth NC, Potter JL, Shephard MK, et al: Assessment of follicular development by ultrasonography and total serum estrogen in human menopausal gonadotropin-stimulated cycles. *J Ultrasound Med* 2:407, 1983.
26. Marrs, RP, Vargyas JM, March CG: Correlation of ultrasonic and endocrinologic measurements in human menopausal gonadotropin therapy. *Am J Obstet Gynecol* 145:417, 1983.
27. O'Herlihy C, Pepperell R, Robinson H: Ultrasound timing of human chorionic gonadotrophin administration in clomiphene stimulated cycles. *Obstet Gynecol* 59:40, 1982.
28. Prins GS, Vogelzang RL: Inherent sources of ultrasound variability in relation to follicular measurements. *J In Vitro Fertil Embryo Trans* 1:221, 1984.
29. Smith B, Porter R, Ahuja K, et al: Ultrasonic assessment of endometrial changes in stimulated cycles in an in vitro fertilization and embryo transfer program. *J In Vitro Fertil Embryo Trans* 1:233, 1984.
30. Brandt T, Levy E, Grant E: Endometrial echo and its significance in female infertility. *Radiology* 157:225, 1985.
31. Fleischer AC, Pittaway DE, Beard LA, et al: Sonographic depiction of endometrial changes occurring with ovulation induction. *J Ultrasound Med* 3:341, 1984.
32. Fleischer AC, Herbert CM, Wentz AC: Sonography of the endometrium in successful and unsuccessful IVP-ET cycles. *Fertil Steril* 1986 (in press).
33. Schenker JG, Weinstein D: Ovarian hyperstimulation syndrome: A current survey. *Fertil Steril* 30:255, 1978.

34. Lenz S: Ultrasonic-guided follicle puncture under local anesthesia. *J In Vitro Fertil Embryo Trans* 1:239, 1984.
35. Feichtinger W, Kemeter P: Laparoscopic or ultrasonically guided follicle aspiration for in vitro fertilization? *J In Vitro Fertil Embryo Trans* 1:244, 1984.
36. Parsons J, Barber M, Goswamy R, et al: Oocyte retrieval for *in vitro* fertilization by ultrasonically guided needle aspiration via the urethra. *Lancet* 1:1076, 1985.
37. Dellenbach P, Nisand I, Fegen B: Transvaginal sonographically controlled follicular puncture for oocyte retrieval. *Fertil Steril* 44:656, 1985.

11 Sonography of Abnormal Early Pregnancy

ARTHUR C. FLEISCHER, M.D., DAVID A. NYBERG, M.D., GLYNIS A. SACKS, M.D.

Sonography has an important role in the evaluation of the early pregnancy of a patient with a history of infertility. Such patients have an increased incidence of complications. In general, the most frequent clinical sign of a complicated early pregnancy is abnormal uterine bleeding. This chapter will emphasize the use of sonography in the evaluation of patients with a complicated or abnormal early pregnancy, particularly as this group of disorders relate to gynecologic infertility. Specifically, the sonographic diagnosis of abortions (incomplete, missed, and complete), blighted ova (anembryonic pregnancy), and ectopic pregnancies will be discussed. Sonographic evaluation of pregnancies occurring with an intrauterine contraceptive device (IUD) will also be mentioned.

Safety Considerations

There have been a large number of investigations concerning the possible adverse bioeffects of ultrasound on the developing embryo and fetus. However, there have, to date, been no substantiated adverse bioeffects on the embryo or fetus at intensities used for diagnosis (1).

As opposed to consideration of bioeffects on the follicle and oocyte, one has to be concerned about the possible bioeffects of ultrasound on the developing embryo and fetus, particularly during periods of organogenesis. Although no bioeffects have been proven at diagnostic levels, the sonographer/sonologist should limit the examination time to only that needed to obtain diagnostically useful information. Similarly, sonography should be performed only when it is clinically indicated. Significant and important information can usually be obtained by sonographic evaluation of the patient with a complicated early pregnancy. New technology has developed scanners that afford improved resolution, explaining, in part, the fact that sonography has gained a vital role in the evaluation of early pregnancy complications.

Instrumentation and Scanning Technique

For early pregnancy, mechanical sector real-time scanners are preferred over linear array scanners since they allow more imaging flexibility and greater resolution. Scanners that utilize phased array transducers also afford excellent resolution.

The highest frequency transducer (usually a 5 MHz) compatible with adequate penetration should be utilized. The sonographer/sonologist should be aware of the focusing characteristics of the transducer that is used so that the images can be optimized.

A moderately distended urinary bladder is necessary for evaluation of the uterus during first trimester. If the fundus of the uterus is not in the focal plane of the transducer, additional maneuvers may be needed to optimize its depiction. For the markedly anteflexed uterus, scanning through a water path such as afforded by an intravenous bag or gel-like standoff pad (Kitekco) held on the abdomen may be helpful for evaluation of the intrauterine contents. A re-

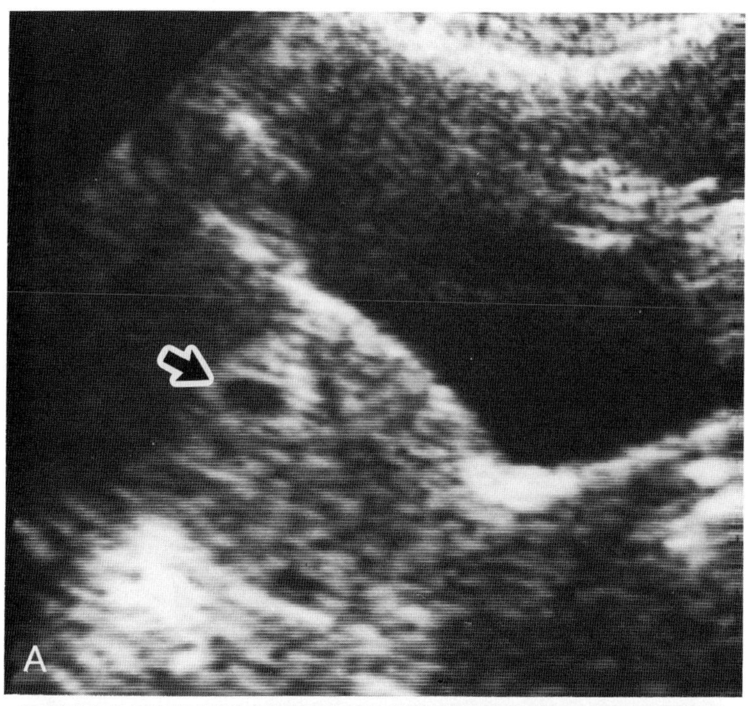

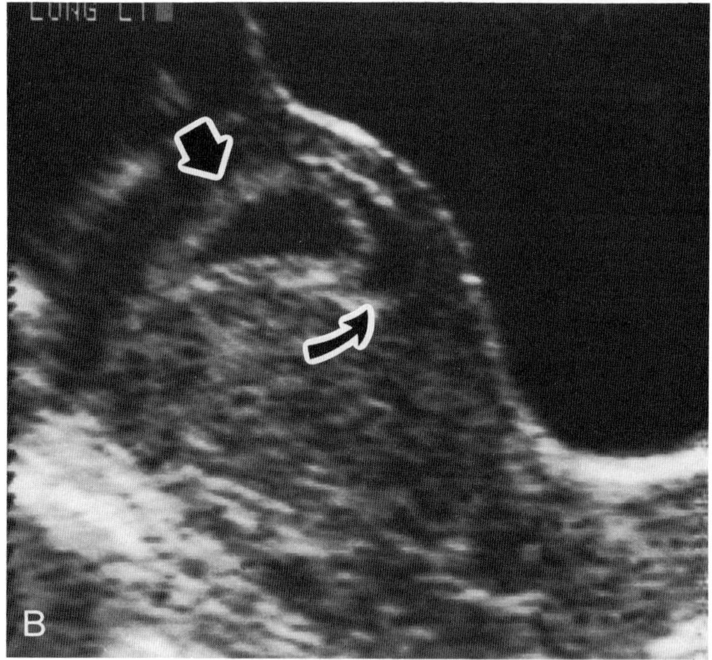

Figure 11.1. Normal early intrauterine pregnancy. *A:* Magnified longitudinal sonogram of 5–6 weeks. The hCG was 2000 mIU/ml. A 7-mm gestational sac (*arrow*) with its two layers of decidua are clearly depicted in the upper portion of the uterus. *B:* Magnified longitudinal real-time scan demonstrating gestational sac (*large arrow*) within upper portion of uterine lumen at 6 weeks menstrual age. There is a hypoechoic space (*curved arrow*) between the inferior aspect of the gestational sac and the decidua vera.

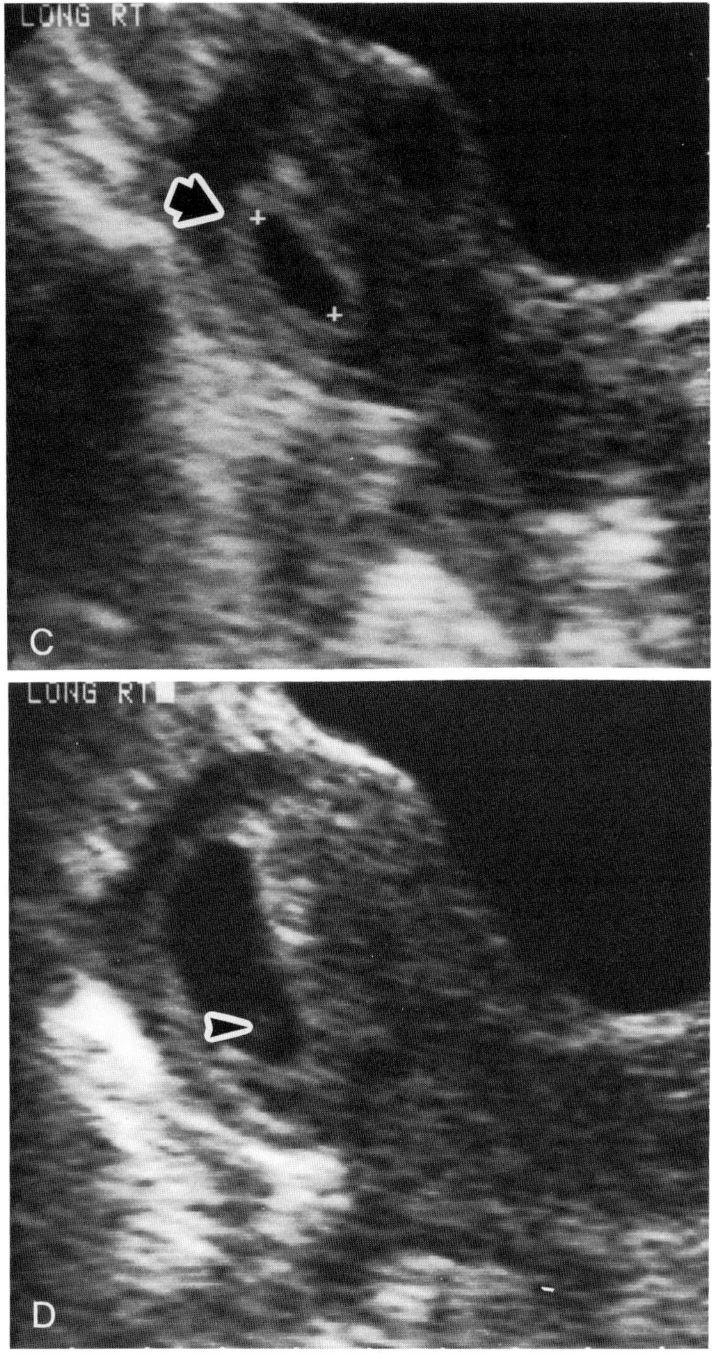

Figure 11.1. *C:* Magnified longitudinal sonogram showing chorion frondosum as localized thickening of chorion (*arrow*). *D:* Yolk sac (*arrow*) within gestational sac at 6 weeks.

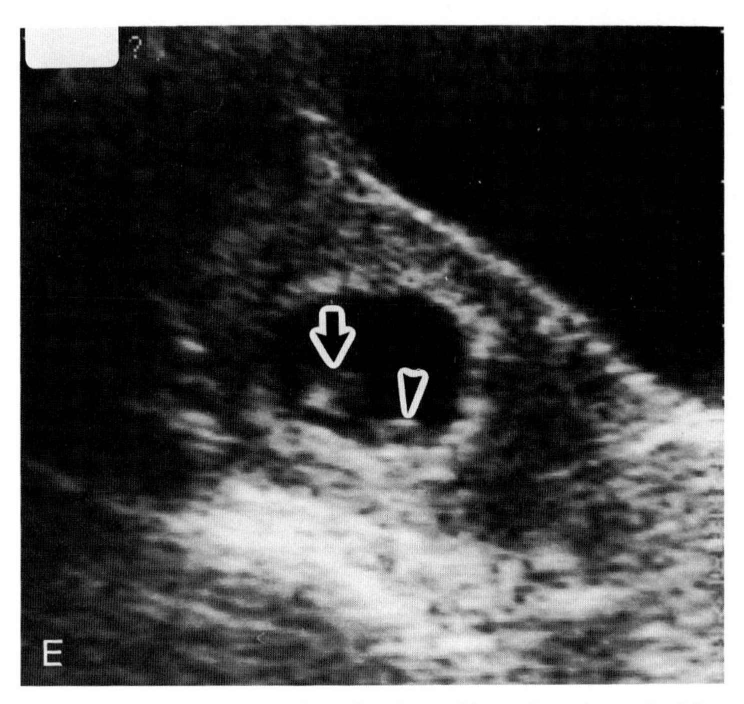

Figure 11.1. *E:* Yolk sac (*arrowhead*) adjacent to embryo (*arrow*) at 7 weeks.

troflexed uterus presents greater difficulty since the fundus may be beyond the focal zone of the transducer. Once the area of interest is displayed on scans obtained with standard field of view, it can be manipulated in various ways to produce a desirable image. A magnified mode is useful for better evaluating the contents of a gestational sac.

The routine sonogram of early pregnancies begins with a longitudinal scan of the uterus (Fig. 11.1A). In this way, the central endometrial interfaces can be examined for evidence of an early gestation. Intraperitoneal fluid can also be identified in the cul-de-sac or superior to the uterine fundus. Transverse scans are useful for further examining an early gestation and its relationship with the uterus, as well as for evaluating the adnexa.

Normal Early Pregnancy

In order to detect the sonographic findings in complicated and/or abnormal early pregnancies, one must first be familiar with the sonographic appearance of normal pregnancy and its variants.

The earliest reliable findings of an intrauterine pregnancy occur at approximately 5 weeks' menstrual age when a 4–7-mm echogenic ring can be delineated within the uterus, within the decidualized endometrium. The central sonolucency represents the chorionic cavity, and the surrounding echogenic ring represents the choriodecidual reaction. On careful inspection, the choriodecidual reaction can actually be seen as two concentric echogenic rings in most cases (2) (Figs. 11.1A and B). The inner ring is thought to represent chorionic villi and the decidua capsularis, whereas the outer layer is believed to be decidua vera (3). There may be a hypoechoic area around the gestational sac itself representing the unobliterated portion of the uterine lumen. This appearance is in distinction to the single decidual layer that is occasionally observed in an ectopic pregnancy due to an intrauterine decidual cast containing blood, and referred to as a "pseudogestational sac" (3,4).

Prior to sonographic depiction of the embryo, the average dimension of the gestational sac can be taken as an approximation of gestational age. This linear measurement is easier to calculate than the gestational sac volume, which has been used as a parameter to assess gestational duration in the first trimester (5). The inner dimensions of the gestational sac should be measured in long, short, and anteroposterior axes. These dimensions should be averaged to obtain the "average gestational sac dimension." The mean dimension of the gestational sac enlarges 1–1.5 mm/day during the first 50–60 days of pregnancy. The gestational age in menstrual days is approximately equal to mean gestational sac dimension + 30 (6).

At 6–7 weeks, a projection from the thickest portion of the choriodecidua (chorion frondosum), which represents the body stalk of the fetus, can be sonographically delineated (Fig. 11.1C). An embryonic pole should be identified in all pregnancies where the mean gestational sac dimension is 2.5 cm or greater (7).

The first structure that can be identified within the gestational sac is a rounded saclike structure that represents the yolk sac (Fig. 11.1D). The yolk sac has an important role in early hematopoiesis; it is approximately 3–5 mm between 5 and 9 weeks' menstrual age (8). After that, it becomes more solid-appearing as it begins to regress.

At 7 to 8 weeks, the embryo can be identified connected to the choriodecidua by a body stalk, and heart motion should be identifiable (Fig. 11.1E). Occasionally, fetal heart motion can be detected as early as 41–43 menstrual days by observing pulsations adjacent to the yolk sac (9). At this stage of development, the yolk sac is in close proximity to the developing embryo and cardiac pulsations will be transmitted to the yolk sac.

Fetal heart motion appears as a rapid, rhythmic pulsation at approximately 120–150 beats/min (9). The frame rate of the real-time scanner utilized should be more than 20 frames/sec in order to capture the rapid heart beat of a developing fetus, thereby confidently distinguishing true heart motion from background electronic flicker.

Recognition of fetal heart motion has important prognostic implications. The vast majority of pregnancies that demonstrate heart motion at 7–8 weeks proceed to term pregnancies, whereas the lack of fetal heart motion is usually associated with missed abortions and nonviable pregnancies (10).

A linear membrane is often depicted within the gestational sac during the 8th to 12th week (Fig. 11.2A). This represents the amnion, which normally fuses with the chorion before 16 menstrual weeks (4). The amnion can be seen separating the embryo from the yolk sac.

Hypoechoic areas around the chorion most likely represent areas of retrochorionic hemorrhage (11). It is not abnormal to have a small (less than 10% of the area of the decidua basalis) hypoechoic area, which probably represents hemorrhage occurring around the chorion as the choriodecidua forms the placenta. However, if this area includes greater than 40% of the basal plate or involves the area immediately beneath the chorion frondosum, the prognosis for completion of pregnancy is somewhat decreased (Fig. 11.3A) (11).

By the 9th week a viable embryo has more specific sonographic features. At this time the crown–rump length is between 25 and 32 mm and the head can be distinguished from the trunk. Trunk motion as well as heart motion can be identified. An intact choriodecidua surrounds the gestational sac.

Abortions

Amazingly, only 30% of fertilized zygotes complete development; the majority undergo atresia and fail to produce a viable fetus. Most of the "fetal and embryonic wastage" can be attributed to chromosomal aberrations (12).

Sonography is helpful for evaluating fetal viability in patients who present with bleeding in the first trimester. Rarely, the bleeding can be secondary to passage of a twin with one being blighted (13). It is felt that most bleeding that occurs in early pregnancy is associated with the process of placentation (14).

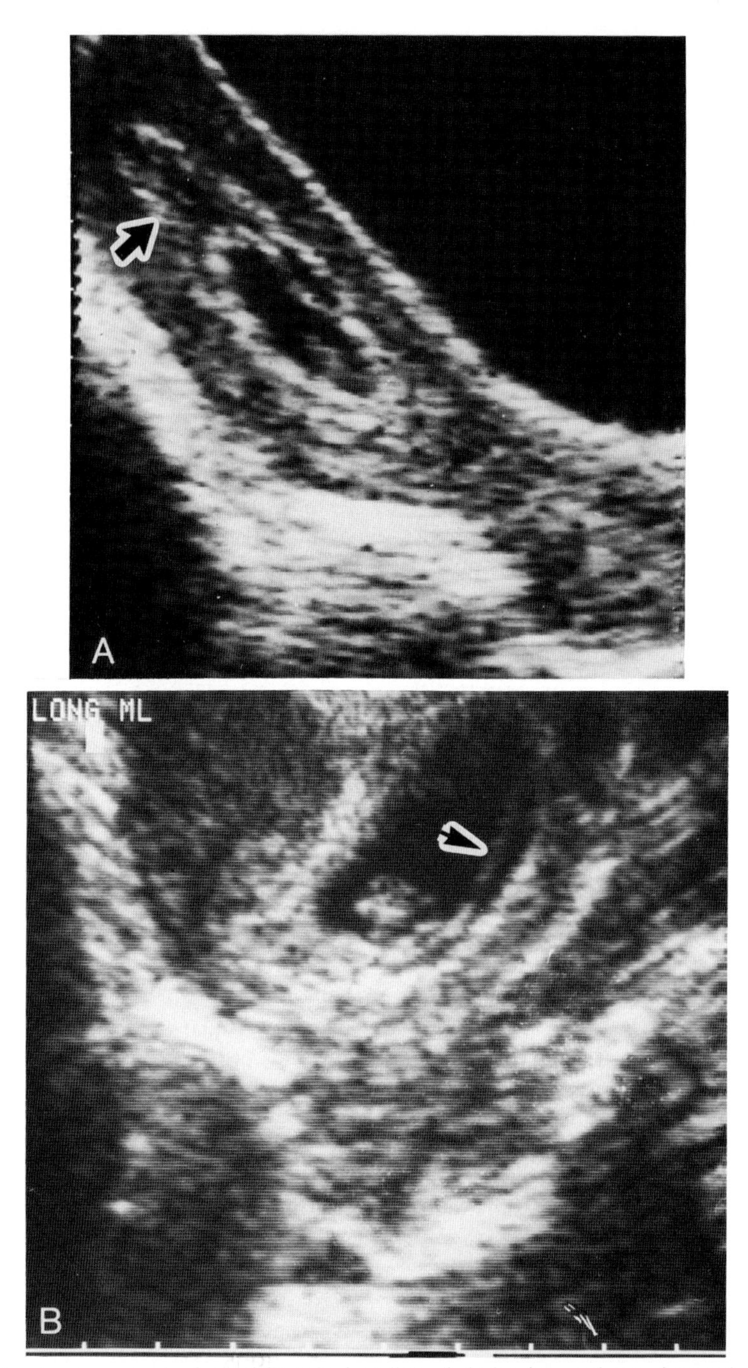

Figure 11.2. Anatomic variants. *A:* Unobliterated portion of uterine lumen (*arrow*) superior to gestational sac. *B:* Linear interface (*arrow*) within gestational sac representing amniotic membrane.

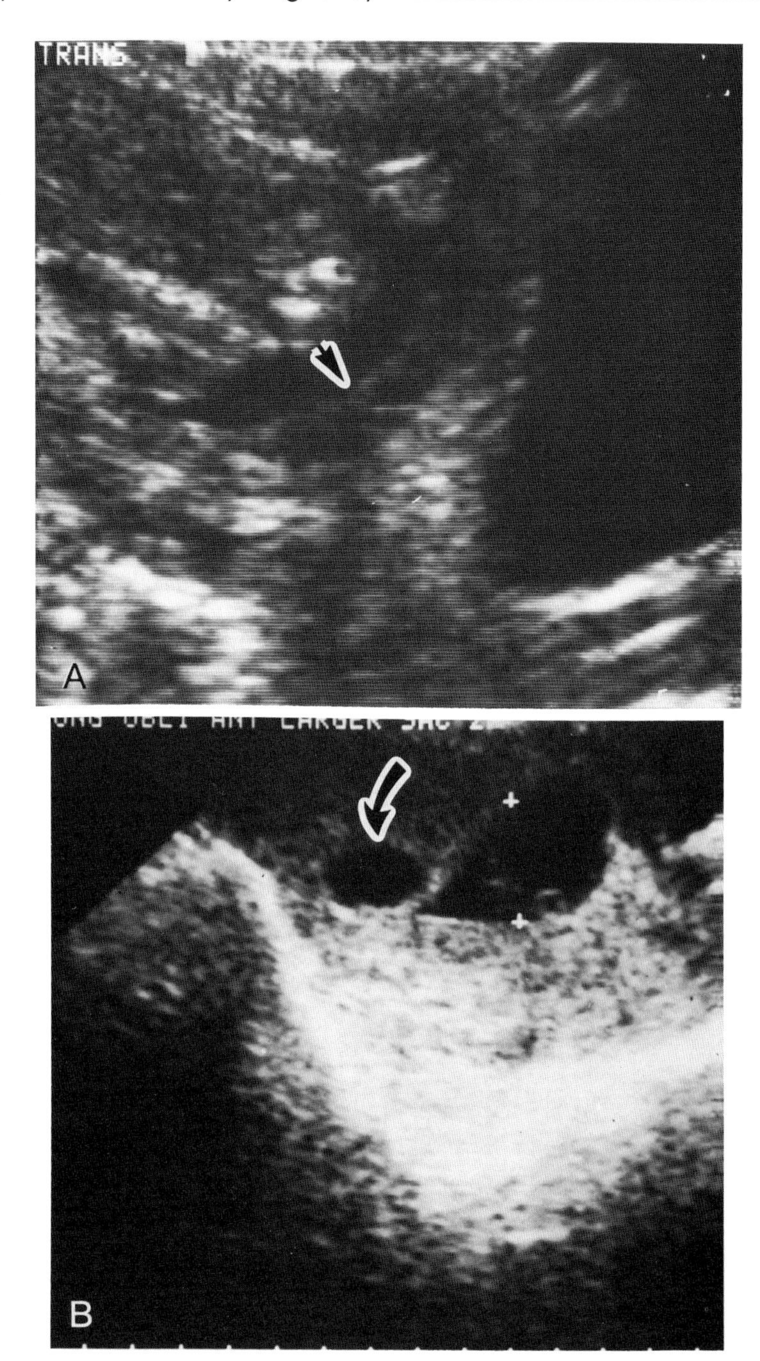

Figure 11.3. Abnormal early pregnancy. *A:* Retrochorionic hemorrhage associated with elevation of chorion (*arrowhead*) from decidua. *B:* Former twin intrauterine pregnancy with one twin blighted, the other twin viable. A previous sonogram demonstrated two viable fetuses. This sonogram, however, shows one "empty sac" (*arrow*).

Sonography is helpful in distinguishing incomplete from missed abortion and confirming the clinical impression of blighted ovum. A recent study has shown that abnormal pregnancies can be differentiated from normal ones by sonographic detection of the gestational sac and its contents (15). Major criteria used to establish abnormal pregnancies included a large (2.5-cm mean sac diameter) gestational sac without an embryo and a distorted sac shape. Minor criteria included weakly echogenic or irregular choriodecidual reaction, absence of a double decidual sac, and low position. The presence of a large sac with a distorted shape that does not contain an embryo was found to be the most accurate prediction of an abnormal pregnancy (15).

In cases of uncertainty, it is wise to "give the fetus the benefit of the doubt." This is particularly true if there is some question as to the accuracy of the menstrual dates. Occasionally, gestations that appear abnormal by sonography may progress to term.

In an incomplete abortion, residual choriodecidua is seen within the uterus (Fig. 11.4A). Retrochorionic hemorrhage may also be present (16). In some patients, the sac itself is abnormally low in position, located in the corpus or cervix.

In missed abortion, a nonviable fetus can be identified within the gestational sac. The amount of amniotic fluid may be markedly decreased. The measured length of the fetus may be less than that expected by menstrual history.

In completed abortion, no choriodecidua or fetus is identified within the uterus (Fig. 11.4B). The endometrial surfaces are closely opposed or coapted (Sacks G, Fleischer A, in preparation). If there is a suspicion of ectopic pregnancy, serial quantitative human chorionic gonadotropin (hCG) assays can be performed to distinguish completed abortion from ectopic pregnancy. In general, completed abortions demonstrate a steady decline in hCG levels while ectopic pregnancies usually show a plateau or transient decline (17).

Sonography has an important role in the evaluation of the infertile patient who may have a higher incidence of having abnormal pregnancies, in particular those associated with blighted ova (12). A blighted ovum can be reliably diagnosed when there is absence of the fetal pole complex with a gestational sac greater than 25 mm in average dimension (Figs. 11.4C,D,E) (7). Other sonographic features of a blighted ovum include distorted sac shape and irregular and/or thin choriodecidual reaction.

Ectopic Pregnancy

Sonography has a significant role in the evaluation of patients suspected of having an ectopic pregnancy. The sonographic information must be correlated with hCG assays (either quantitative or qualitative) in order to accurately determine the possibilities of an ectopic pregnancy (18).

The sonographic findings in patients with ectopic pregnancy vary depending on several factors. These include the stage of gestation in which the patient presents, the location of the ectopic implantation, and whether or not rupture of the tube has occurred.

Although not infallible, the combination of sonography and hCG can detect most ectopic pregnancies. In one recent study of 223 patients suspected of having an ectopic pregnancy, 93% were prospectively diagnosed correctly with hCG and sonography (19). Most authorities agree that sonography is most useful for suggesting an ectopic gestation by demonstrating absence of an intrauterine pregnancy, assuming a hCG pregnancy test is positive. Other possibilities for this combination of findings are an early intrauterine pregnancy (before 5 weeks) and a recent spontaneous abortion. Findings that favor an ectopic gestation are the presence of an adnexal mass and/or a moderate to large amount of fluid in the cul-de-sac. However, it is important to realize that up to 20% of confirmed ectopic pregnancies may not demonstrate any abnormality (20).

The adnexae should be accurately evaluated in patients with suspected ectopic pregnancies. In a recent study, careful inspection revealed a noncystic mass in the majority of cases with ectopic pregnancy

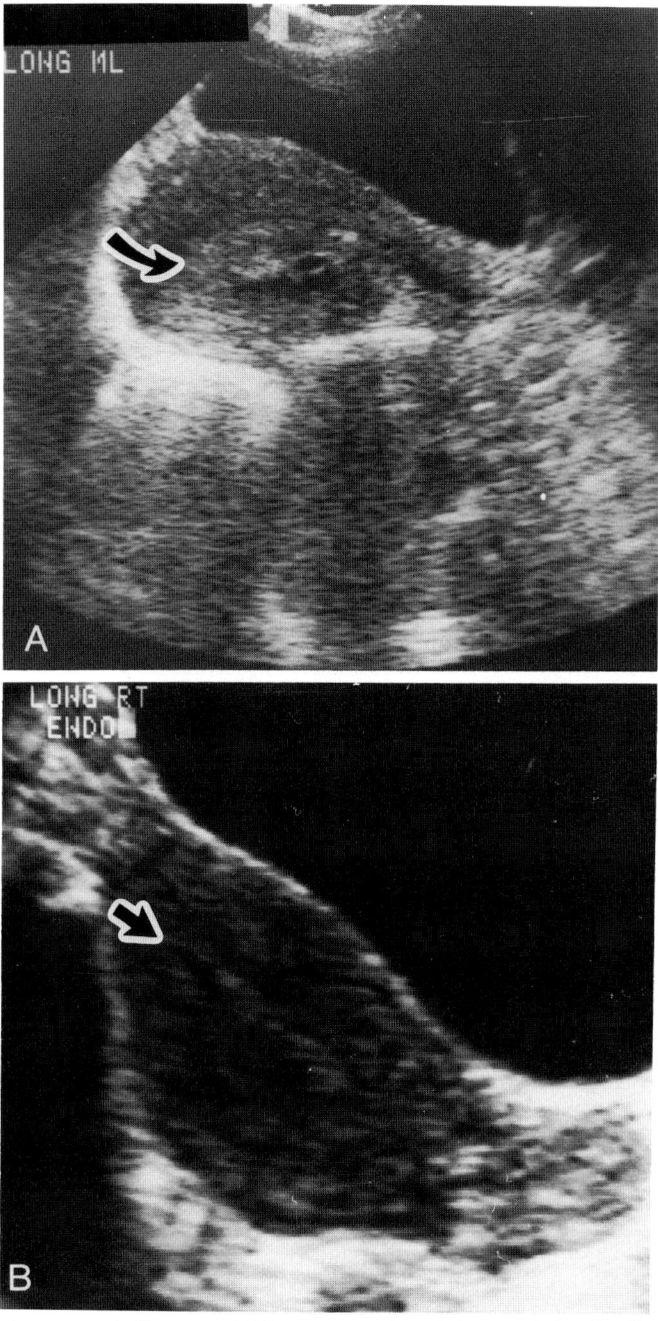

Figure 11.4. Miscarriages and other nonviable pregnancies. *A:* Incomplete abortion showing retained choriodecidua (*arrow*). *B:* Completed abortion showing endometrial interfaces (*arrow*) to be totally coapted.

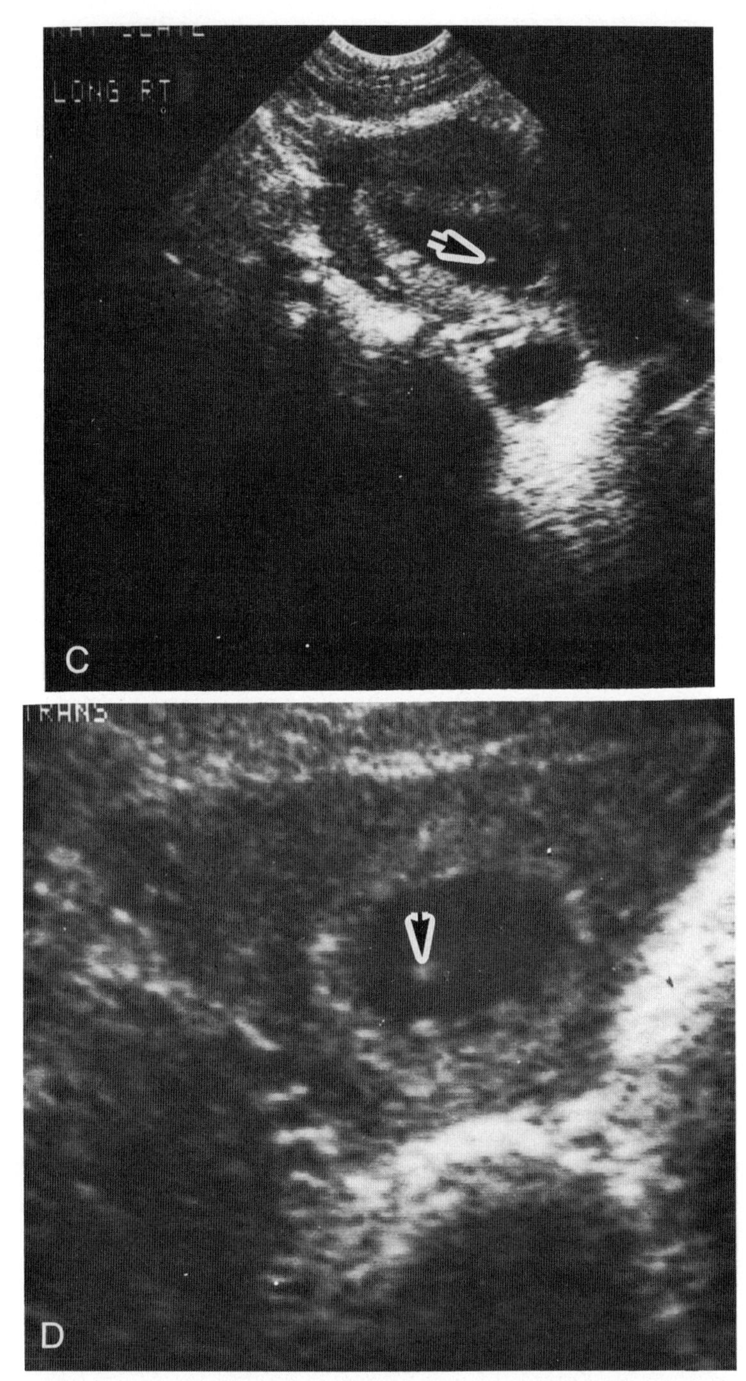

Figure 11.4. *C:* Initial scan at 8 weeks showing discrepancy between gestational sac size and size of "embryonic pole" (*arrow*). *D:* Transverse magnified sonogram showing abnormally small structure within gestational sac (*arrow*).

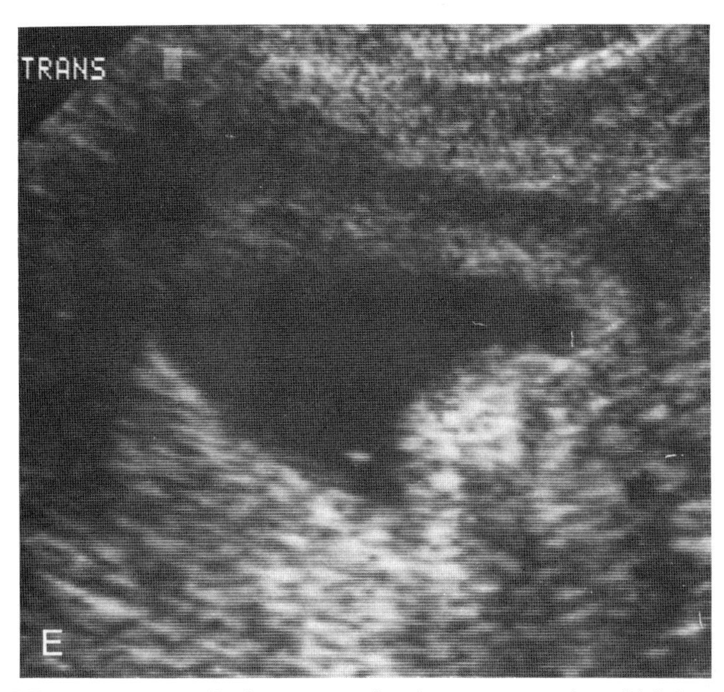

Figure 11.4. *E:* Transverse, magnified sonogram showing no progression of fetal growth in 1 week.

(Fig. 11.5) (20). In advanced tubal ectopic pregnancies, the mass may measure up to 5 cm. In about 10% of cases, a living extrauterine embryo can be identified, which is, of course, a diagnostic finding (21).

A corpus luteum cyst is frequently identified in both ectopic and intrauterine pregnancies and it should not be confused with an ectopic gestation itself. It appears as a rounded, hypoechoic structure that can usually be delineated with the ovary. Occasionally, a corpus luteum cyst that has hemorrhaged may appear indistinguishable from an ectopic gestation (Fig. 11.6A). In approximately one-third of cases, the ectopic gestation may be on the side opposite from the corpus luteum, perhaps as the result of transperitoneal migration of the ovum (22).

Ectopic implantation within the tube usually is associated with some degree of hematosalpinx and intraperitoneal blood collection. If the bleeding is extensive, collections of fluid within the cul-de-sac and lower peritoneal cavity can be identified. The fluid collections associated with ectopic pregnancies are usually greater than those encountered during ovulation, which are rarely greater than 5 ml.

In about one out of 10 advanced tubal ectopic pregnancies (8–10 weeks gestation) the formation of a decidual cast can mimic some features of a true gestational sac. However, the decidual cast associated with an ectopic pregnancy has only a single decidual layer. In comparison, a recently published study found two concentric layers of decidua in 87% of proven early (5–6 weeks) intrauterine pregnancies, prior to sonographic depiction of the embryo (6). In some rare instances, the decidual cast of ectopic pregnancy may separate from the myometrium and mimic the double-layer decidua of an intrauterine pregnancy.

A method that has recently been used for diagnosing ectopic pregnancy as well as nonviable intrauterine pregnancies is correlating sonographic findings with quantitative hCG (Figs. 11.7 and 11.8) (6). One should realize that there is a range of normal for these values and that research is ongoing to define accurately these ranges. Recently it has been found that a normal intrauterine gestational sac is always de-

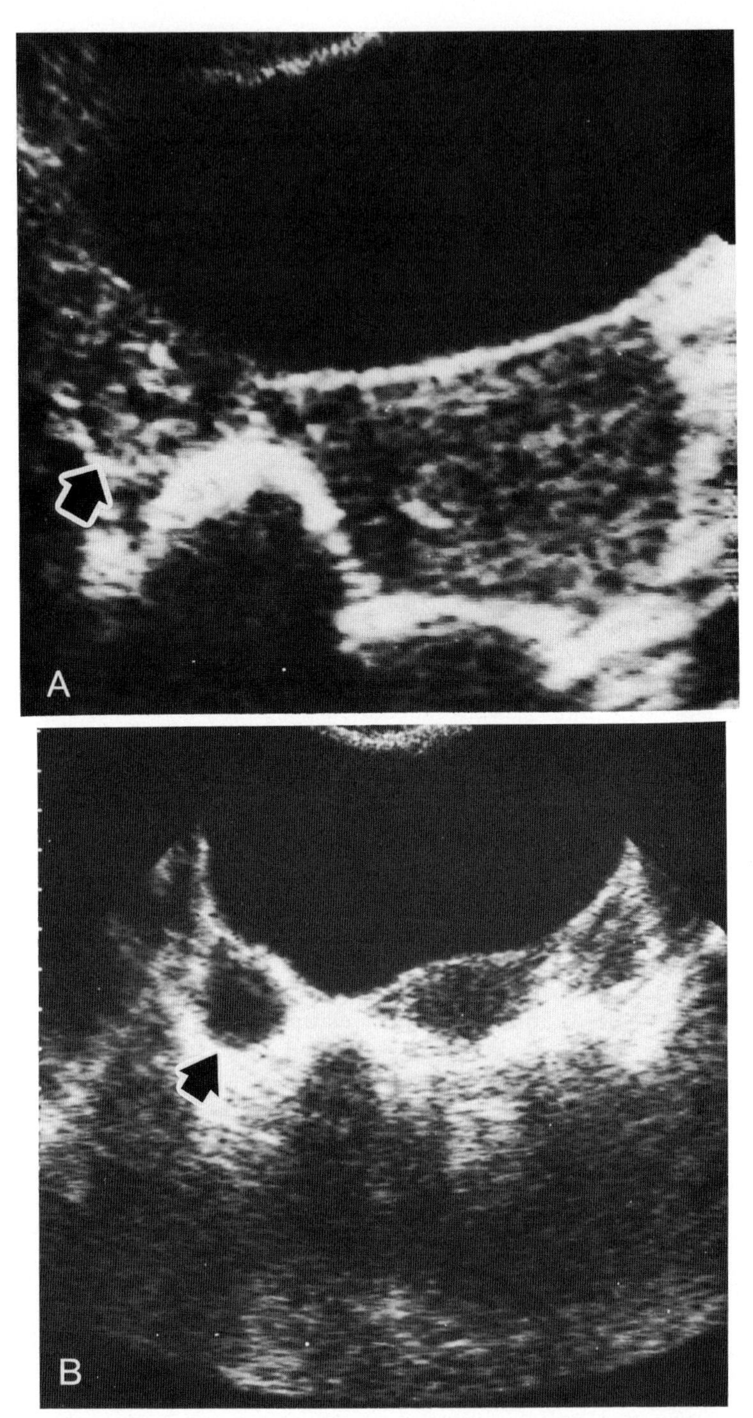

Figure 11.5. Transverse magnified sonogram showing ectopic pregnancy. *A:* Unruptured ectopic pregnancy with an echogenic rim and anechoic center (*arrow*) in isthmic portion of tube appearing as an adnexal mass. (Courtesy of G. Thieme, M.D.) *B:* Inferior to the ectopic pregnancy is a rounded cystic area within right ovary representing a corpus luteum.

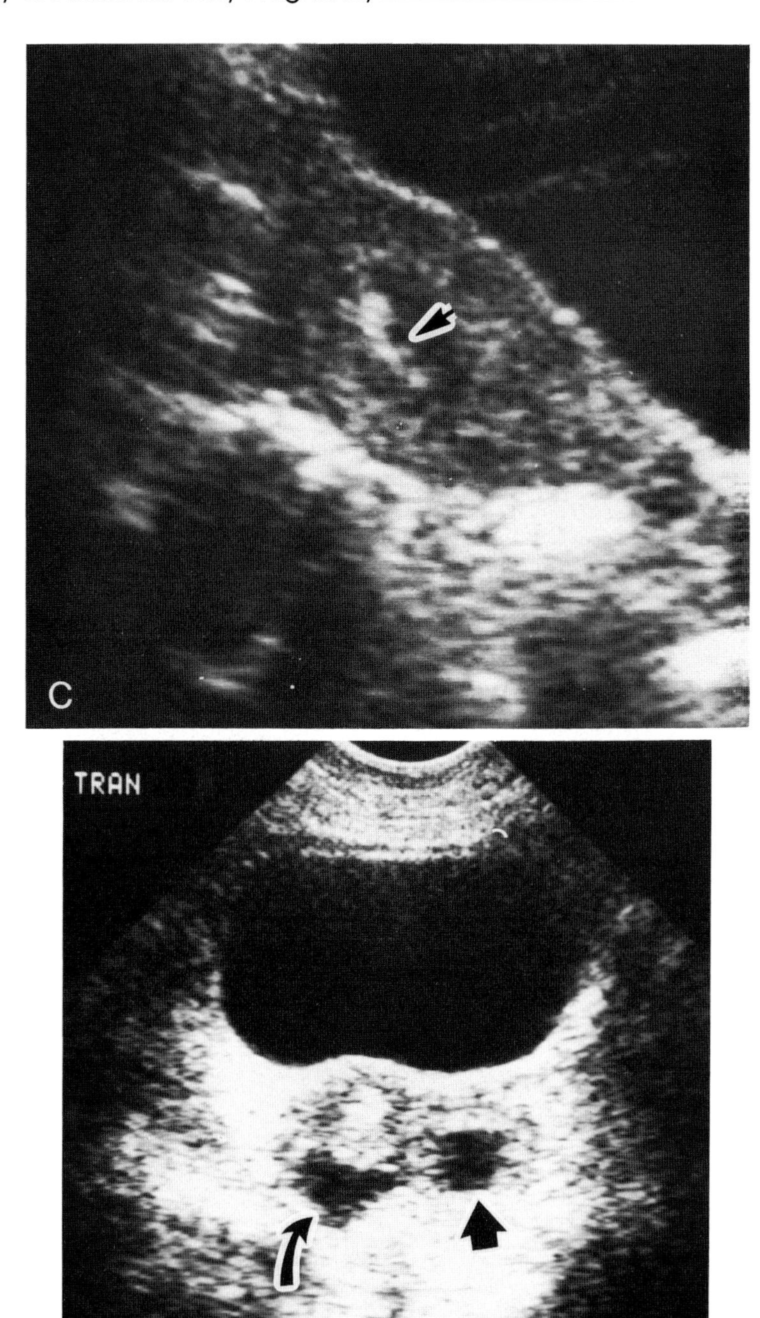

Figure 11.5. *C:* The endometrium (*arrowhead*) has become decidualized but has an appearance similar to secretory phase endometrium. *D:* Ruptured ectopic pregnancy (*arrow*) associated with blood in the cul-de-sac (*curved arrow*).

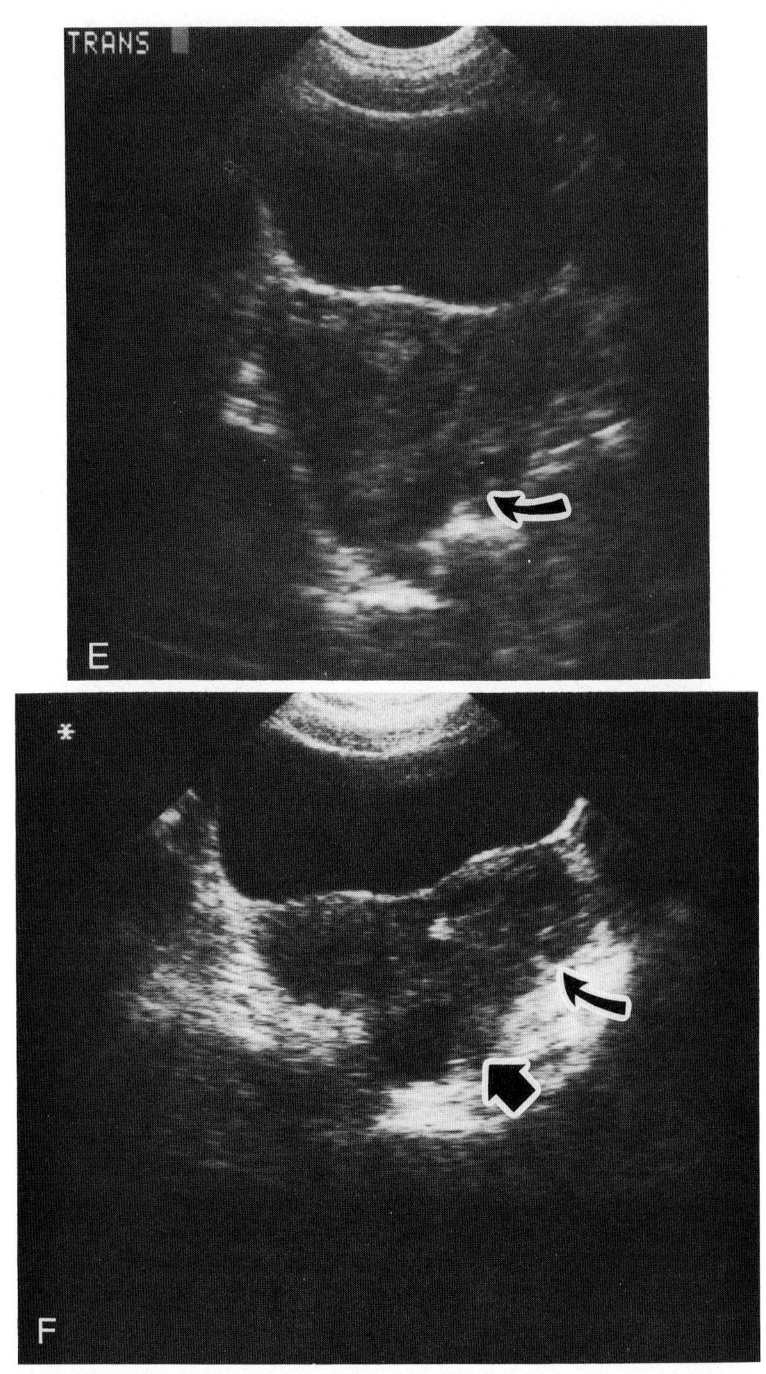

Figure 11.5. *E:* Irregularly shaped solid left adnexal mass representing ruptured ectopic pregnancy with surrounding organized hematoma (*curved arrow*). *F:* Chronic ectopic pregnancy (*arrow*) between uterus and hemorrhagic corpus luteum cyst (*curved arrow*).

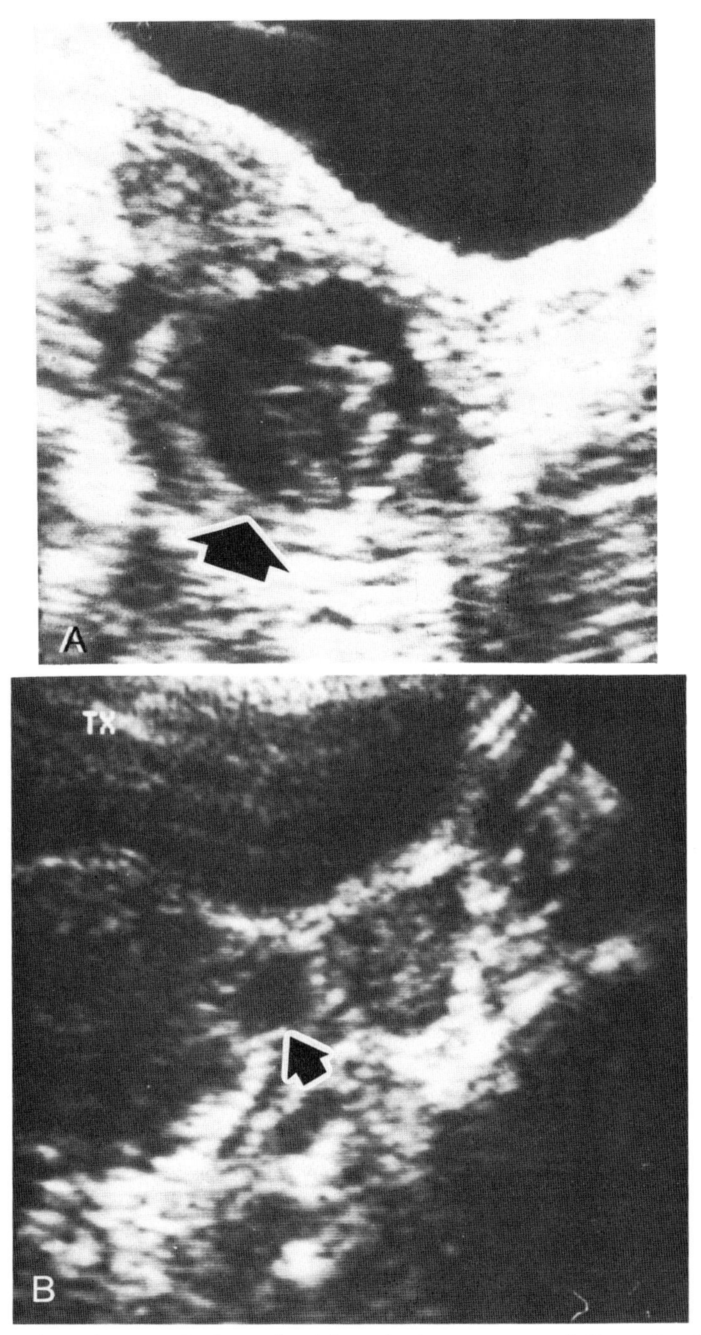

Figure 11.6. Other adnexal masses that can simulate ectopic pregnancy. *A:* Hemorrhagic corpus luteum appearing as complex adnexal mass (*arrow*). The internal echoes represent clotted blood within the cyst. *B:* Endometrioma (*arrow*) between uterus and left ovary.

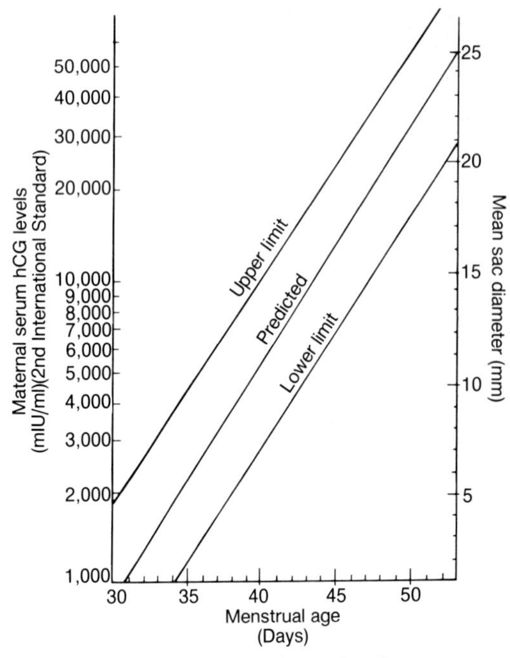

Figure 11.7. Correlation of menstrual age, serum hCG, and gestational sac size.

tected when a simultaneous hCG level exceeds 1800 mIU/ml. Furthermore, the size of the gestational sac correlates well with the hCG level. However, there is apparently some variation from institution to institution regarding the hCG threshold level at which a gestational sac is recognized (4,23). This depends on the accuracy of the hCG assay, the calibration standard used for the assay, the experience of the sonographer/sonologist, and the resolution of the ultrasound equipment. Correlating gestational sac size with hCG levels might represent a way of comparing differences in sonographic and hCG assays between institutions.

At our institution and others, a gestational sac can be identified when the hCG level is more than 2000 mIU/ml (6,24). Absence of a detectable gestational sac at this hCG level is evidence for an ectopic pregnancy. An hCG level less than 2000 mIU/ml may result from an early intrauterine pregnancy or nonviable (incomplete, missed, or completed abortion) intrauterine preg-

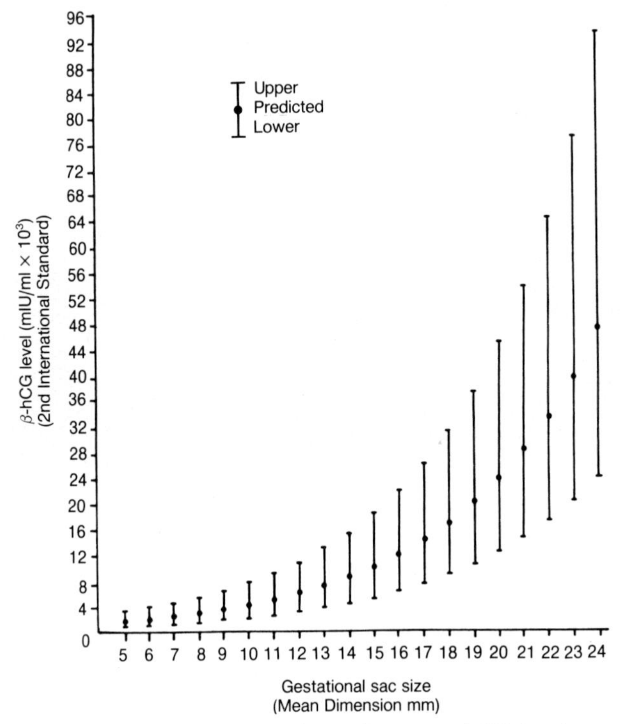

Figure 11.8. Gestational sac size versus hCG level.

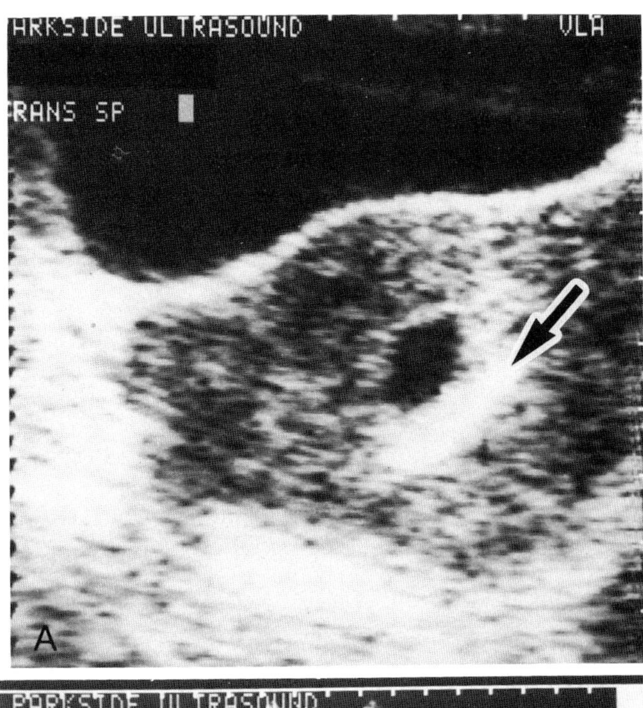

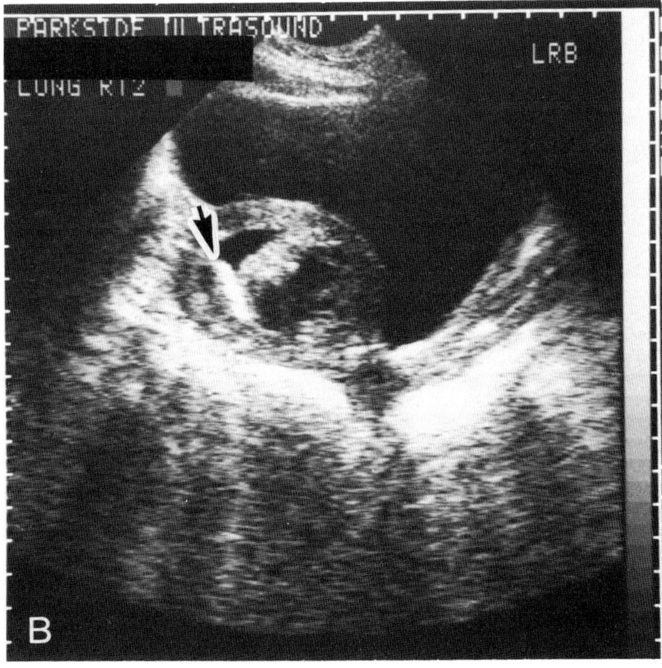

Figure 11.9. Intrauterine pregnancy coexistent with IUD. *A:* Cu-7 IUD (*arrow*) located adjacent to gestational sac. *B:* Long axis view with Cu-7 (*arrowhead*) with retrochorionic hemorrhage superior to the gestational sac.

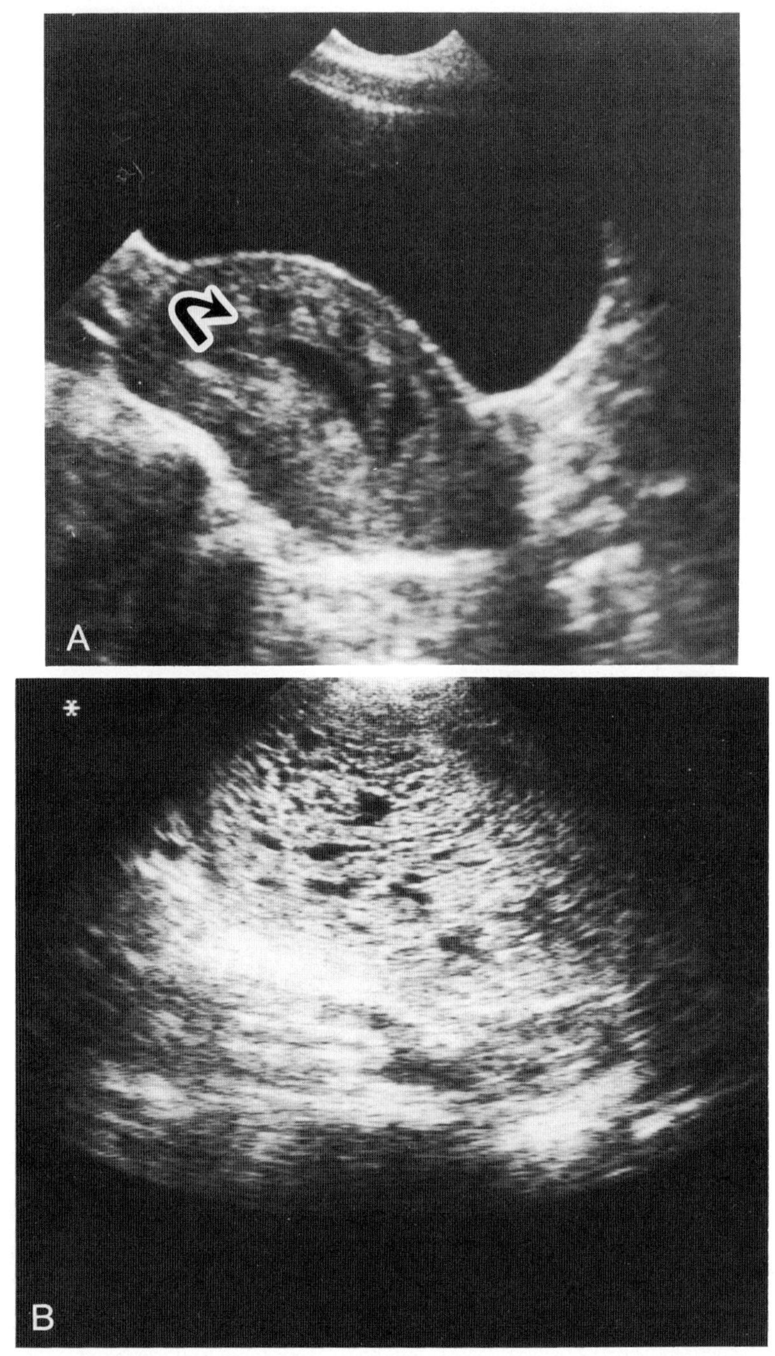

Figure 11.10. Molar pregnancy. *A:* Early molar pregnancy appearing as echogenic tissue with punctate hypoechoic areas within the uterine lumen. *B:* Typical appearance of hydatidiform mole consisting of echogenic tissue that contains multiple cystic spaces that represent hydropic villi.

nancy as well as an ectopic pregnancy. In this situation, serial hCG levels are very helpful since normal intrauterine pregnancies produce doubling times of 2–3 days. A follow-up sonogram performed 7–10 days after the initial study may also be useful in the clinically stable patient.

Other Disorders

Sonography is helpful in evaluation of the location and complications associated with IUDs. The optimal location of the IUD is in the upper portion of the uterine lumen. Metal-containing devices can be recognized by thin echogenic appearance. Plastic IUDs appear as moderately echogenic linear interfaces within the uterus.

Sonography is very useful in detecting pregnancies coexistent with an IUD (Fig. 11.9A). Usually, the IUD is extrachorionic. Occasionally, in patients who experience bleeding, a hypoechoic area can be detected and represents retrochorionic hemorrhage (Fig. 11.9B).

The sonographic appearance of an early molar pregnancy is difficult to distinguish from that associated with an incomplete or missed abortion (Fig. 11.10). Occasionally, echogenic material that corresponds to trophoblastic tissue within the uterus can be identified. Correlating sonographic findings with a quantitative hCG level may also be helpful for distinguishing molar pregnancies, which have disproportionately high hCG levels, from incomplete abortions, which have low hCG levels.

Summary

Sonography has an important role in evaluation of the complicated pregnancy, particularly as it relates to the causes of repeated nonviable pregnancies (6). In these cases, sonography can assess the degree of normal development and occasionally can establish the etiology of the recurrent pregnancy failures.

REFERENCES

1. Biological effects of ultrasound, in Ziskin M (ed): *Clinical Diagnostic Ultrasound*. New York, Churchill-Livingston, 1985.
2. Nyberg DA, Laing FC, Filly RA, et al: Ultrasonographic differentiation of the gestational sac of early intrauterine pregnancy from the pseudogestational sac of ectopic pregnancy. *Radiology* 146:755, 1983.
3. Bradley W, Fiske C, Filly R: The double sac sign of early intrauterine pregnancy: Use in exclusion of ectopic pregnancy (work in progress). *Radiology* 143:223, 1982.
4. Batzer FR, Weiner S, Corson SL, Schlaff S, Otis C: Landmarks during the first forty-two days of gestation demonstrated by the β-subunit of human chorionic gonadotropin and ultrasound. *Am J Obstet Gynecol* 146:973, 1982.
5. Robinson H: "Gestation sac" volumes as determined by sonar in the first trimester of pregnancy. *Br J Obstet Gynaecol* 82:100, 1975.
6. Nyberg D, Filly R, Mahoney B, et al: Early gestation: Correlation of hCG levels and sonographic identification. *AJR* 144:951, 1985.
7. Bernard KG, Cooperberg PL: Sonographic differentiation between blighted ovum and early viable pregnancy. *AJR* 144:597, 1985.
8. Sauerbrei E, Cooperberg P, Poulan B: Ultrasound demonstration of the normal fetal yolk sac. *J Clin Ultrasound* 8:217, 1980.
9. Cadkin AV, McAlpin J: Detection of fetal cardiac activity between 41 and 43 days of gestation. *J Ultrasound Med* 3:499, 1984.
10. Anderson S: Management of threatened abortion with real-time sonography. *Obstet Gynecol* 55:259, 1981.
11. Goldstein S, Subyamnyam B, Ragnavendra BN: Subchorion bleeding in threatened abortion. *AJR* 141:975, 1983.
12. Biggers JD: In vitro fertilization and embryo transfer in human beings. *N Engl J Med* 304:336, 1981.
13. Finberg HJ, Birnholz JC: Ultrasound observations in multiple gestation with first trimester bleeding: The blighted twin. *Radiology* 132:137, 1979.
14. Kaufman A, Fleischer A, Thieme G, et al: Separated chorioamnion and elevated chorion: Sonographic features and clinical significance. *J Ultrasound Med* 4:119, 1985.
15. Nyberg D, Laing F, Filly R: Sonographic distinction of normal from abnormal gestational sacs in threatened abortion. *Radiology* 158:397, 1986.
16. Jouppila P: Clinical consequences after ultrasonic diagnosis of intrauterine hematoma in threatened abortion. *J Clin Ultrasound* 13:107, 1985.
17. Steier J, Berasjo P, Myking O: Human chorionic gonadotropin in maternal plasma after induced abortion, spontaneous abortion, and removed ectopic pregnancy. *Obstet Gynecol* 64:391, 1985.
18. Fleischer A, Cartwright P, dePietro D: Sonographic evaluation of ectopic pregnancy, in Sanders R, James AE Jr (eds): *Principles and Practice of Ultrasonography in Obstetrics and Gynecology*, ed 3. New York, Appleton-Century-Crofts, 1984.

19. Weckstein D, Boucher M, Tacher H: Accurate diagnosis of early ectopic pregnancy. *Obstet Gynecol* 65:393, 1985.
20. Mahoney B, Filly R, Nyberg D, Callen P: Sonographic evaluation of ectopic pregnancy. *J Ultrasound Med* 4:221, 1985.
21. Coleman B, Baron R, Arger P: Ectopic embryo detection using real-time sonography. *J Clin Ultrasound* 13:545, 1985.
22. Berry S, Coulam C, Hill L, et al: Evidence of contralateral ovulation in ectopic pregnancy. *J Ultrasound Med* 4:293, 1985.
23. Kadar N, DeVore G, Romero R: Discriminatory hCG zone: Its use in the sonographic evaluation for ectopic pregnancy. *Obstet Gynecol* 58:256, 1981.
24. Pittaway D, Wentz AC, Maxson W, Herbert C, Daniel J, Fleischer A: The efficacy of early pregnancy monitoring with serial chorionic gonadotrophin: Determination and real-time sonography in an infertility population. *Fertil Steril* 44:190, 1985.

12 Male Infertility

FRED K. KIRCHNER, M.D.
MURRAY MAZER, M.D.

When evaluating large numbers of couples for infertility a male factor alone will be found in approximately 30% of the cases. In another 20%, abnormalities will be found in both partners. Therefore, it is vitally important to screen the male partner even if abnormalities are initially suspected in the woman.

Clinical Evaluation of the Infertile Male

A brief summary of the evaluation of the infertile male is in order so that the role of radiologic procedures can be put into perspective. Central to the evaluation is the semen analysis.

SEMEN ANALYSIS

In general, the minimal parameters that are considered acceptable for potential fertility are an ejaculate volume of 1.5–5 ml containing a sperm density of greater than 20 million/ml. The sperm should show at least 60% motility with reasonable forward progression and a morphology of at least 60% normal forms. Additionally, there should be no significant sperm agglutination, pyospermia, or hyperviscosity. If the patient is azoospermatic, the semen specimen is evaluated for the presence or absence of fructose. Fructose is produced in the seminal vesicles and its absence in the semen implies either congenital absence of the seminal vesicles or obstruction of the ejaculatory ducts.

The semen analysis can occasionally be quite variable on different occasions in any one individual, and therefore a minimum of two semen analyses should be obtained to define the parameters of any individual patient. Also, because spermatogenesis takes approximately 72 days, any potential toxic influence, such as a fever, may result in a suboptimal semen analysis 2 to 3 months after the episode. Therefore, in a man with such a history who has a suboptimal semen analysis the study should be repeated 2 to 3 months later.

If it is determined that the man does indeed have a suboptimal semen analysis, then a careful history and physical examination is indicated. The following is a brief overview of the evaluation of the subfertile male. For a more detailed account the reader is referred to any standard text of infertility, such as Lipshultz and Howards' *Infertility in the Male* (1).

HISTORY

The past history should include inquiries with respect to childhood events. Unilateral cryptorchidism has an adverse effect on a man's potential fertility even if orchiopexy is performed at an early age. Postpubertal mumps may lead to mumps orchitis with devastating testicular damage. Any history of bladder neck surgery, either open or endoscopic, may be a clue as to possible retrograde ejaculation.

A history of exposure to toxic substances or radiation should also be elicited as well as a medication history. Certain medications such as sulfasalazine, cimetidine, and nitrofurantoin have all been implicated as potential spermatotoxic agents.

Men with a history of testicular cancer have been known for some time to have special problems with relation to fertility. These problems have centered around those patients who have had retroperitoneal lymph node dissections with subsequent risk of decreased transport of sperm and, in some instances, retrograde ejaculation due to surgical interruption of neural pathways. Also, these patients are at risk from the toxic effects of either radiotherapy or chemotherapy.

More recently it has become appreciated that many of these men are subfertile prior to the initiation of any therapy (2). Whether this is a bilateral phenomenon representing subnormal testicles with respect to both fertility potential and risk of tumor or simply represents a toxic effect of the tumor on the contralateral testis is not known.

Along similar lines, it has been recently shown in animal studies that testicular torsion that is not rapidly surgically corrected may have a noxious effect on the contralateral testicle. There is experimental evidence to show that early (less than 8 hr) detorsion or orchiectomy can lessen the chance of damage to the opposite testicle (3). Whether these observations apply to the human situation is not known.

Even a slight increase in temperature can have an adverse effect on spermatogenesis, and therefore a history of the patient's work environment should be elicited. Occasionally patients who are in the habit of taking long, hot tub baths on a daily basis can improve their semen analysis parameters by switching their bathing habits to cool showers.

PHYSICAL EXAMINATION

A general physical examination with emphasis on evaluation of the genitalia should be carried out. The general body habitus should be noted, and if eunuchoid proportions are suspected then the arm span should be measured. The arm span should not exceed the patient's height by more than 2 inches. The presence and distribution of body hair should also be noted as well as the presence or absence of gynecomastia.

With respect to the genitalia, adequate size of the penis, retractability of the prepuce and location of the urethral meatus should be noted. The scrotal examination should be performed with the patient standing, to better assess the presence or absence of a varicocele. Testicles can be measured in a variety of ways, and their size is important with respect to fertility. Approximately 85% of the testicular volume is involved with sperm production and any decrease in volume will probably be reflected by reduced sperm production. The testicles can be measured either with a tape or calipers, but most urologists find it handy to have a set of graded prostheses that they can hold up near the testicle for assessment of the volume. The normal testicular volume should be greater than 15 ml. Evidence of acute or chronic inflammatory disease in the epididymis is also important to evaluate. Any nodularity or tenderness may indicate problems in this area. Careful palpation of the prostate completes the exam.

OTHER TESTS

At this point in the evaluation further tests are done depending upon the past history and the findings of physical examination. If the patient is azoospermatic, a determination of semen fructose should be carried out. As mentioned above, if fructose is present, then obstruction of the ejaculatory duct or maldevelopment of the seminal vesicles is essentially ruled out. Blood is often drawn for follicle-stimulating hormone (FSH), luteinizing hormone (LH), and testosterone determinations. Usually the latter two will be normal. If the FSH is greater than twice normal, this is indicative of end-organ (testicular) failure and the evaluation can be terminated. If, on the other hand, these values are normal, then testicular biopsy and seminal vesiculograms are indicated.

A large number of patients will fall into the category of "idiopathic oligospermia." These patients should have an endocrine evaluation, although in virtually all cases the findings will be normal. In recent years the sperm penetration assay has been used to aid in prognosis (4). Occasionally evaluation for the presence or absence of antisperm antibodies may be indicated, although their exact role in infertility has yet to be completely elucidated.

Radiographic Procedures

As can be seen above, radiographic procedures do not play as central a role in the evaluation of male infertility as they do in female infertility. However, in selected patients, spermatic venography or seminal vesiculography can be quite helpful in the diagnostic process. In the former instance, therapy (occlusion of the spermatic vein) may be instituted at the time of the diagnostic study.

SPERMATIC VENOGRAPHY

A varicocele is a scrotal venous tumefaction caused by variceal dilatation of the veins of the pampiniform plexus secondary to obstruction of venous drainage and/or valve incompetence, and is the most common diagnosis made in male patients attending an infertility clinic. The presence of a varicocele appears to be virtually nonexistent in the prepubertal years, and the incidence increases through the teens (5, 6). An incidence of approximately 15% has been described in groups of teenage boys (5, 6). On the other hand, in the infertility setting, the number of men found to have varicoceles ranges from 21% to 41% (7–10). If these individuals undergo ligation of their spermatic vein, the majority, but not all, will show an improved semen quality, although in most series the resulting impregnation rate is less than in those showing improvement in semen parameters (10–13). Those men who fail to improve their semen analysis after spermatic vein ligation may represent either failure of the procedure itself, or a subpopulation of men whose infertility is secondary to other unknown causes but who also happen to have a varicocele.

The diagnosis of a varicocele is often straightforward. When the patient is examined in the upright position, a characteristic "bag of worms" can be felt above the testicle. Sometimes the varicocele is large enough to be seen on visual examination. The significance of the so-called subclinical varicocele has yet to be demonstrated by any prospective study. Some urologists feel that a more diligent search for a very small, impalpable varicocele should be carried out in a man with abnormal semen parameters. The diagnostic maneuvers include: having the patient Valsalva in the upright position while palpating the spermatic cord, Doppler stethoscope examination (14), scrotal thermography (15), scrotal ultrasound (16), isotopic studies (17), and venography. Renal venography (on the left) or cavography (for assessment of the right gonadal vein) should not produce significant filling of the gonadal veins if the valves are competent. Retrograde flow down the gonadal veins is a sign of incompetent valves from congenital or acquired causes. If a varicocele is present, the gonadal vein is distended and the pampiniform plexus engorged (Fig. 12.1).

The pathophysiology of the deleterious effect of a varicocele on fertility has yet to be fully elucidated. Numerous hypotheses, including reflux of adrenal or renal metabolites, hormonal imbalance, and hypoxia secondary to venous stasis have all been proposed but not definitely proven. It has been appreciated for a long time that even a slight increase in testicular temperature will have an adverse effect on spermatogenesis. Animal studies have demonstrated that elevated testicular temperature in an experimentally induced varicocele may be secondary to elevated testicular arterial flow. Both of these abnormalities return to normal after ligation of the varicocele (18).

The finding of an incidental varicocele in the noninfertility setting poses a dilemma to both the physician and the patient. There is no way to predict prospectively whether a varicocele has adverse fertility implications. If, in addition to the varicocele, the associated testicle is smaller and softer, then prophylactic spermatic vein ligation can be recommended. It is often not feasible to ask a teenage boy to produce a semen specimen for analysis, and so the physician is left with counseling the patient and possibly his parents with the above facts and letting them help make the decision as to whether an elective spermatic varicocele ligation should be done.

As stated above, varicoceles are much more common on the left side, presumably

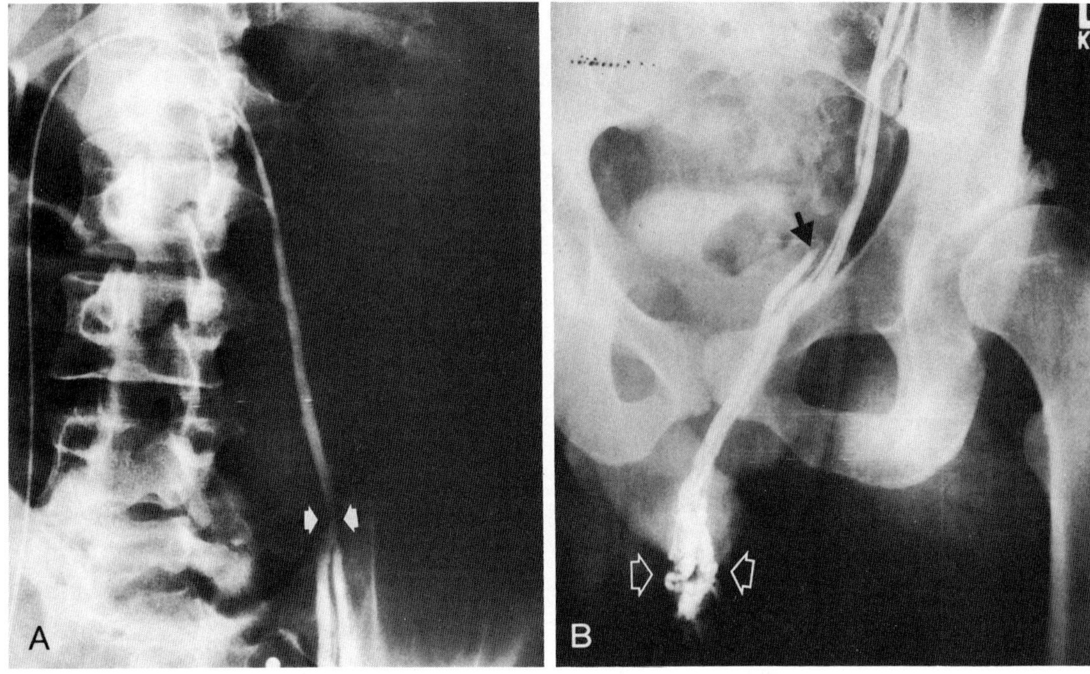

Figure 12.1. Varicocele: Left gonadal venography. *A*: Selective catheterization of the gonadal vein with injection of contrast results in retrograde filling of the spermatic vein and its tributaries, reflecting absent or incompetent valves. Collateral venous pathways are exemplified by the demonstration of two large venous channels that unite at the L5 level into a single major vein (*paired arrows*). *B*: Lower segment of same venogram as in A. Numerous venous channels (*large arrow*) are seen draining an engorged and dilated pampiniform plexus (*open arrows*).

because the left internal spermatic vein has a longer course than its counterpart on the right. Also, the left internal spermatic vein enters the left renal vein at a right angle. This is in contradistinction to the oblique entrance of the right vein into the vena cava. Increased hydrostatic pressure in the left internal spermatic vein may also result from the so-called nutcracker phenomenon. The left renal vein passes between the superior mesenteric artery and aorta, and an area of narrowing at this point with proximal venous dilatation can often be seen in patients examined with ultrasound or computerized tomography (19).

It has been stated that in 90% of varicocele patients, only a left varicocele is found at clinical examination. However, 8–9% of patients are found to have bilateral varicoceles, and in a few patients (1–2%) only a right varicocele is detected (20). Some authors have claimed that as many as 25–66% of patients with only a palpable left varicocele also have an incompetent right internal spermatic vein (21, 22). However, a recent cadaver study (23) has shown that valves in the right internal spermatic vein commonly occur at or within 0.5 cm of the orifice. The act of inserting the catheter tip into this orifice may bypass these very proximal valves, leading to a false-positive diagnosis of varicocele.

The traditional therapy for the symptomatic varicocele is high ligation of the internal spermatic vein. Recently, spermatic venography has been popularized because this procedure is not only diagnostic but can be therapeutic. Using a femoral or jugular venous catheter approach, if a varicocele is demonstrated, the internal spermatic vein can be occluded by a variety of techniques such as insertion of a coil (24) (Fig. 12.2), a sclerosing agent (25, 26), or a detachable balloon (27). Vital to the complete success of this procedure is a meticulous evaluation of collateral veins that often par-

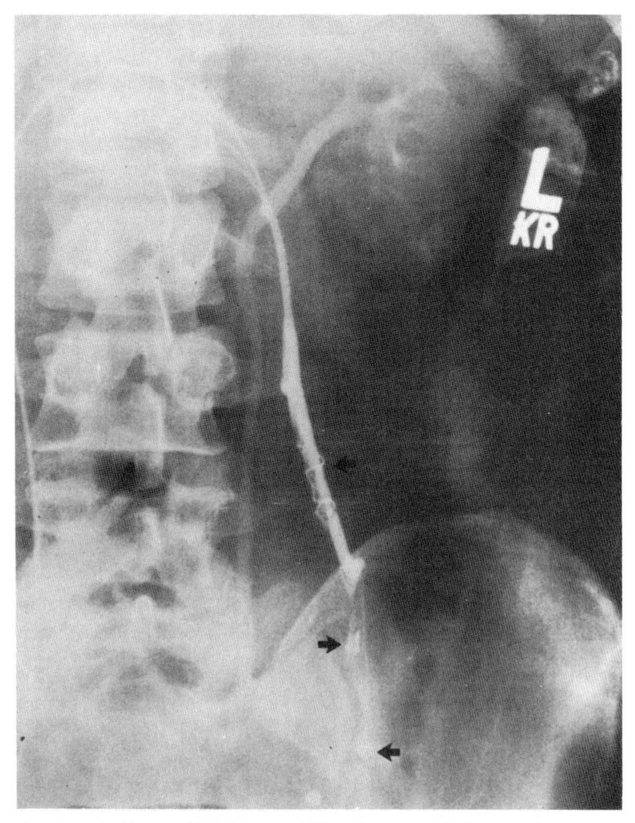

Figure 12.2. Same patient as in Figure 12.1. Several Gianturco coils (*arrows*) are now seen in the gonadal vein.

allel the main spermatic vein above or below the iliac crest to re-enter the upper spermatic vein or the renal vein separately or drain into the retroperitoneum, renal capsular veins, iliac veins, or the contralateral scrotal spermatic system. By placing the embolizing agent as peripheral as possible in the spermatic vein close to the inguinal ligament, and occasionally, when necessary, more central in the spermatic vein close to the renal vein, the chance of varicocele recurrence is diminished (28). The use of the detachable balloon technique is felt to be particularly successful in thoroughly evaluating the smallest of collateral veins that could become clinical problems if not recognized. The balloon can be inflated as far down the course of the spermatic vein as possible without being detached while the more proximal introducer catheter is used to inject contrast without inducing spasm. Contrast is thus diverted into the most subtle collateral pathways at their upper ends while the balloon is occluding the main spermatic vein channel in the lowermost position. After this more thorough evaluation of all potential collaterals the balloon is detached in an optimal position along the course of the spermatic vein.

Whether the above technique will supplant traditional surgical ligation of the internal spermatic vein remains to be seen. Both procedures can be performed on an outpatient basis. Usually the surgical procedure is done under light general anesthesia, but can be done with local anesthesia, while the radiologic procedure is generally done under local anesthesia with or without sedation. Success rates with regard to achievement of pregnancy appear to be comparable for both procedures. The cost of the surgical procedure is, in general, slightly higher and the patients will miss a few days of work, whereas after the radiologic procedure he should be able to return to full activity the next day. Except in the most experienced hands, the time of the surgical procedure is shorter. Surgery also obviates the need for radiation exposure to the patient. Finally, there is a small but definite risk of wire or balloon embolization with the percutaneous approach.

Recurrence rates for the two procedures appear to be comparable. In a large series from Johns Hopkins Hospital, a 6-year evaluation of outpatient balloon embolotherapy in 300 varicoceles (left-sided in 78%, right-sided in 4%, bilateral in 18%) was conducted (29). Of these, 86% were performed for infertility and 14% performed for scrotal discomfort. Long-term follow-up demonstrated a decrease in varicocele recurrence rate from 11% in the first 70 patients to 4% in the last 230 patients, thus emphasizing that experience with this procedure and careful evaluation of potential collaterals is necessary to achieve optimal results.

Approximately 5% of men treated with spermatic vein ligation will have recurrent varicocele (30), and it is perhaps in this clinical situation that venography assumes a more important role. Other patients who might be candidates for venography are those whose semen analysis fails to improve after spermatic vein ligation. The pattern of collaterals seen on venography after an obliterative procedure may vary depending upon the method used to occlude the vein (surgical vs. percutaneous venous occlusion) (31).

Two particular and unusual anatomic variations are worth mentioning separately. Three percent of men have transcrotal collaterals (21). Two confirmed cases of disappearance of the left varicocele after right internal spermatic vein embolizations in the Johns Hopkins series emphasize the importance of this pathway (32).

There is also a proposed theoretical mechanism for varicocele formation in which an obstruction of the left common iliac vein (probably due to pressure by the right common iliac artery as it crosses the vein) would produce collateral flow via the ductus deferens vein, a branch of the internal iliac vein, through the varicocele and up the internal spermatic vein to the renal vein (33). In this situation, occlusion of the internal spermatic vein might actually result in worsening of the varicocele. Left internal iliac venography and measurement of a pressure gradient across the left iliac vein could be performed in patients with recurrence if no other sources of collateralization are visualized. Embolization of the branch of the internal iliac vein that leads to the varicocele could potentially obliterate the varicocele in this theoretical situation.

Finally, the clinical significance and management of the so-called subclinical varicocele remains controversial. A recent retrospective evaluation of venographic findings and improvement of fertility after embolization in 46 clinical and 38 subclinical varicoceles is available (34). A major difference between the two groups was the greater degree of reflux in the clinical varicocele patients. However, in terms of semen analysis improvement and pregnancy rates, embolization treatment appeared to be equally effective in clinical and subclinical varicoceles (34). The issue of patient selection for evaluation of subclinical right internal spermatic vein varicocele also remains controversial. Because incompetence of the right internal spermatic vein is an uncommon finding, some choose to perform venography on the right only in those patients who have palpable right varicoceles or who have left varicoceles but no evidence of reflux on the left (32).

SEMINAL VESICULOGRAMS

Seminal vesiculography is reserved for those men who have azoospermia or severe oligospermia (sperm density less than 1 million/ml) and who have no evidence of pituitary insufficiency (hypogonadotropic hypogonadism) or "end-organ failure" as reflected by an elevated FSH. Generally, testicular biopsy with frozen section evaluation is done at the same time as the proposed radiographic study. If the biopsy shows severe germinal hypoplasia or aplasia then the seminal vesiculogram is not done. The procedure is generally done on an outpatient basis under general anes-

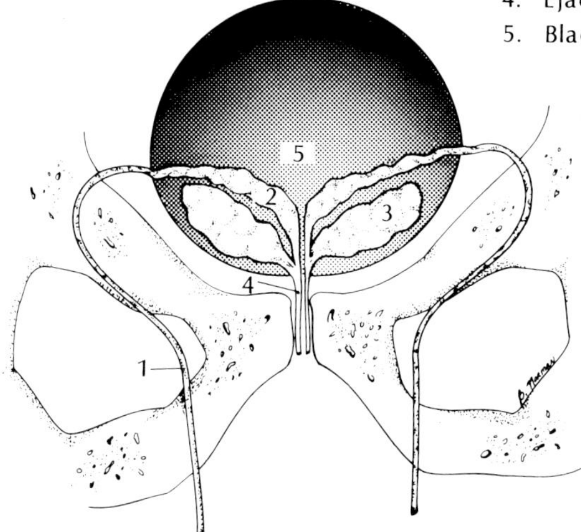

Figure 12.3. A diagrammatic representation of a bilateral seminal vesiculogram and the anatomy that should be visualized.

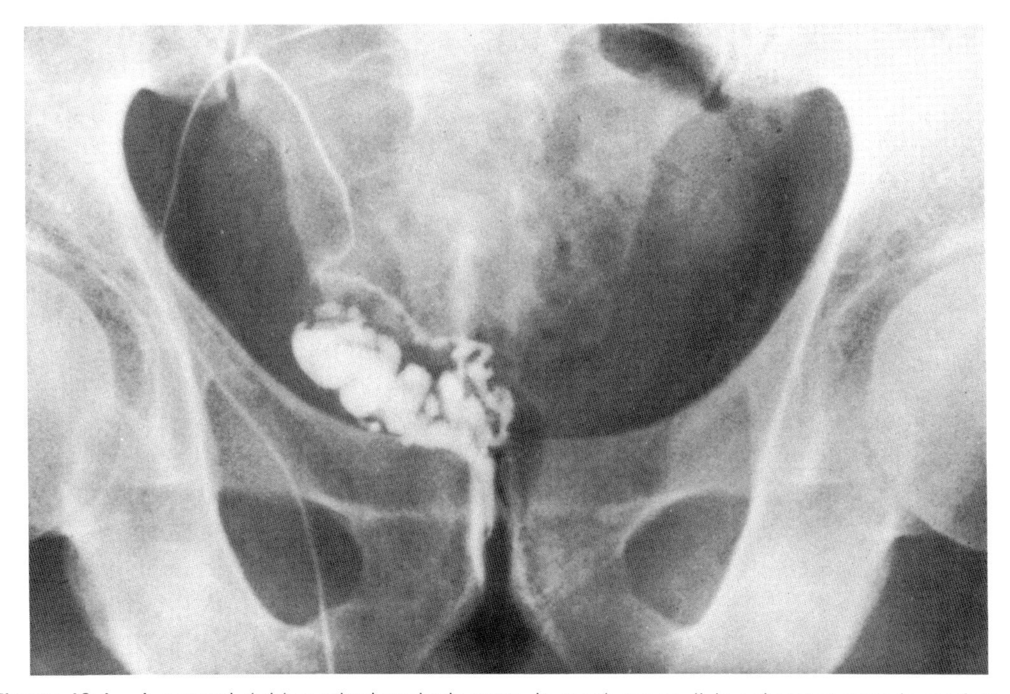

Figure 12.4. A normal right seminal vesiculogram. In most cases, it is only necessary to perform a unilateral seminal vesiculogram. (Courtesy of N. Reed Dunnick, M.D., Durham, NC.)

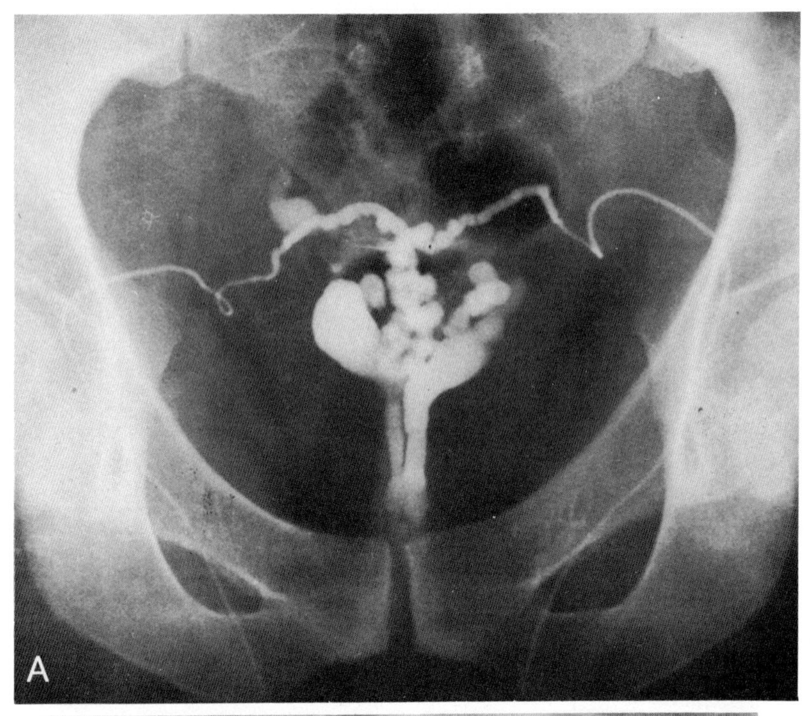

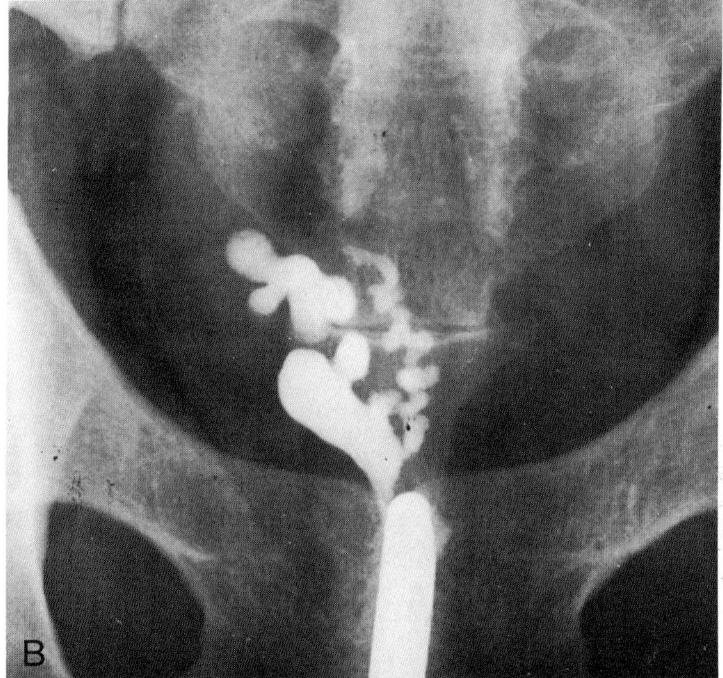

Figure 12.5. *A*: Operative vasogram showing dilated ejaculatory ducts and nonfilling of prostatic urethra or bladder. *B*: Right ejaculatory duct catheterized following transurethral resection and demonstrating patency. (Reprinted by permission from Porch PP Jr: Aspermia owing to obstruction of distal ejaculatory duct and treatment by transurethral resection. *J Urol* 119:141, 1978. © by Williams & Wilkins, 1978.)

thesia, although local anesthesia can be used. There have been concerns raised that this study may induce an inflammatory response and subsequent obstruction at the injection site; however, it is felt that with meticulous technique, using a fine needle (23-gauge butterfly), this should be of minimal concern (35). Techniques of injection that involve incising the vas probably should not be done. After injection of 2.5–5 ml of dilute water-soluble iodinated contrast material, the vas, ampulla of the vas, seminal vesicles, and ejaculatory ducts should be easily visualized (Figs. 12.3 and 12.4). Patency of the duct as it enters the prostatic urethra is confirmed by seeing the contrast either in the urethra or bladder. This is an important observation, and the study can be supplemented by mixing a small amount of methylene blue with the contrast agent. By placing a urethral catheter during the study, demonstration that the contrast has reached the urethra can be observed.

This study is often normal, an especially frustrating finding if the patient should also have a normal testicular biopsy (36). Perhaps the most easily reparable problem that is found is obstruction of the distal ejaculatory ducts, which can be treated by judicious transurethral resection of the area of the prostate where the ejaculatory duct enters (37) (Fig. 12.5). The finding of diverticula-like structures in the ampulla of the vas may be a normal variant (38) or may represent inflammatory changes (39).

Finally, if at the time of the scrotal exploration the biopsy shows spermatogenesis and the seminal vesiculograms show no obstruction, the epididymis should be carefully inspected and palpated, looking for evidence of dilated epididymal tubules. If obstruction of the epididymis is encountered then an epididymovastostomy should be performed at the same sitting.

Summary

The evaluation of the infertile male requires a systematic and thorough investigation carried out in a logical and expeditious manner. It should be stressed that infertility is a "couple problem" even if the primary pathophysiology appears to reside in one individual. Careful explanations of the rationale for the various tests and procedures should be made not only to the patient but also to his spouse. Such words as "abnormal" should be avoided, and the physician caring for such couples must be aware of the great amount of stress that this condition engenders and be prepared to offer sympathetic counseling.

REFERENCES

1. Lipschultz LI, Howards SS: *Infertility in the Male.* New York, Churchill-Livingston, 1983.
2. Bracken RB, Smith KD: Is semen cryopreservation helpful in testicular cancer? *Urology* 15:581, 1980.
3. Nagler HM, deVere White R: The effect of testicular torsion on the contralateral testis. *J Urol* 128:1343, 1982.
4. Rogers BJ: The sperm penetration assay: Its usefulness reevaluated. *Fertil Steril* 43:821, 1985.
5. Oster J: Varicocele in children and adolescents. *Scand J Urol Nephrol* 5:27, 1971.
6. Steeno O, Knops J, Declerck A, Adimoelja A, Van de Voorde H: Prevention of fertility disorders by detection and treatment of varicocele at school and college age. *Andrologia* 8:47, 1976.
7. Hendry WF, Sommerville IF, Hall RR, Pugh RCB: Investigation and treatment of the subfertile male. *Br J Urol* 45:684, 1973.
8. Dubin L, Amelar RD: Etiologic factors in 1294 consecutive cases of male infertility. *Fertil Steril* 22:469, 1971.
9. Stewart BH: Varicocele in infertility: Incidence and results of surgical therapy. *J Urol* 112:222, 1974.
10. Cockett ATK, Urry RL, Dougherty KA: The varicocele and semen characteristics. *J Urol* 121:435, 1979.
11. MacLeod J: Further observations on the role of varicocele in human male infertility. *Fertil Steril* 20:545, 1969.
12. Brown JS: Varicocelectomy in the subfertile male: A ten-year experience with 295 cases. *Fertil Steril* 27:1046, 1976.
13. Dubin L, Amelar RD: Varicocelectomy: 986 cases in a twelve-year study. *Urology* 10:446, 1977.
14. Greenberg SH, Lipschultz LI, Morganroth J, Wein AJ: The use of the Doppler stethoscope in the evaluation of varicoceles. *J Urol* 117:296, 1977.
15. Lewis RW, Harrison RM: Contact scrotal thermography: Application to problems of infertility. *J Urol* 122:40, 1979.
16. McClure RD, Hricak, H: Scrotal ultrasound in the infertile man: Detection of subclinical unilateral and bilateral varicoceles. *J Urol* 135:711, 1986.
17. Wheatley JK, Fajman WA, Witten FR: Clinical experience with the radioisotopic varicocele scan as a screening method for detection of subclinical varicoceles. *J Urol* 128:57, 1982.
18. Green KF, Turner TT, Howards SS: Varicocele: Reversal of the testicular blood flow and temperature effects by varicocele repair. *J Urol* 131:1208, 1984.

19. Buschi AJ, Harrison RB, Brenbridge ANAG, Williamson BRJ, Gentry RR, Cole R: Distended left renal vein: CT/sonographic normal variant. *AJR* 135:339, 1980.
20. Tjia TT, Rumping WJM, Landman GHM, et al: Phlebography of the internal spermatic vein (and the ovarian vein). *Diagn Imaging* 51:8, 1982.
21. Bigot JM, Chatel A: The value of retrograde spermatic phlebography in varicocele. *Eur Urol* 6:301, 1980.
22. Comhaire F, Kunnen M, Nahoum C: Radiological anatomy of the internal spermatic vein(s) in 200 retrograde venograms. *Int J Androl* 4:379, 1981.
23. Nadel SN, Hutchins GM, Albertson PC, et al: Valves of the internal spermatic vein: Potential for misdiagnosis of varicocele by venography. *Fertil Steril* 4:479, 1984.
24. Gonzales R, Narayan P, Castaneda-Zuniga WR, Amplatz K: Transvenous embolization of the internal spermatic veins for the treatment of varicocele scroti. *Urol Clin North Am* 9:177, 1982.
25. Lima SS, Castro MP, Costa OF: A new method for the treatment of varicocele. *Andrologia* 10:103, 1978.
26. Seyferth W, Jecht E, Zeitler E: Percutaneous sclerotherapy of varicocele. *Radiology* 139:335, 1981.
27. White RI Jr, Kaufman SL, Barth KH, Kadir S, Smyth JW, Walsh PC: Occlusion of varicoceles with detachable balloons. *Radiology* 139:327, 1981.
28. Kaufman SL, Kadir S, Barth KH, et al: Mechanisms of recurrent varicocele after balloon occlusion or surgical ligation of the internal spermatic vein. *Radiology* 147:435, 1983.
29. Mitchell SE, White RI Jr, Chang R, et al: Long-term results of outpatient balloon embolotherapy in 300 varicoceles. Abstract #216, 71st Scientific Assembly of the RSNA, November 1985.
30. Sayfan J, Adam YG, Soffer Y: A natural "venous bypass" causing postoperative recurrence of a varicocele. *J Androl* 2:108, 1981.
31. Murray RR Jr, Mitchell SE, Kadir S, Kaufman SL, Chang R, Kinnison ML, Smyth JW, White RI Jr: Comparison of recurrent varicocele anatomy following surgery and percutaneous balloon occlusion. *J Urol* 135:286, 1986.
32. Shuman L, White RI Jr, Mitchell SE, et al: Right-sided varicocele technique and clinical results of balloon embolotherapy from the femoral approach. *Radiology* 158:787, 1986.
33. Coolsaet BLRA: The varicocele syndrome: Venography determining the optimal level for surgical treatment. *J Urol* 124:833, 1980.
34. Marsman JWP: Clinical vs. subclinical varicocele: Venographic findings and improvement of fertility after embolization. *Radiology* 155:635, 1985.
35. Gordon JA, Clahassey EB: Evaluation of stricture formation as a complication of vasopuncture and vasography in the guinea pig. *Fertil Steril* 29:180, 1978.
36. Ford K, Carson CC III, Dunnick NR, Osborne D, Paulson DF: The role of seminal vesiculography in the evaluation of male infertility. *Fertil Steril* 37:552, 1982.
37. Porch PP Jr: Aspermia owing to obstruction of distal ejaculatory duct and treatment by transurethral resection. *J Urol* 119:141, 1978.
38. Banner MP, Hassler R: The normal seminal vesiculogram. *Radiology* 128:339, 1978.
39. Dunnick NR, Ford K, Osborne D, Carson CC III, Paulson DF: Seminal vesiculography: Limited value in vesiculitis. *Urology* 20:454, 1982.

Index

Page numbers in *italics* denote figures; those followed by "t" denote tables.